MW00845661

Mixtures

WILEY SERIES IN PROBABILITY AND STATISTICS

Established by WALTER A. SHEWHART and SAMUEL S. WILKS

A complete list of the titles in this series can be found on http://www.wiley.com/WileyCDA/Section/id-300611.html.

Mixtures

Estimation and Applications

Edited by

Kerrie L. Mengersen

Queensland University of Technology, Australia

Christian P. Robert

Université Paris-Dauphine, France

D. Michael Titterington

University of Glasgow, UK

A John Wiley & Sons, Ltd., Publication

Registered office
John Wiley & Sons Ltd, The Atrium, Southern Gate, Chichester, West Sussex, PO19 8SQ, United Kingdom

For details of our global editorial offices, for customer services and for information about how to apply for permission to reuse the copyright material in this book please see our website at www.wiley.com.

Library of Congress Cataloging-in-Publication Data

Mengersen, Kerrie L.
Mixtures: estimation and applications / Kerrie L. Mengersen, Christian P. Robert, D. Michael Titterington.
p. cm.
Includes bibliographical references and index.
ISBN 978-1-119-99389-6 (cloth)
1. Mixture distributions (Probability theory) I. Robert, Christian P., 1961- II. Titterington, D. M. III. Title.
QA273.6.M46 2011
519.2'4–dc22

2010053469

A catalogue record for this book is available from the British Library.

Print ISBN: 978-1-119-99389-6
ePDF ISBN: 978-1-119-99568-5
oBook ISBN: 978-1-119-99567-8
ePub ISBN: 978-1-119-99844-0
Mobi ISBN: 978-1-119-99845-7

Typeset in 10.25/12pt Times Roman by Thomson Digital, Noida, India
Printed and bound in Great Britain by TJ International Ltd, Padstow, Cornwall

Contents

Preface xiii

Acknowledgements xv

List of contributors xvii

1 The EM algorithm, variational approximations and expectation propagation for mixtures 1
D. Michael Titterington

1.1 Preamble 1
1.2 The EM algorithm 2
1.2.1 Introduction to the algorithm 2
1.2.2 The E-step and the M-step for the mixing weights 3
1.2.3 The M-step for mixtures of univariate Gaussian distributions 4
1.2.4 M-step for mixtures of regular exponential family distributions formulated in terms of the natural parameters 6
1.2.5 Application to other mixtures 7
1.2.6 EM as a double expectation 8
1.3 Variational approximations 8
1.3.1 Preamble 8
1.3.2 Introduction to variational approximations 9
1.3.3 Application of variational Bayes to mixture problems 11
1.3.4 Application to other mixture problems 16
1.3.5 Recursive variational approximations 18
1.3.6 Asymptotic results 19
1.4 Expectation–propagation 20
1.4.1 Introduction 20
1.4.2 Overview of the recursive approach to be adopted 22
1.4.3 Finite Gaussian mixtures with an unknown mean parameter 23
1.4.4 Mixture of two known distributions 24
1.4.5 Discussion 26
Acknowledgements 27
References 27

2 Online expectation maximisation **31**
Olivier Cappé

2.1 Introduction 31
2.2 Model and assumptions 33
2.3 The EM algorithm and the limiting EM recursion 36
2.3.1 The batch EM algorithm 36
2.3.2 The limiting EM recursion 37
2.3.3 Limitations of batch EM for long data records 38
2.4 Online expectation maximisation 40
2.4.1 The algorithm 40
2.4.2 Convergence properties 41
2.4.3 Application to finite mixtures 45
2.4.4 Use for batch maximum-likelihood estimation 47
2.5 Discussion 50
References 52

3 The limiting distribution of the EM test of the order of a finite mixture **55**
Jiahua Chen and Pengfei Li

3.1 Introduction 55
3.2 The method and theory of the EM test 56
3.2.1 The definition of the EM test statistic 57
3.2.2 The limiting distribution of the EM test statistic 58
3.3 Proofs 60
3.4 Discussion 74
References 74

4 Comparing Wald and likelihood regions applied to locally identifiable mixture models **77**
Daeyoung Kim and Bruce G. Lindsay

4.1 Introduction 77
4.2 Background on likelihood confidence regions 79
4.2.1 Likelihood regions 79
4.2.2 Profile likelihood regions 80
4.2.3 Alternative methods 81
4.3 Background on simulation and visualisation of the likelihood regions 82
4.3.1 Modal simulation method 82
4.3.2 Illustrative example 84
4.4 Comparison between the likelihood regions and the Wald regions 85
4.4.1 Volume/volume error of the confidence regions 86
4.4.2 Differences in univariate intervals via worst case analysis 87
4.4.3 Illustrative example (revisited) 88

4.5 Application to a finite mixture model 89
4.5.1 Nonidentifiabilities and likelihood regions for the mixture parameters 90
4.5.2 Mixture likelihood region simulation and visualisation 93
4.5.3 Adequacy of using the Wald confidence region 94
4.6 Data analysis 95
4.7 Discussion 98
References 99

5 Mixture of experts modelling with social science applications 101
Isobel Claire Gormley and Thomas Brendan Murphy

5.1 Introduction 101
5.2 Motivating examples 102
5.2.1 Voting blocs 102
5.2.2 Social and organisational structure 103
5.3 Mixture models 103
5.4 Mixture of experts models 104
5.5 A mixture of experts model for ranked preference data 107
5.5.1 Examining the clustering structure 111
5.6 A mixture of experts latent position cluster model 112
5.7 Discussion 118
Acknowledgements 118
References 118

6 Modelling conditional densities using finite smooth mixtures 123
Feng Li, Mattias Villani and Robert Kohn

6.1 Introduction 123
6.2 The model and prior 125
6.2.1 Smooth mixtures 125
6.2.2 The component models 126
6.2.3 The prior 128
6.3 Inference methodology 129
6.3.1 The general MCMC scheme 129
6.3.2 Updating β and $\mathcal{I}$ using variable-dimension finite-step Newton proposals 130
6.3.3 Model comparison 132
6.4 Applications 133
6.4.1 A small simulation study 133
6.4.2 LIDAR data 135
6.4.3 Electricity expenditure data 137
6.5 Conclusions 141
Acknowledgements 142
Appendix: Implementation details for the gamma and log-normal models 142
References 143

7 Nonparametric mixed membership modelling using the IBP compound Dirichlet process 145
Sinead Williamson, Chong Wang, Katherine A. Heller and David M. Blei

7.1 Introduction 145
7.2 Mixed membership models 146
7.2.1 Latent Dirichlet allocation 147
7.2.2 Nonparametric mixed membership models 147
7.3 Motivation 148
7.4 Decorrelating prevalence and proportion 150
7.4.1 Indian buffet process 150
7.4.2 The IBP compound Dirichlet process 150
7.4.3 An application of the ICD: focused topic models 153
7.4.4 Inference 154
7.5 Related models 155
7.6 Empirical studies 156
7.7 Discussion 158
References 159

8 Discovering nonbinary hierarchical structures with Bayesian rose trees 161
Charles Blundell, Yee Whye Teh and Katherine A. Heller

8.1 Introduction 161
8.2 Prior work 163
8.3 Rose trees, partitions and mixtures 165
8.4 Avoiding needless cascades 169
8.4.1 Cluster models 171
8.5 Greedy construction of Bayesian rose tree mixtures 172
8.5.1 Prediction 175
8.5.2 Hyperparameter optimisation 176
8.6 Bayesian hierarchical clustering, Dirichlet process models and product partition models 176
8.6.1 Mixture models and product partition models 177
8.6.2 PCluster and Bayesian hierarchical clustering 178
8.7 Results 179
8.7.1 Optimality of tree structure 179
8.7.2 Hierarchy likelihoods 180
8.7.3 Partially observed data 182
8.7.4 Psychological hierarchies 182
8.7.5 Hierarchies of Gaussian process experts 182
8.8 Discussion 183
References 185

9 Mixtures of factor analysers for the analysis of high-dimensional data 189
Geoffrey J. McLachlan, Jangsun Baek and Suren I. Rathnayake

9.1 Introduction 189
9.2 Single-factor analysis model 191
9.3 Mixtures of factor analysers 192
9.4 Mixtures of common factor analysers (MCFA) 193
9.5 Some related approaches 196
9.6 Fitting of factor-analytic models 197
9.7 Choice of the number of factors q 199
9.8 Example 199
9.9 Low-dimensional plots via MCFA approach 200
9.10 Multivariate t-factor analysers 202
9.11 Discussion 205
Appendix 206
References 210

10 Dealing with label switching under model uncertainty 213
Sylvia Frühwirth-Schnatter

10.1 Introduction 213
10.2 Labelling through clustering in the point-process representation 214
10.2.1 The point-process representation of a finite mixture model 214
10.2.2 Identification through clustering in the point-process representation 217
10.3 Identifying mixtures when the number of components is unknown 220
10.3.1 The role of Dirichlet priors in overfitting mixtures 221
10.3.2 The meaning of K for overfitting mixtures 222
10.3.3 The point-process representation of overfitting mixtures 224
10.3.4 Examples 227
10.4 Overfitting heterogeneity of component-specific parameters 231
10.4.1 Overfitting heterogeneity 231
10.4.2 Using shrinkage priors on the component-specific location parameters 233
10.5 Concluding remarks 237
References 237

11 Exact Bayesian analysis of mixtures 241
Christian P. Robert and Kerrie L. Mengersen

11.1 Introduction 241
11.2 Formal derivation of the posterior distribution 242
11.2.1 Locally conjugate priors 242
11.2.2 True posterior distributions 244

11.2.3 Poisson mixture 246
11.2.4 Multinomial mixtures 250
11.2.5 Normal mixtures 252
References 254

12 Manifold MCMC for mixtures 255
Vassilios Stathopoulos and Mark Girolami

12.1 Introduction 255
12.2 Markov chain Monte Carlo Methods 257
12.2.1 Metropolis-Hastings 257
12.2.2 Gibbs sampling 257
12.2.3 Manifold Metropolis adjusted Langevin algorithm 258
12.2.4 Manifold Hamiltonian Monte Carlo 259
12.3 Finite Gaussian mixture models 259
12.3.1 Gibbs sampler for mixtures of univariate Gaussians 261
12.3.2 Manifold MCMC for mixtures of univariate Gaussians 261
12.3.3 Metric tensor 262
12.3.4 An illustrative example 263
12.4 Experiments 266
12.5 Discussion 272
Acknowledgements 273
Appendix 273
References 274

13 How many components in a finite mixture? 277
Murray Aitkin

13.1 Introduction 277
13.2 The galaxy data 278
13.3 The normal mixture model 279
13.4 Bayesian analyses 280
13.4.1 Escobar and West 281
13.4.2 Phillips and Smith 281
13.4.3 Roeder and Wasserman 282
13.4.4 Richardson and Green 282
13.4.5 Stephens 283
13.5 Posterior distributions for K (for flat prior) 283
13.6 Conclusions from the Bayesian analyses 285
13.7 Posterior distributions of the model deviances 286
13.8 Asymptotic distributions 286
13.9 Posterior deviances for the galaxy data 287
13.10 Conclusions 291
References 291

14 Bayesian mixture models: a blood-free dissection of a sheep **293**
Clair L. Alston, Kerrie L. Mengersen and Graham E. Gardner

14.1 Introduction 293
14.2 Mixture models 294
14.2.1 Hierarchical normal mixture 294
14.3 Altering dimensions of the mixture model 296
14.4 Bayesian mixture model incorporating spatial information 298
14.4.1 Results 301
14.5 Volume calculation 302
14.6 Discussion 307
References 307

Index **309**

Preface

This edited volume was stimulated by a workshop entitled 'Mixture Estimation and Applications' held at the International Centre for Mathematical Science (ICMS) in Edinburgh on 3–5 March 2010. With the exception of the chapters written by the editors, all chapters were presented during this workshop.

Statistical mixture distributions are used to model scenarios in which certain variables are measured but a categorical variable is missing. For example, although clinical data on a patient may be available their disease category may not be, and this adds significant degrees of complication to the statistical analysis. The above situation characterises the simplest mixture-type scenario; variations include, among others, hidden Markov models, in which the missing variable follows a Markov chain model, and latent structure models, in which the missing variable or variables represent model-enriching devices rather than real physical entities. Finally, mixture models can simply be employed as a more flexible parametric or nonparametric description of data, processes and systems. In the title of the workshop the term 'mixture' was taken to include these and other variations along with the simple mixture. The motivating factors for this three-day workshop were that research on inference and computational techniques for mixture-type models is currently experiencing major advances and that simultaneously the application of mixture modelling to many fields in science and elsewhere has never been so rich. The workshop thus assembled leading statisticians and computer scientists, in both methodological research and applied inference, at this fertile interface. The methodological component encompassed both Bayesian and non-Bayesian contributions, with biology and economics featuring strongly among the application areas addressed.

In addition to the lectures per se, there were two special lectures, given by Michael Jordan and Kerrie Mengersen. Michael Jordan gave a wide-reaching lecture on 'Applied Bayesian nonparametrics' as part of the Edinburgh Informatics Distinguished Lecture series. Kerrie Mengersen presented an evening public lecture at the ICMS on 'Where are they and what do they look like? Discovering patterns in data using statistical mixture models'. Both lectures were highly successful in attracting large audiences and were very well received by those audiences but they do not appear in this volume.

The workshop itself was attended by 70 participants, who had all contributed to the high quality of both the presentations and the corresponding exchanges. The meeting started with a session dedicated to label-switching, with John Geweke and Sylvia Frühwirth-Schnatter (Chapter 10) presenting their views on this difficult issue and Gilles Celeux, Agostino Nobile and Christian Robert discussing the presentations. Other Bayesian talks included Murray Aitkin's (Chapter 13) views on the estimation of the number of components, Clair Alston's (Chapter 14) application to sheep dissection, Kim-Anh Do's application to translational cancer research, Richard Gerlach's smooth transition mixture GARCH models, Chris Holmes's investigations in variable selection, Robert Kohn's (Chapter 6) Modelling conditional densities using finite smooth mixtures, Peter Müller's semiparametric mixture models with covariate-dependent weights and Michael Newton's gamma-based clustering with application to gene-expression analysis. Work on the asymptotics of mixtures was represented by Jiahua Chen's (Chapter 3) lecture on testing the order of finite mixture models by the EM test and Bruce Lindsay's (Chapter 4) talk on mixture analysis in many dimensions, a topic related to Geoff McLachlan's (Chapter 9) modelling of high-dimensional data. Similarly, there was a wide range of talks at the interface between nonparametric Bayes and machine learning, introduced by Michael Jordan's overview and followed by Mark Girolami's (Chapter 12) on spectral mixture component inference, Katherine Heller's (Chapter 7) IBP compound Dirichlet process, Iain Murray's sampling in latent variable models, Yee Whye Teh's (Chapter 8) presentation on hierarchical clustering and Chris Williams's greedy learning of binary latent trees. Brendan Murphy (Chapter 5) covered the mixture-of-experts model from a clustering perspective. Talks with a more computational emphasis were also presented during the workshop, with Christophe Andrieu's approximations of MCMC algorithms, Olivier Cappé's (Chapter 2) work on online EM and Paul Fearnhead's lecture on sequential Monte Carlo.

We believe that this collection of chapters represents the state of the art in mixture modelling, inference and computation. It is our hope that the compilation of our current understanding of this important field will be useful and profitable to active researchers and practitioners in the field as well as to newcomers.

Kerrie L. Mengersen
Christian P. Robert
D. Michael Titterington
Brisbane, Sceaux and Glasgow, 18 November 2010

Acknowledgements

This book consists of chapters contributed by invited speakers at the meeting 'Mixture Estimation and Applications' organised by the International Centre for Mathematical Science (ICMS) in Edinburgh on 3–5 March 2010. The editors are very grateful for the exemplary organisational efficiency of the ICMS staff and for funding provided, through the ICMS, by the UK Engineering and Physical Sciences Research Council, the London Mathematical Society and the Glasgow Mathematical Journal Trust, as well as by the Royal Statistical Society, the Australian Research Council and l'Agence Nationale de la Recherche.

The editors are also most grateful to the contributors for contributing a chapter and for their help during the preparation stage. The support of John Wiley through the encouragement of Kathryn Sharples was most welcome. Part of the editing of the book was done during a visit by the second editor to the Wharton School of Business, University of Pennsylvania, whose support and welcome he gratefully acknowledges.

Acknowledgements

[illegible]

List of contributors

Murray Aitkin
Department of Mathematics and Statistics
University of Melbourne, Australia

Clair L. Alston
School of Mathematical Sciences
Queensland University of Technology
Australia

Jangsun Baek
Department of Statistics
Chonnam National University
Gwangju, Korea

David M. Blei
Computer Science Department
Princeton University, New Jersey
USA

Charles Blundell
Gatsby Computational Neuroscience Unit
University College London, UK

Olivier Cappé
LTCI, Telecom ParisTech, Paris
France

Jiahua Chen
Department of Statistics
University of British Columbia
Vancouver, Canada

Sylvia Frühwirth-Schnatter
Department of Applied Statistics and Econometrics
Johannes Kepler Universität Linz
Austria

Graham E. Gardner
School of Veterinary and Biomedical Sciences
Murdoch University, Australia

Mark Girolami
Department of Statistical Science
University College London, UK

Isobel Claire Gormley
School of Mathematical Sciences
University College Dublin
Ireland

Katherine A. Heller
Department of Engineering
University of Cambridge, UK

Robert Kohn
Australian School of Business
University of New South Wales, Sydney
Australia

Daeyoung Kim
Department of Mathematics
and Statistics
University of Massachusetts, Amherst
Massachusetts, USA

Feng Li
Department of Statistics
Stockholm University, Sweden

Pengfei Li
Department of Mathematical and Statistical Sciences
University of Alberta, Edmonton
Canada

Bruce G. Lindsay
Department of Statistics
Pennsylvania State University
Pennsylvania, USA

Geoffrey J. McLachlan
Department of Mathematics and
Institute for Molecular Bioscience
University of Queensland, St Lucia
Australia

Kerrie L. Mengersen
School of Mathematical Sciences
Queensland University of Technology
Australia

Thomas Brendan Murphy
School of Mathematical Sciences
University College Dublin, Ireland

Suren I. Rathnayake
Department of Mathematics and
Institute for Molecular Bioscience
University of Queensland, St Lucia
Australia

Christian P. Robert
Université Paris-Dauphine
CEREMADE, Paris, France

Vassilios Stathopoulos
Department of Statistical Science
University College London, UK

Yee Whye Teh
Gatsby Computational Neuroscience
Unit
University College London, UK

D. Michael Titterington
University of Glasgow
Glasgow, UK

Mattias Villani
Sveriges Riksbank
Stockholm, Sweden

Chong Wang
Computer Science Department
Princeton University, New Jersey
USA

Sinead Williamson
Department of Engineering
University of Cambridge, UK

1

The EM algorithm, variational approximations and expectation propagation for mixtures

D. Michael Titterington

1.1 Preamble

The material in this chapter is largely tutorial in nature. The main goal is to review two types of deterministic approximation, variational approximations and the expectation propagation approach, which have been developed mainly in the computer science literature, but with some statistical antecedents, to assist approximate Bayesian inference. However, we believe that it is helpful to preface discussion of these methods with an elementary reminder of the EM algorithm as a way of computing posterior modes. All three approaches have now been applied to many model types, but we shall just mention them in the context of mixtures, and only a very small number of types of mixture at that.

Mixtures: Estimation and Applications, First Edition. Edited by Kerrie L. Mengersen, Christian P. Robert and D. Michael Titterington.

1.2 The EM algorithm

1.2.1 Introduction to the algorithm

Parameter estimation in mixture models often goes hand-in-hand with a discussion of the EM algorithm. This is especially so if the objective is maximum likelihood estimation, but the algorithm is also relevant in the Bayesian approach if maximum a posteriori estimates are required. If we have a set of data D from a parametric model, with parameter θ, probably multidimensional, and with likelihood function $p(D|\theta)$ and prior density $p(\theta)$, then the posterior density for θ is

$$p(\theta|D) \propto p(D|\theta)p(\theta),$$

and therefore the posterior mode $\hat{\theta}_{MAP}$ is the maximiser of $p(D|\theta)p(\theta) = p(D, \theta)$. Of course, if the prior density is uniform then the posterior mode is the same as the maximum likelihood estimate. If explicit formulae for the posterior mode do not exist then recourse has to be made to numerical methods, and the EM algorithm is a popular general method in contexts that involve incomplete data, either explicitly or by construct. Mixture data fall into this category, with the component membership indicators $\mathbf{z}$ regarded as missing values.

The EM algorithm is as follows. With data D and initial guess $\theta^{(0)}$ for $\hat{\theta}_{MAP}$, a sequence $\{\theta^{(m)}\}$ of values are generated from the following double-step that creates $\theta^{(m+1)}$ from $\theta^{(m)}$.

- **E-step**: Evaluate

$$\begin{aligned} Q(\theta; \theta^{(m)}) &= E\{\log[p(D, \mathbf{z}, \theta)]|D, \theta^{(m)}\} \\ &= \sum_{\mathbf{z}} \log[p(D, \mathbf{z}, \theta)]\ p(\mathbf{z}|D, \theta^{(m)}) \\ &= \sum_{\mathbf{z}} \log[p(D, \mathbf{z}|\theta)]\ p(\mathbf{z}|D, \theta^{(m)}) + p(\theta). \end{aligned}$$

- **M-step**: Find $\theta = \theta^{(m+1)}$ to maximise $Q(\theta; \theta^{(m)})$ with respect to θ.

Remarks

1. In many other incomplete data problems the missing values $\mathbf{z}$ are continuous, in which case the summation is replaced by an integration in the above.
2. Not surprisingly, $Q(\theta; \theta^{(m)})$ is usually very like a complete-data log-posterior, apart from a constant that is independent of θ, so that the M-step is easy or difficult according as calculation of the complete-data posterior mode is easy or difficult.
3. The usual monotonicity proof of the EM algorithm in the maximum-likelihood context can be used, with minimal adaptation, to show that

$$\log[p(D, \theta^{(m+1)})] \geq \log[p(D, \theta^{(m)})].$$

Thus, the EM algorithm 'improves' $p(D, \theta)$ at each stage and, provided the posterior density for θ is locally bounded, the values of $p(D, \theta^{(m)})$ should converge to a local maximum of $p(D, \theta)$. The corresponding sequence $\{\theta^{(m)}\}$ will also often converge, one hopes to $\hat{\theta}_{MAP}$, but convergence may not occur if, for instance, $p(D, \theta)$ contains a ridge. The niceties of convergence properties are discussed in detail, for maximum likelihood, in Chapter 3 of McLachlan and Krishnan (1997).

1.2.2 The E-step and the M-step for the mixing weights

Suppose now that the data are a random sample $D = \{y_1, \ldots, y_n\}$ from a distribution with probability density function

$$p(y|\theta) = \sum_{j=1}^{k} \lambda_j f_j(y|\phi_j),$$

where $\lambda = (\lambda_1, \ldots, \lambda_k)$ are the mixing weights, the f_j are the component densities, each corresponding to a subpopulation, and k is finite. The density f_j is parameterised by ϕ_j and the set of all these is to be called ϕ. Often we shall assume that the component densities are of the same type, in which case we shall omit the subscript j from f_j. The complete set of parameters is $\theta = (\lambda, \phi)$.

The complete-data joint distribution can be conveniently written as

$$p(D, \mathbf{z}, \theta) = \prod_{i=1}^{n} \left\{ \prod_{j=1}^{k} [\lambda_j f_j(y_i|\phi_j)]^{z_{ij}} \right\} p(\theta),$$

with the help of the indicator notation, where $z_{ij} = 1$ if the ith observation comes from component j and is zero otherwise. Thus

$$\log[p(D, \mathbf{z}, \theta)] = \sum_{i=1}^{n} \sum_{j=1}^{k} z_{ij} \log[\lambda_j f_j(y_i|\phi_j)] + \log p(\theta).$$

For the E-step of the EM algorithm all that we have to compute are the expectations of the indicator variables. Given $\theta^{(m)} = (\lambda^{(m)}, \phi^{(m)})$, we obtain

$$\begin{aligned} z_{ij}^{(m)} &= P(z_{ij} = 1|y_i, \theta^{(m)}) \\ &= P(i\text{th observation belongs to component } j|y_i, \theta^{(m)}) \\ &= \frac{\lambda_j^{(m)} f_j(y_i|\phi_j^{(m)})}{\sum_l \lambda_l^{(m)} f_l(y_i|\phi_l^{(m)})}, \end{aligned}$$

for each i, j. We now have

$$\begin{aligned} Q(\theta;\theta^{(m)}) &= \sum_{i=1}^{n}\sum_{j=1}^{k} z_{ij}^{(m)} \log[\lambda_j f_j(y_i|\phi_j)] + \log p(\theta) \\ &= \sum_{j=1}^{k} n_j^{(m)} \log \lambda_j + \sum_{i=1}^{n}\sum_{j=1}^{k} z_{ij}^{(m)} \log f_j(y_i|\phi_j) + \log p(\theta) \\ &= Q_1(\lambda) + Q_2(\phi) + \log p(\theta), \end{aligned}$$

say, where $n_j^{(m)} = \sum_{i=1}^{n} z_{ij}^{(m)}$, a 'pseudo' sample size associated with subpopulation j. (In the case of complete data, $n_j = \sum_{i=1}^{n} z_{ij}$ is precisely the sample size for subpopulation j.)

We now consider the M-step for the mixing weights λ. Before this we make some assumptions about the prior distributions. In particular, we assume that λ and ϕ are a priori independent and that the prior for λ takes a convenient form, namely that which would be conjugate were the data complete. Thus, we assume that, a priori, $\lambda \sim \text{Dir}(a^{(0)})$; that is,

$$p(\lambda) \propto \prod_{j=1}^{k} \lambda_j^{a_j^{(0)}-1},$$

for prescribed hyperparameters $a^{(0)} = \{a_j^{(0)}\}$. Clearly, $\lambda^{(m+1)}$ must maximise

$$\sum_{j=1}^{k} (n_j^{(m)} + a_j^{(0)} - 1) \log \lambda_j,$$

which is essentially a log-posterior associated with multinomial data. Thus

$$\lambda_j^{(m+1)} = (n_j^{(m)} + a_j^{(0)} - 1)/(n + a_.^{(0)} - k),$$

where $a_.^{(0)} = \sum_j a_j^{(0)}$.

Example 1: Mixture of two known densities

In this simplest case with $k = 2$, write $\lambda_1 = \lambda = 1 - \lambda_2$. Then the iteration is

$$\lambda^{(r+1)} = (n_1^{(m)} + a_1^{(0)} - 1)/(n + a_.^{(0)} - 2). \qquad (1.1)$$

In examples involving mixtures of *known* densities, this completes the analysis. Otherwise the nature of the M-step for ϕ depends on the model used for the component densities.

1.2.3 The M-step for mixtures of univariate Gaussian distributions

This is by far the most commonly used finite mixture in practice. Here, for $j = 1, \ldots, k$, and using the natural notation for means and variances of

Gaussian distributions,

$$\begin{aligned} f_j(y|\phi_j) &= f(y|\mu_j, \sigma_j^2) \\ &= (2\pi\sigma_j^2)^{-1/2}\exp[-(y-\mu_j)^2/(2\sigma_j^2)]. \end{aligned}$$

Thus

$$Q_2(\phi) = \text{const.} - \frac{1}{2}\sum_{j=1}^{k}\left(n_j^{(m)}\log\sigma_j^2 + \frac{1}{\sigma_j^2}\sum_{i=1}^{n} z_{ij}^{(m)}(y_i-\mu_j)^2\right),$$

which is very like a sum of k Gaussian log-likelihoods, with the $z_{ij}^{(m)}$ included.

For the prior $p(\phi)$ we make the convenient assumptions that the parameters corresponding to the different components are a priori independent, with densities that belong to the appropriate conjugate families. Thus

$$p(\phi) = \prod_{j=1}^{k}[p(\mu_j|\tau_j)p(\tau_j)],$$

in which $\tau_j = 1/\sigma_j^2$ denotes the jth component's precision,

$$\begin{aligned} p(\mu_j|\tau_j) &= N[\mu_j; \rho_j^{(0)}, (b_j^{(0)}\tau_j)^{-1}], \\ p(\tau_j) &= \text{Ga}(\tau_j; c_j^{(0)}, d_j^{(0)}), \end{aligned}$$

where $N[\mu_j; \rho_j^{(0)}, (b_j^{(0)}\tau_j)^{-1}]$ denotes the density of the $N[\rho_j^{(0)}, (b_j^{(0)}\tau_j)^{-1}]$ distribution and $\text{Ga}(\tau_j; c_j^{(0)}, d_j^{(0)})$ denotes the density of the $\text{Ga}(c_j^{(0)}, d_j^{(0)})$ distribution. Thus, for a given j, $\mu_j^{(m+1)}$ and $\sigma_j^{2(m+1)}$ are the joint maximisers of

$$-\frac{1}{2}n_j^{(m)}\log\sigma_j^2 - \frac{1}{2\sigma_j^2}\sum_{i=1}^{n} z_{ij}^{(m)}(y_i-\mu_j)^2 - (c_j^{(0)}-1)\log\sigma_j^2 - d_j^{(0)}/\sigma_j^2$$

$$-\frac{1}{2}\log\sigma_j^2 - \frac{b_j^{(0)}}{2\sigma_j^2}(\rho_j^{(0)}-\mu_j)^2.$$

Straightforward calculus gives

$$\begin{aligned} \mu_j^{(m+1)} &= \frac{\sum_{i=1}^{n} z_{ij}^{(m)}y_i + b_j^{(0)}\rho_j^{(0)}}{n_j^{(m)} + b_j^{(0)}}, \\ \sigma_j^{2(m+1)} &= \frac{\sum_{i=1}^{n} z_{ij}^{(m)}(y_i-\mu_j^{(m+1)})^2 + 2d_j^{(0)} + b_j^{(0)}(\rho_j^{(0)}-\mu_j^{(m+1)})^2}{n_j^{(m)} + 2c_j^{(0)} - 1}. \end{aligned}$$

As usual, note the similarity in structure with formulae for posterior modes for Gaussian parameters given complete data.

1.2.4 M-step for mixtures of regular exponential family distributions formulated in terms of the natural parameters

Here $\log f_j(y|\phi_j) = \text{const.} + t(y)^\top \phi_j - C(\phi_j)$ and therefore

$$Q_2(\phi) = \text{const.} + \sum_i \sum_j z_{ij}^{(m)} [t(y_i)^\top \phi_j - C(\phi_j)]$$

$$= \text{const.} + \sum_j \left\{ \left[\sum_i z_{ij}^{(m)} t(y_i) \right]^\top \phi_j - n_j^{(m)} C(\phi_j) \right\}.$$

Then an appropriate prior for ϕ_j has the form

$$\log p(\phi_j) = d_j^{(0)\top} \phi_j - c_j^{(0)} C(\phi_j) + \text{const},$$

and therefore $\phi^{(m+1)}$ is the maximiser of

$$\left[\sum_i z_{ij}^{(m)} t(y_i) + d_j^{(0)} \right]^\top \phi_j - (n_j^{(m)} + c_j^{(0)}) C(\phi_j).$$

Differentiation with respect to ϕ leads to the following equation satisfied by $\phi_j = \phi_j^{(m+1)}$:

$$\sum_i z_{ij}^{(m)} t(y_i) + d_j^{(0)} = (n_j^{(m)} + c_j^{(0)}) \frac{\partial C(\phi_j)}{\partial \phi_j}.$$

However, if we parameterise by

$$\psi_j := E[t(Y)],$$

in which the expectation is over Y assumed to belong to the jth subpopulation, then

$$\psi_j = \frac{\partial C(\phi_j)}{\partial \phi_j},$$

so the M-step is just

$$\psi_j^{(m+1)} = \frac{\sum_{i=1}^n z_{ij}^{(m)} t(y_i) + d_j^{(0)}}{n_j^{(m)} + c_j^{(0)}}.$$

Example 2: Mixture of Poisson distributions

In this example the jth component density can be written as

$$f_j(y|\phi_j) \propto e^{-\phi_j} \phi_j^y,$$

so that an appropriate conjugate prior for ϕ_j is a $\mathrm{Ga}(c_j^{(0)}, d_j^{(0)})$ distribution. Then it is easy to see that the M-step of the EM algorithm leads to

$$\phi_j^{(m+1)} = \frac{\sum_{i=1}^n z_{ij}^{(m)} y_i + c_j^{(0)} - 1}{n_j^{(m)} + d_j^{(0)}}.$$

The terms in $Q(\theta; \theta^{(m)})$ that involve ϕ_j constitute, up to an additive constant, the logarithm of a $\mathrm{Ga}(c_j^{(m+1)}, d_j^{(m+1)})$ density function, with

$$c_j^{(m+1)} = \sum_{i=1}^n z_{ij}^{(m)} y_i + c_j^{(0)},$$
$$d_j^{(m+1)} = n_j^{(m)} + d_j^{(0)}.$$

Example 3: Mixture of exponential distributions

In the 'natural' parameterisation,

$$f_j(y|\phi_j) = \phi_j \exp(-\phi_j y),$$

so that $t(y) = -y$, $C(\phi_j) = -\log \phi_j$ and $\psi_j = -1/\phi_j$. To fit in with the previous notation, we assume that the prior for ϕ_j is a $\mathrm{Ga}(c_j^{(0)}, d_j^{(0)})$ distribution. Thus the M-step is

$$\psi_j^{(m+1)} = -\frac{\sum_{i=1}^n z_{ij}^{(m)} y_i + d_j^{(0)}}{n_j^{(m)} + c_j^{(0)} - 1},$$

from which $\phi_j^{(m+1)} = -1/\psi_j^{(m+1)}$ can be obtained.

1.2.5 Application to other mixtures

Application of EM to various other specific mixture models is discussed in monographs such as Titterington *et al.* (1985) and McLachlan and Peel (2000), not to mention many individual papers. Included in these special cases is that of hidden Markov models, more precisely called hidden Markov chain models. In a mixture sample the complete data corresponding to the ith observation consist of observed data y_i together with the component-membership indicators $z_i = \{z_{ij}, j = 1, \ldots, k\}$. In a mixture sample the $\{z_i\}$ are missing or 'hidden', it is assumed that the ys for different observations are independent, given the zs, and also that the zs are themselves independent. In the hidden Markov model, this second assumption is modified; instead, it is assumed that the $\{z_i, i = 1, \ldots, n\}$ come from a homogeneous, first-order Markov chain with states corresponding to the component subpopulations. Thus, dependence is assumed, but is of the simplest, one-dimensional kind. This model is very popular, in areas such as ecology, speech modelling and DNA sequencing, and associated methodology has been developed based on both maximum likelihood (Rabiner, 1989) and the Bayesian approach

(Robert *et al.*, 1993). The dependence among the hidden variables leads to additional complications, in comparison to the case of mixture data, but typically not to a severe degree. More precisely, the E-step in the EM algorithm for finding a posterior mode is not explicit, but requires a (terminating) so-called forwards–backwards algorithm.

1.2.6 EM as a double expectation

There is an appealing interpretation of EM as a double maximisation rather than as an expectation followed by a maximisation. The idea goes back at least as far as Csiszár and Tusnády (1984) but has more recently been set out clearly, in the context of maximum likelihood estimation, by Neal and Hinton (1999). The version corresponding to calculation of a Bayesian posterior mode with our notation goes as follows. Define

$$F(q, \theta) = \sum_{\mathbf{z}} q(\mathbf{z}) \log \left[\frac{p(D, \mathbf{z}, \theta)}{q(\mathbf{z})} \right],$$

where q is a density function, and suppose that we are at stage m in the algorithm.

- The first step is to choose $q = q^{(m)}$ to maximise $F(q, \theta^{(m)})$. Since we can write

$$F(q, \theta^{(m)}) = \sum_{\mathbf{z}} q(\mathbf{z}) \log \left[\frac{p(\mathbf{z}|D, \theta^{(m)})}{q(\mathbf{z})} \right] + \log p(D, \theta^{(m)}),$$

  the solution is to take $q^{(m)}(\mathbf{z}) = p(\mathbf{z}|D, \theta^{(m)})$.
- The second step is to choose $\theta = \theta^{(m+1)}$ to maximise $F(q^{(m)}, \theta)$. This amounts to maximising $Q(\theta; \theta^{(m)})$, which is just the EM algorithm.

It is easy to see that

$$F(q^{(m)}, \theta^{(m)}) = \log p(D, \theta^{(m)}),$$

and therefore the above *alternating maximisation* technique leads to monotonicity:

$$\begin{aligned} \log p(D, \theta^{(m)}) &= F(q^{(m)}, \theta^{(m)}) \leq F(q^{(m)}, \theta^{(m+1)}) \leq F(q^{(m+1)}, \theta^{(m+1)}) \\ &= \log p(D, \theta^{(m+1)}). \end{aligned} \tag{1.2}$$

1.3 Variational approximations

1.3.1 Preamble

Exact Bayesian analysis of mixture data is complicated, from a computational point of view, because the likelihood function, in expanded form, consists of the sum of a large number of terms. In practice, the use of some form of approximation is inevitable, and much space has been devoted to the idea of approximating the posterior density or predictive density of interest stochastically, in the form of a set of realisations from the distribution, typically created by an MCMC procedure. In principle the resulting inferences will be asymptotically 'correct', in that the empirical distribution associated with a very large set of realisations should be very similar to the target distribution. However, practical difficulties may arise, especially for problems involving many observations and many parameters, because of storage costs, computation time, the need to confirm convergence and so on. These considerations have led to the development of deterministic approximations to complicated distributions. In the next two sections we describe two such approaches, based respectively on *variational approximations* and on *expectation propagation*. We shall see in this section that, unlike with MCMC methods, the resulting variational approximations are not exact, even asymptotically as $n \to \infty$, but they are less unwieldy and this may be sufficiently attractive to outweigh technical considerations. In fact some attractive asymptotic properties do hold, as is discussed in Section 1.3.6.

1.3.2 Introduction to variational approximations

The fundamental idea is very natural, namely to identify a best approximating density q to a target density p, where 'best' means the minimisation of some 'distance' between q and p. There are many measures of distance between density functions, but the one generally used in this context is the Kullback–Leibler directed divergence $\text{KL}(q, p)$, defined by

$$\text{KL}(q, p) = \int q \log(q/p). \tag{1.3}$$

Of course, $\text{KL}(q, p)$ is not symmetric in its arguments and is therefore not a metric, but it is nonnegative, and zero only if q and p are the same except on a set of measure zero. Without further constraints, the optimal q is of course p, but in practice the whole point is that the p in question is very complicated and the optimisation is carried out subject to approximating but facilitating constraints being imposed on q. In many applications the target, p, is $p_D = p(\cdot|D)$, the conditional density of a set of 'unobservables', $\mathbf{u}$, conditional on a set of observed data, D, and q is chosen to minimise

$$\text{KL}(q, p_D) = \int q(\mathbf{u}) \log \left[q(\mathbf{u})/p(\mathbf{u}|D)\right] d\mathbf{u},$$

subject to simplifying constraints on q.

The same solution yields a lower bound on $p(D)$, the (marginal) probability of the data. This follows because

$$\log p(D) = \int q(\mathbf{u})\log[p(D, \mathbf{u})/q(\mathbf{u})]d\mathbf{u} + \text{KL}(q, p_D),$$

as can be seen by combining the two terms on the right-hand side. The properties of the Kullback–Leibler divergence then both provide the desired lower bound, in that then

$$\log p(D) \geq \int q(\mathbf{u})\log[p(D, \mathbf{u})/q(\mathbf{u})]\mathrm{d}\mathbf{u} = \mathcal{F}(q), \tag{1.4}$$

say, and demonstrate that a q that minimises $\text{KL}(q, p_D)$ provides the best lower bound for $\log p(D)$. The right-hand side of (1.4), $\mathcal{F}(q)$, is known in statistical physics as the *free energy* associated with q. The optimum q can be interpreted as the maximiser of the free energy, and it is this interpretation that stimulates the methodology described in this chapter.

In a non-Bayesian context these results permit the approximation of a likelihood function corresponding to a missing-data problem. Here $\mathbf{u}$ represents the missing data and $(D, \mathbf{u})$ the complete data. Thus, the target for q is the conditional density of the missing data, given the observed data, evaluated at a specified value θ for the parameters, and (1.4) provides a lower bound for the observed-data loglikelihood, again for the specified value of θ. More discussion of this use of variational approximations, with references, is given in Section 3.1 of Titterington (2004).

In the Bayesian context, the unobservables include the parameters themselves, as well as any missing values. Thus $\mathbf{u} = (\theta, \mathbf{z})$, where $\mathbf{z}$ denotes the missing data, which for us are the mixture component indicators. The target for q is therefore $p(\theta, \mathbf{z}|D)$, the posterior density of the parameters and the missing values, and (1.4) provides a lower bound for the marginal loglikelihood for the observed data. This marginal likelihood is called the 'evidence' by MacKay (1992) or the Type-II likelihood, and it is a key component of Bayes factors in model-comparison contexts. As we have said, constraints have to be imposed on q in order to create a workable procedure, and the standard approximation is to assume a factorised form,

$$q(\theta, \mathbf{z}) = q^{(\theta)}(\theta)q^{(\mathbf{z})}(\mathbf{z}),$$

for q. We are therefore imposing a (posterior) assumption of independence between the parameters and the missing values. This is clearly a substantive concession, and it is crucial to assess the degree to which the extra computational feasibility counteracts the loss of accuracy. One consequence of the assumption is that the factor $q^{(\theta)}(\theta)$ represents the variational approximation to the 'true' posterior, $p(\theta|D)$. The lower bound in (1.4) is

$$\log p(D) \geq \int \sum_{\mathbf{z}} q^{(\theta)}(\theta)q^{(\mathbf{z})}(\mathbf{z})\log[p(D, \mathbf{z}, \theta)/q^{(\theta)}(\theta)q^{(\mathbf{z})}(\mathbf{z})]\mathrm{d}\theta = \mathcal{F}(q^{(\theta)}, q^{(\mathbf{z})}), \tag{1.5}$$

say, and can also be written

$$\log p(D) \geq \int q^{(\theta)}(\theta)\left\{\sum_{\mathbf{z}} q^{(\mathbf{z})}(\mathbf{z})\log[p(D, \mathbf{z}|\theta)/q^{(\mathbf{z})}(\mathbf{z})] + \log[p(\theta)/q^{(\theta)}(\theta)]\right\}\mathrm{d}\theta.$$

The optimal $q^{(\theta)}$ and $q^{(\mathbf{z})}$ maximise the right-hand side of (1.5). Thus, the two factors respectively maximise the coupled formulae

$$\int q^{(\theta)}(\theta) \log \left(\frac{\exp\{\sum q^{(\mathbf{z})}(\mathbf{z})\log[p(D, \mathbf{z}, \theta)]\}}{q^{(\theta)}(\theta)} \right) \mathrm{d}\theta, \tag{1.6}$$

$$\sum q^{(\mathbf{z})}(\mathbf{z}) \log \left(\frac{\exp\{\int q^{(\theta)}(\theta)\log[p(D, \mathbf{z}, \theta)]\mathrm{d}\theta\}}{q^{(\mathbf{z})}(\mathbf{z})} \right). \tag{1.7}$$

It follows that the optimum $q^{(\theta)}$ and $q^{(\mathbf{z})}$ satisfy

$$q^{(\theta)}(\theta) \propto \exp\left\{\sum q^{(\mathbf{z})}(\mathbf{z})\log[p(D, \mathbf{z}, \theta)]\right\} \tag{1.8}$$

and

$$q^{(\mathbf{z})}(\mathbf{z}) \propto \exp\left\{\int q^{(\theta)}(\theta)\log[p(D, \mathbf{z}, \theta)]\mathrm{d}\theta\right\}. \tag{1.9}$$

Explicit solution of these equations will not be possible. However, writing the equations in the shorthand form $q^{(\theta)} = T^{(\theta)}(q^{(\mathbf{z})})$, $q^{(\mathbf{z})} = T^{(\mathbf{z})}(q^{(\theta)})$ suggests the following iterative algorithm for computing $q^{(\theta)}$ and $q^{(\mathbf{z})}$ from an initial $q^{(\mathbf{z})} = q^{(\mathbf{z})(0)}$: for $m = 0, 1, \ldots$, calculate

$$\begin{aligned} q^{(\theta)(m+1)} &= T^{(\theta)}(q^{(\mathbf{z})(m)}), \\ q^{(\mathbf{z})(m+1)} &= T^{(\mathbf{z})}(q^{(\theta)(m+1)}). \end{aligned} \tag{1.10}$$

The construction of the algorithm is such that, so far as the 'free energy' $\mathcal{F}(q^{(\theta)}, q^{(\mathbf{z})})$ in (1.5) is concerned,

$$\mathcal{F}(q^{(\theta)(m)}, q^{(\mathbf{z})(m)}) \le \mathcal{F}(q^{(\theta)(m+1)}, q^{(\mathbf{z})(m)}) \le \mathcal{F}(q^{(\theta)(m+1)}, q^{(\mathbf{z})(m+1)}), \tag{1.11}$$

for $m = 1, 2, \ldots$, so that the free energy is monotonically increasing and bounded above by $\log\, p(D)$, and therefore the sequence of free energies converges. This behaviour is a direct parallel of that of the EM algorithm; compare (1.11) with (1.2). Whether or not there are multiple local maxima, and so on, is a different issue; in fact the existence of multiple local maxima, or multiple fixed points of the algorithm, is a very common phenomenon.

Again the reader is referred to Titterington (2004), and to Bishop (2006), for general discussions of variational Bayes approaches to missing-data problems and a substantial body of references.

1.3.3 Application of variational Bayes to mixture problems

We shall consider some of the same examples as those used in section 1.2 about the EM algorithm, with the intention of revealing strong methodological similarities between the two approaches.

Example 4: Mixture of k known densities

Recall that here we assume that the observed data are a random sample $D = \{y_1, \ldots, y_n\}$ from a distribution with probability density function

$$p(y|\theta) = \sum_{j=1}^{k} \lambda_j f_j(y),$$

where the f_js are known so that θ just consists of the λ_js. In this case,

$$p(D, \mathbf{z}, \theta) = \prod_{i=1}^{n} \prod_{j=1}^{k} (\lambda^{z_{ij}} f_{ij}^{z_{ij}}) p(\theta),$$

where $z_{ij} = 1$ if the ith observation comes from component j and is zero otherwise, $f_{ij} = f_j(y_i)$ and $p(\theta)$ is the prior density for θ. If we assume a $\text{Dir}(a^{(0)})$ prior for θ, then

$$p(D, \mathbf{z}, \theta) \propto \prod_{j=1}^{k} \lambda_j^{\sum_i z_{ij} + a_j^{(0)} - 1},$$

so that

$$\sum_{\mathbf{z}} q^{(\mathbf{z})}(\mathbf{z}) \log[p(D, \mathbf{z}, \theta)] = \sum_{j=1}^{k} \left(\sum_i q_{ij}^{(\mathbf{z})} + a_j^{(0)} - 1 \right) \log \lambda_j + \text{const},$$

where 'const.' does not depend on θ and $q_{ij}^{(\mathbf{z})}$ is the marginal probability that $z_{ij} = 1$, according to the distribution $q^{(\mathbf{z})}(\mathbf{z})$. From (1.8), therefore, the optimal $q^{(\theta)}(\theta)$ density is that of the $\text{Dir}(a)$ distribution, where $a_j = \sum_i q_{ij}^{(\mathbf{z})} + a_j^{(0)}$, for $j = 1, \ldots, k$.

Next we identify the optimal $q^{(\mathbf{z})}(\mathbf{z})$ as a function of $q^{(\theta)}(\theta)$. We have

$$\int q^{(\theta)}(\theta) \log[p(D, \mathbf{z}, \theta)] d\theta = \sum_{i=1}^{n} \sum_{j=1}^{k} z_{ij} [\log(\phi_j^{(q)}) + \log f_{ij}] + \text{const},$$

where $\phi_j^{(q)} = \exp[E_{q^{(\theta)}} \log(\lambda_j)]$ and now the 'const.' is independent of the $\{z_{ij}\}$. If we substitute this on the right-hand side of (1.9) we see that the optimal $q_{\mathbf{z}}$ takes a factorised form, with one factor for each observation, and that the optimal factors are defined by

$$q_{ij}^{(\mathbf{z})} \propto f_{ij} \phi_j^{(q)},$$

for $j = 1, \ldots k$, subject to $\sum_j q_{ij}^{(\mathbf{z})} = 1$, for each i. Properties of the $\text{Dir}(a)$ distribution imply that $E_{q^{(\theta)}} \log(\lambda_j) = \Psi(a_j) - \Psi(a_.)$, where Ψ denotes the digamma function and $a_. = \sum_j a_j$. Thus, the equations for $q^{(\mathbf{z})}$ are

$$q_{ij}^{(\mathbf{z})} = f_{ij}\exp[\Psi(a_j) - \Psi(a_.)] / \sum_{r=1}^{k} f_{ir}\exp[\Psi(a_r) - \Psi(a_.)],$$

for each i and j. As predicted earlier, clearly the equations for the $\{a_j\}$, which define the Dirichlet distribution that represents the variational approximation to the posterior distribution of θ, and the $\{q_{ij}^{(\mathbf{z})}\}$ cannot be solved explicitly, but the version of the iterative algorithm (1.10) works as follows. Initialise, for example by choosing $q_{ij}^{(\mathbf{z})(0)} = f_{ij}a_j^{(0)}/(\sum_{r=1}^{k} f_{ir}a_r^{(0)})$, for each i and j, and carry out the following steps, for $m = 0, 1, \ldots$.

- *Step 1*: For $j = 1, \ldots, k$, calculate

$$a_j^{(m+1)} = \sum_i q_{ij}^{(\mathbf{z})(m)} + a_j^{(0)}. \tag{1.12}$$

- *Step 2*: For $i = 1, \ldots, n$ and $j = 1, 2$, calculate

$$q_{ij}^{(\mathbf{z})(m+1)} = f_{ij}\exp\left[\Psi(a_j^{(m+1)}) - \Psi(a_.^{(m+1)})\right] / \sum_{r=1}^{k} f_{ir}\exp\left[\Psi(a_r^{(m+1)}) - \Psi(a_.^{(m+1)})\right]. \tag{1.13}$$

Note the strong similarity between Step 1 and an EM algorithm M-step and between Step 2 and an EM-algorithm E-step, the major difference being the replacement of $z_{ij}^{(m)}$ by $q_{ij}^{(\mathbf{z})(m)}$. Indeed, in some of the literature these algorithms are called 'variational EM'.

Example 5: Mixture of k univariate Gaussian densities

In this case

$$p(y|\theta) = \sum_{j=1}^{k} \lambda_j N(y; \mu_j, \sigma_j^2),$$

where $\lambda = \{\lambda_j\}$ are the mixing weights, $\mu = \{\mu_j\}$ are the component means and $\sigma = \{\sigma_j^2\}$ are the component variances, so that $\theta = (\lambda, \mu, \sigma)$. Given data $D = \{y_i, i = 1, \ldots, n\}$, we have

$$p(D, \mathbf{z}, \theta) = \prod_{i=1}^{n} \prod_{j=1}^{k} \left[\lambda_j N(y_i; \mu_j, \sigma_j^2)\right]^{z_{ij}} p(\theta).$$

At this point it is revealing to return to Example 4 for a couple of remarks.

1. The variational posterior for θ there took the form of a Dirichlet distribution, as a result of the factorisation assumption about $q(\theta, \mathbf{z})$ and of the choice of a Dirichlet prior. In other words, having chosen the traditional complete-data conjugate family of priors, conjugacy was obtained for the variational method.
2. The joint variational approximation for the distribution of $\mathbf{z}$ took a factorised form, essentially because $p(D, \mathbf{z}, \theta)$ took the form of a product over i; in other words the (y_i, z_i) are independent over i, where $z_i = \{z_{ij}, j = 1, \ldots .k\}$.

The second remark holds again here, so that we shall have

$$q_{ij}^{(\mathbf{z})} \propto \exp\left\{\int q^{(\theta)}(\theta)\log[p(y_i, z_{ij} = 1, \theta)]\mathrm{d}\theta\right\}, \tag{1.14}$$

for each i and j, normalised so that $\sum_j q_{ij}^{(\mathbf{z})} = 1$, for each i.

Also, if we choose a complete-data conjugate prior for θ then the optimal $q^{(\theta)}$ will be a member of that family. The appropriate hyperparameters will satisfy equations interlinked with those for the $\{q_{ij}^{(\mathbf{z})}\}$ that can be solved iteratively, by alternately updating the hyperparameters and the $\{q_{ij}^{(\mathbf{z})}\}$. This structure will clearly obtain for a wide range of other examples, but we concentrate on the details for the univariate Gaussian mixture, for which the appropriate prior density takes the form mentioned before, namely

$$p(\theta) = p(\lambda, \mu, \sigma) = p(\lambda)\prod_{j=1}^{k}[p(\mu_j|\tau_j)p(\tau_j)],$$

in which $\tau_j = 1/\sigma_j^2$ denotes the jth component precision,

$$\begin{aligned} p(\lambda) &= \mathrm{Dir}(\lambda; a^{(0)}), \\ p(\mu_j|\tau_j) &= N[\mu_j; \rho_j^{(0)}, (b_j^{(0)}\tau_j)^{-1}], \\ p(\tau_j) &= \mathrm{Ga}(\tau_j; c_j^{(0)}, d_j^{(0)}), \end{aligned}$$

where $\mathrm{Dir}(\lambda; a^{(0)})$ denotes the density of the Dirichlet distribution with parameters $a^{(0)} = \{a_j^{(0)}, j = 1, \ldots, k\}$, and the other notation has been defined already.

Often the priors will be exchangeable, so that all the $c_j^{(0)}$s will be the same, and so on, but we shall work through the more general case. The optimal variational approximation $q^{(\theta)}(\theta)$ then takes the form

$$q^{(\theta)}(\theta) = q^{(\lambda)}(\lambda)\prod_j[q^{(\mu_j|\tau_j)}(\mu_j|\tau_j)q^{(\tau_j)}(\tau_j)],$$

within which the factors are given by

$$\begin{aligned} q^{(\lambda)}(\lambda) &= \mathrm{Dir}(\lambda; a), \\ q^{(\mu_j|\tau_j)}(\mu_j|\tau_j) &= N[\mu_j; \rho_j, (b_j\tau_j)^{-1}], \\ q^{(\tau_j)}(\tau_j) &= \mathrm{Ga}(\tau_j; c_j, d_j), \end{aligned}$$

where the hyperparameters satisfy

$$a_j = \sum_{i=1}^n q_{ij}^{(\mathbf{z})} + a_j^{(0)},$$

$$\rho_j = \left(\sum_{i=1}^n q_{ij}^{(\mathbf{z})} y_i + b_j^{(0)} \rho_j^{(0)}\right) / \left(\sum_{i=1}^n q_{ij}^{(\mathbf{z})} + b_j^{(0)}\right),$$

$$b_j = \sum_{i=1}^n q_{ij}^{(\mathbf{z})} + b_j^{(0)},$$

$$c_j = \frac{1}{2} \sum_{i=1}^n q_{ij}^{(\mathbf{z})} + c_j^{(0)},$$

$$d_j = \frac{1}{2} \left[\sum_{i=1}^n q_{ij}^{(\mathbf{z})} (y_i - \bar{\mu}_j)^2 + b_j^{(0)} \left(\sum_{i=1}^n q_{ij}^{(\mathbf{z})}\right) (\bar{\mu}_j - \rho_j^{(0)})^2 / \left(\sum_{i=1}^n q_{ij}^{(\mathbf{z})} + b_j^{(0)}\right)\right] + d_j^{(0)},$$

in which $\bar{\mu}_j = \sum_{i=1}^n q_{ij}^{(\mathbf{z})} y_i / \sum_{i=1}^n q_{ij}^{(\mathbf{z})}$, for each j, representing a pseudo sample mean for component j in the same way that $\sum_i q_{ij}^{(\mathbf{z})}$ represents a pseudo sample size.

If we write the total set of hyperparameters as h, then the above equations take the form

$$h = G_1(q^{(\mathbf{z})}). \tag{1.15}$$

We now have to identify the optimal $q^{(\mathbf{z})}(\mathbf{z})$ as a function of $q^{(\theta)}(\theta)$ or, to be more specific, as a function of the associated hyperparameters h. As we have seen, for each i we have

$$\begin{aligned} q_{ij}^{(\mathbf{z})} &\propto \exp\left\{\int q^{(\theta)}(\theta) \log[p(y_i, z_{ij} = 1, \theta)] \mathrm{d}\theta\right\} \\ &= \exp\left\{\int q^{(\theta)}(\theta) \sum_j z_{ij} [\log(\lambda_j) + \frac{1}{2}\log(\tau_j) - \frac{1}{2}\tau_j (y_i - \mu_j)^2] \mathrm{d}\theta\right\} + \text{const} \\ &= \exp\left(\sum_j z_{ij} \{E_{q^{(\theta)}} \log(\lambda_j) + \frac{1}{2} E_{q^{(\theta)}} \log(\tau_j) - \frac{1}{2} E_{q^{(\theta)}} [\tau_j (y_i - \mu_j)^2]\}\right) + \text{const} \end{aligned}$$

From Example 1 we know that

$$E_{q^{(\theta)}} \log(\lambda_j) = \Psi(a_j) - \Psi(a_.).$$

Also, from properties of the gamma distribution, we have

$$E_{q^{(\theta)}} \log(\tau_j) = \Psi(c_j) - \log(d_j),$$

and, averaging out first conditionally on τ_j and then over τ_j, we obtain

$$E_{q^{(\theta)}} [\tau_j (y_i - \mu_j)^2] = c_j d_j^{-1} (y_i - \rho_j)^2 + b_j^{-1}.$$

Thus

$$q_{ij}^{(\mathbf{z})} \propto \exp\left(\Psi(a_j) + \frac{1}{2}[\Psi(c_j) - \log(d_j)] - \frac{1}{2}[c_j d_j^{-1}(y_i - \rho_j)^2 + b_j^{-1}]\right),$$

with the normalisation carried out over j for each i. The set of relationships can be represented concisely in the form

$$q^{(\mathbf{z})} = G_2(h). \tag{1.16}$$

The obvious algorithm for obtaining the variational approximations involves initialising, by setting $q^{(\mathbf{z})}$ to some $q^{(\mathbf{z})(0)}$, and then calculating $h^{(m+1)} = G_1(q^{(\mathbf{z})(m)})$ and $q^{(\mathbf{z})(m+1)} = G_2(h^{(m+1)})$, for $m = 0, 1, \ldots$. Again, this is the form of (1.10) corresponding to this example. Since the component densities are unknown, initialisation is not so straightforward, especially if exchangeable priors are assumed for the parameters in the component densities. An ad hoc approach that seemed to be adequate in practice is to perform a cluster analysis of the data into k clusters and to base the initial $q^{(\mathbf{z})}$ on that. For univariate data the crude method of using sample quantiles to determine the clusters can be adequate, and for multivariate data a k-means clustering can be effective enough.

1.3.4 Application to other mixture problems

As already mentioned, Titterington (2004) and Bishop (2006) list many references to particular cases of variational Bayes. The case of mixtures of multivariate Gaussian distributions has been treated in a number of places. The natural analogue of the univariate case as discussed in Example 5 is considered by McGrory (2005) and McGrory and Titterington (2007). The usual conjugate prior is used for the parameters of the component distributions, namely independent Wishart distributions for the inverses of the covariance matrices and independent Gaussian priors for the mean vectors, conditionally on the inverse covariance matrices. The version of the variational posterior distribution, $q^{(\theta)}(\theta)$, naturally has the same structure.

The same mixture model was investigated by Corduneanu and Bishop (2001), but with a different structure for the prior. The prior distributions for the component means were independent Gaussians with zero means, but with covariance matrices that were βI, where I is the identity matrix and β is a positive scalar; this is clearly different from the usual conjugate prior structure. Furthermore, they did not assume any prior distribution for the mixing weights. Instead, they kept the mixing weights λ as fixed parameters and chose $q^{(\theta)}(\theta)q^{(\mathbf{z})}(\mathbf{z})$, where θ consists of the component mean vectors and precision matrices, so as to maximise

$$\sum_{\mathbf{z}} \int q^{(\theta)}(\theta) \log[(D, \mathbf{z}, \theta|\lambda)/q^{(\theta)}(\theta)q^{(\mathbf{z})}(\mathbf{z})] \mathrm{d}\theta,$$

which is a lower bound $\mathcal{F}_\lambda(q)$ for the 'marginal loglikelihood' $\log p(D|\lambda)$. Corduneanu and Bishop (2001) alternately calculate the optimal q for the current value of λ and then obtain λ so as to maximise $\mathcal{F}_\lambda(q)$ for that q. In fact they only perform one iteration of the algorithm for the calculation of the optimal q before moving on to obtain a new λ. As the algorithm progresses the values of $\mathcal{F}_\lambda$ increase, reflecting the fact that the algorithm is similar to the generalised EM algorithm (Dempster *et al.*, 1977). Hyperparameters such as β are pre-set, although explicit procedures are not stated in the paper.

The method of Corduneanu and Bishop (2001) is unconventional in statistical terms in not using the usual structure for the prior distributions for the component mean vectors and in not treating the mixing weights in a fully Bayesian way, as was done in McGrory (2005) and McGrory and Titterington (2007). The more traditional structure is also followed by Ueda and Ghahramani (2002), who consider a more complicated mixture model known as the mixture-of-experts model (Jordan and Jacobs, 1994). This model assumes the existence of covariates, the mixing weights depend on the covariates and the component distributions correspond to regression models, usually Gaussian, again depending on the covariates. The calculations for the variational approach involve much careful detail but are again evocative of complete-data conjugate Bayesian analysis.

Ueda and Ghahramani (2002) emphasise other issues beyond the calculation of approximate posterior distributions, and in particular they cover prediction and model choice. So far as prediction is concerned, in the case of 'ordinary' Gaussian mixtures, the calculation of the predictive density of a new observation is straightforward, being a mixture of the corresponding component predictive densities, with mixing weights given by the means of the variational posterior distribution for the mixing weights. In the univariate case this gives a mixture of Student's t distributions. In the case of mixture-of-experts models the complexity of the mixing weights requires a mild amount of approximation in order to obtain a closed-form predictive density. Ueda and Ghahramani deal with model choice by assuming a specified finite class of models, with associated prior probabilities, incorporating a factor $q^{(M)}(m)$ in the formula to represent the variational posterior distribution on the class of models, and basing model choice on the resulting values of $q^{(M)}(m)$. In their empirical work on Gaussian mixtures, Corduneanu and Bishop (2001) observed automatic model selection taking place, when analysing data that were actually simulated from mixture distributions. If a model were fitted that contained more components than the true model, then, as the 'marginal likelihood' maximisation progressed, at least one of the estimates of the mixing weights would become smaller and smaller. At that stage that component was dropped from the model, the algorithm proceeded and this automatic pruning stopped only when the estimated mixture had the same number of components as the true model. If the fitted model did not have as many components as the true model then no pruning occurred. The same phenonemon occurred in the fully Bayesian approach of McGrory and Titterington (2007) in that, if the fitted model was too rich, then, for some j, $\sum_i q_{ij}^{(\mathbf{z})}$ became very small, corresponding to a component for which the other parameters were very similar to those of a second component. As we have said, $\sum_i q_{ij}^{(\mathbf{z})}$ is a pseudo sample size of those observations nominally assigned to the jth component, and if that number fell much below 1 the component was dropped from the fitted model. There does not seem to be any formal explanation for this intriguing yet helpful behaviour.

Application to hidden Markov models is described in MacKay (1997) and Humphreys and Titterington (2001). Software for variational Bayes developed by J. Winn is available at http://vibes.sourceforge.net/.

1.3.5 Recursive variational approximations

Example 1 (Revisited): Mixture of two known densities

We introduce the idea of recursive methods by returning to the very simple case of a mixture of two known densities, with λ denoting the mixing weight, which is the sole unknown parameter. The variational posterior turned out to be a beta distribution, provided that a beta prior was proposed. Alternative ways of deriving beta approximations to the true posterior are given by recursive approximations, in which the observations are processed one by one and the posterior is updated each time, within the beta class. If, after the ith observation has been dealt with, the approximate posterior is the $\text{Be}(a_1^{(i)}, a_2^{(i)})$ distribution, then the recursion computes hyperparameters $(a_1^{(i+1)}, a_2^{(i+1)})$ on the basis of $(a_1^{(i)}, a_2^{(i)})$ and y_{i+1}. Nonvariational versions of these recursions are motivated by the exact Bayesian step for incorporating observation $i+1$, given a 'current' $\text{Be}(a_1^{(i)}, a_2^{(i)})$ prior. The correct posterior is the mixture

$$p(\lambda|y_{i+1}) = w_{i+1,1}\text{Be}(\lambda; a_1^{(i)} + 1, a_2^{(i)}) + w_{i+1,2}\text{Be}(\lambda; a_1^{(i)}, a_2^{(i)} + 1),$$

where, for $j = 1, 2$, $w_{i+1,j} = f_{i+1,j}a_j^{(i)}/(f_{i+1,1}a_1^{(i)} + f_{i+1,2}a_2^{(i)})$ and $\text{Be}(\lambda; a_1, a_2)$ denotes the $\text{Be}(a_1, a_2)$ density.

Recursions such as these were discussed in detail in Chapter 6 of Titterington *et al.* (1985). For this simple, one-parameter problem, two particular versions are the quasi-Bayes method (Makov and Smith, 1977; Smith and Makov, 1978, 1981), in which $a_j^{(i+1)} = a_j^{(i)} + w_{i+1,j}$, for $j = 1, 2$, and the probabilistic editor (Athans *et al.*, 1977; Makov, 1983), in which $a_1^{(i+1)}$ and $a_2^{(i+1)}$ are calculated to ensure that the first two moments of the beta approximation match the 'correct' moments corresponding to the mixture. This moment-matching is possible beause there are two hyperparameters.

It is easy to define a sequence of beta variational approximations $\{\text{Be}(a_1^{(i)}, a_2^{(i)})\}$. Given an approximating $\text{Be}(a_1^{(i)}, a_2^{(i)})$ at stage i, choose $q_{i+1,j}^{(\mathbf{z})(0)} = f_{i+1,j}a_j^{(i)}/\{\sum_{r=1}^{2}(f_{i+1,r}a_r^{(i)})\}$, for each $j = 1, 2$, and carry out the following steps, for $m = 0, 1, \ldots$.

- *Step 1*: For $j = 1, 2$, calculate

$$a_j^{(i+1)(m+1)} = q_{i+1,j}^{(\mathbf{z})(m)} + a_j^{(i)}.$$

- *Step 2*: For $j = 1, 2$, calculate

$$q_{i+1,j}^{(\mathbf{z})(m+1)} = \frac{f_{i+1,j}\exp[\Psi(a_j^{(i+1)(m+1)}) - \Psi(a_.^{(i+1)(m+1)})]}{\sum_{r=1}^{2} f_{i+1,r}\exp[\Psi(a_r^{(i+1)(m+1)}) - \Psi(a_.^{(i+1)(m+1)})]}.$$

Note the obvious similarity to Equations (1.12) and (1.13).

Clearly, at each recursive stage, the iterative procedure may take some time, but for large i very few iterations should be necessary, and even terminating at $m = 1$ may eventually be adequate. One important factor is that the results obtained will depend on the order in which the observations are incorporated, as tends to be the case with all recursive methods.

Humphreys and Titterington (2000) report an experiment based on a sample of size 50 from a mixture of the $N(3, 1)$ and $N(5, 1)$ densities with mixing weight $\lambda = 0.65$, starting from a Un(0,1) prior, so that $a_1^{(0)} = a_2^{(0)} = 1$. So far as the posterior variances are concerned, the values that were obtained empirically from the nonrecursive variational method, the quasi-Bayes and the recursive variational methods were 0.0043, 0.0044 and 0.0043, respectively. In contrast, the values obtained from Gibbs sampling and the Probabilistic Editor (PE) were 0.0088 and 0.0091, respectively.

Of particular note is the fact that only the probabilistic editor leads to a posterior variance for λ that is very similar to the 'correct' value provided by the Gibbs sampling result; all the other methods, including the nonrecursive variational method, 'underestimate' the posterior variability, although they produce a reasonable mean. Further detailed elucidation of this behaviour is provided in Section 1.4.4.

As remarked above, the recursive results are influenced by the ordering of the data. In the case of a hidden Markov chain, a natural ordering does exist, and indeed recursive (online) analysis could be of genuine practical importance. Humphreys and Titterington (2001) develop versions of the quasi-Bayes, probabilistic editor and recursive variational methods for this problem.

1.3.6 Asymptotic results

Variational Bayes approximations have been developed for many particular scenarios, including a number involving mixture models, and have led to useful empirical results in applications. However, it is important to try to reinforce these pragmatic advantages with consideration of the underlying theoretical properties. What aspects, if any, of the variational posteriors match those of the true, potentially highly complex, posteriors, at least asymptotically? Attias (1999, 2000) and Penny and Roberts (2000) claim that, in certain contexts, the variational Bayes estimator, given by the variational posterior mean, approaches the maximum likelihood estimator in the large sample limit, but no detailed proof was given.

More recently, some more detailed investigations have been carried out. In Wang and Titterington (2004a) the case of a mixture of k known densities was considered, for which the variational approximation is a Dirichlet distribution whereas the true posterior corresponds to a mixture of Dirichlets. However, it was proved that the variational Bayes estimator (the variational posterior mean) of the mixing weights converges locally to the maximum likelihood estimator. Wang and Titterington (2005a) extended the treatment of this mixture example by proving asymptotic normality of the variational posterior distribution of the parameters. In terms of asymptotic mean and distribution type, therefore, the variational approximation behaved satisfactorily. However, the paper went on to examine the asymptotic covariance matrix and showed that it was 'too small' compared with that of the maximum likelihood estimators. As a result, interval estimates based on the variational approximation will be unrealistically narrow.

This confirms, through theory, intuition inspired by the fact that the variational approximation is a 'pure' Dirichlet rather than a complicated mixture of Dirichlets, as well as reinforcing the empirical results described in the previous section. For the case of $k = 2$, the asymptotic variational posterior variance of the mixing weight λ is $\lambda(1 - \lambda)/n$, in other words the same as the variance of the complete-data maximum likelihood estimator. For the example in Humphreys and Titterington (2000), mentioned in the previous section, in which $\lambda = 0.65$ and $n = 50$, this variance is 0.00455, which is rather similar to some of the estimates reported in Section 1.3.5. As already mentioned, we return to this in Section 1.4.4.

In Wang and Titterington (2005b) the demonstration of this unsatisfactory nature of posterior second moments was extended to the case of mixtures of multivariate normal distributions with all parameters unknown. Also in the context of this type of mixture, Wang and Titterington (2006) allowed more flexibility in the algorithm for computing the hyperparameters in $q^{(\theta)}$ by introducing a variable step-length ϵ reminiscent of the extension of the EM algorithm in Peters and Walker (1978); the standard algorithm corresponds to $\epsilon = 1$. It was proved that, for $0 < \epsilon < 2$, the algorithm converges and that the variational Bayes estimators of the parameters converge to the maximum likelihood estimators. Wang and Titterington (2004b) considered the more general context of data from exponential family models with missing values, establishing local convergence of the iterative algorithm for calculating the variational approximation and proving asymptotic normality of the estimator.

In the context of hidden Markov chains, the use of a fully factorised version of $q^{(\mathbf{z})}$ does not reflect the correct solution of the variational problem and this shows up in unsatisfactory empirical results. Wang and Titterington (2004c) reinforced this more formally by showing that, in the context of a simple state-space model, which is a continuous-time analogue of the hidden Markov chain model, the variational Bayes estimators could be asymptotically biased; the factorisation constraints imposed on $q^{(\mathbf{z})}$ are inconsistent with the essential correlations between the hidden states that constitute $\mathbf{z}$.

1.4 Expectation–propagation

1.4.1 Introduction

In this section we describe, in the mixtures context, another deterministic approximation to Bayesian analysis, as an alternative to the variational approximation of the previous section and as before stimulated by the fact that exact calculation of posterior and predictive distributions is not feasible when there are latent or missing variables.

Suppose that we have data of the form $D := \{y_1, \ldots, y_n\}$. The posterior distribution of interest is given formally by

$$p(\theta|D) \propto p(D|\theta)\, p(\theta),$$

where $p(\theta)$ on the right-hand side is the prior density for θ. In this account we shall assume that the data are independently distributed, so that we can write

$$p(\theta|D) \propto t_0(\theta) \prod_{i=1}^{n} t_i(\theta),$$

in which $t_0(\theta) = p(\theta)$ and $t_i(\theta) = p(y_i|\theta)$, the probability density or mass function for the ith observation, for $i = 1, \ldots, n$.

The alternative class of deterministic approximations is that provided by the method of expectation propagation (Minka, 2001a, 2001b). In this approach it is assumed that the posterior distribution for θ is approximated by a product of terms,

$$q_\theta(\theta) = \prod_{i=0}^{n} \tilde{t}_i(\theta),$$

where the $i = 0$ term corresponds to the prior density and, typically, if the prior comes from a conjugate family then, as functions of θ, all the other terms take the same form so that $q_\theta(\theta)$ does also. The explicit form of the approximation is calculated iteratively.

- *Step 1*: From an initial or current proposal for $q_\theta(\theta)$ the ith factor $\tilde{t}_i$ is discarded and the resulting form is renormalised, giving $q_{\theta,\backslash i}(\theta)$.
- *Step 2*: This $q_{\theta,\backslash i}(\theta)$ is then combined with the 'correct' ith factor $t_i(\theta)$ (implying that the correct posterior does consist of a product) and a new $q_\theta(\theta)$ of the conjugate form is selected that gives an 'optimal' fit with the new product, $p_i(\theta)$. In Minka (2001b) optimality is determined by Kullback–Leibler divergence, in that, with

  $$p_i(\theta) \propto q_{\theta,\backslash i}(\theta) t_i(\theta),$$

  then the new $q_\theta(\theta)$ minimises

  $$\text{KL}(p_i, q) = \int p_i \log(p_i/q). \tag{1.17}$$

  However, in some cases, simpler solutions are obtained if moment-matching is used instead; if the conjugate family is Gaussian then the two approaches are equivalent.
- *Step 3*: Finally, from $q_{\theta,\backslash i}(\theta)$ and the new $q_\theta(\theta)$, a new factor $\tilde{t}_i(\theta)$ can be obtained such that

  $$q_\theta(\theta) \propto q_{\theta,\backslash i}(\theta) \tilde{t}_i(\theta).$$

This procedure is iterated over choices of i repeatedly until convergence is attained.

We note in passing that the position of q among the arguments of the Kullback–Leibler divergence in (1.17) is different from that in (1.3).

As with the variational Bayes approach, an approximate posterior is obtained that is a member of the complete-data conjugate family. The important question is to what extent the approximation improves on the variational approximation. Much empirical evidence suggests that it is indeed better, but it is important to investigate this issue at a deeper level. Does expectation propagation achieve what variational Bayes achieves in terms of Gaussianity and asymptotically correct mean and does it in addition manage to behave appropriately, or at least better, in terms of asymptotic second posterior moments?

This section takes a few first steps by discussing some very simple scenarios and treating them without complete rigour. In some cases positive results are obtained, but the method is shown not to be uniformly successful in the above terms.

1.4.2 Overview of the recursive approach to be adopted

As Minka (2001b) explains, expectation propagation (EP) was motivated by a recursive version of the approach known as assumed density filtering (ADF) (Maybeck, 1982). In ADF an approximation to the posterior density is created by incorporating the data one by one, carrying forward a conjugate-family approximation to the true posterior and updating it observation-by-observation using the sort of Kullback–Leibler-based strategy described above in Step 2 of the EP method; the same approach was studied by Bernardo and Girón (1988) and Stephens (1997, Chapter 5) and there are some similarities with Titterington (1976). The difference from EP is that the data are run through only once, in a particular order, and therefore the final result is order-dependent. However, recursive procedures such as these often have desirable asymptotic properties, sometimes going under the name of stochastic approximations. There is also a close relationship with the probabilistic editor. Indeed, our analysis concentrates on the recursive step defined by Step 2 in the EP approach, which is essentially the probabilistic editor if moment-matching is used to update.

We shall summarise results for two one-parameter scenarios, namely normal mixtures with an unknown mean parameter, for which the conjugate prior family is Gaussian, and mixtures with an unknown mixing weight, for which the conjugate prior family is beta. Full details will appear in a paper co-authored with N. L. Hjort.

We shall suppose that the conjugate family takes the form $q(\theta|a)$, where a represents a set of hyperparameters and θ is a scalar. For simplicity we shall omit the subscript θ attached to q. Of interest will be the relationship between consecutive sets of hyperparameters, $a^{(n-1)}$ and $a^{(n)}$, corresponding to the situation before and after the nth observation, y_n, is incorporated. Thus, $q(\theta|a^{(n)})$ is the member of the conjugate family that is closest, in terms of Kullback–Leibler divergence, or perhaps of moment-matching, to the density that is given by

$$q(\theta|a^{(n-1)})t_n(\theta) = q(\theta|a^{(n-1)})f(y_n|\theta),$$

suitably normalised. If E_n and V_n are the functions of $a^{(n)}$ that represent the mean and the variance corresponding to $q(\theta|a^{(n)})$, then we would want E_n and V_n asymptotically to be indistinguishable from the corresponding values for the correct posterior.

We would also expect asymptotic Gaussianity. So far as E_n is concerned, it is a matter of showing that it converges in some sense to the true value of θ. The correct asymptotic variance is essentially given by the asymptotic variance of the maximum likelihood estimator, with the prior density having negligible effect, asymptotically. Equivalently, the increase in precision associated with the addition of y_n, $V_n^{-1} - V_{n-1}^{-1}$, should be asymptotically the same, in some sense, as the negative of the second derivative with respect to θ of $\log f(y_n|\theta)$, again by analogy with maximum likelihood theory.

1.4.3 Finite Gaussian mixtures with an unknown mean parameter

The assumption is that the observed data are a random sample from a univariate mixture of J Gaussian distributions, with means and variances $\{c_j\mu, \sigma_j^2; j = 1, \ldots, J\}$ and with mixing weights $\{v_j; j = 1, \ldots, J\}$. The $\{c_j, \sigma_j^2, v_j\}$ are assumed known, so that μ is the only unknown parameter. For observation y, therefore, we have

$$p(y|\mu) \propto \sum_{j=1}^{J} \frac{v_j}{\sigma_j} \exp\left[-\frac{1}{2\sigma_j^2}(y - c_j\mu)^2\right].$$

In the case of a Gaussian distribution the obvious hyperparameters are the mean and the variance themselves. For comparative simplicity of notation we shall write those hyperparameters before treatment of the nth observation as (a, b), the hyperparameters afterwards as (A, B) and the nth observation itself as y.

In the recursive step the hyperparameters A and B are chosen to match the moments of the density for μ that is proportional to

$$b^{-1/2}\exp\left[-\frac{1}{2b}(\mu - a)^2\right] \sum_{j=1}^{J} \frac{v_j}{\sigma_j} \exp\left[-\frac{1}{2\sigma_j^2}(y - c_j\mu)^2\right].$$

Detailed calculation, described in Hjort and Titterington (2010), shows that the changes in mean and precision satisfy, respectively,

$$A - a = b\sum_j R_j S_j T_j / \left[\sum_{j'} T_{j'} + o(b)\right] \tag{1.18}$$

and

$$B^{-1} - b^{-1} = \frac{\sum_j R_j^2 T_j}{\sum_j T_j} - \frac{\sum_j T_j R_j^2 S_j^2}{\sum_j T_j} + \frac{(\sum_j T_j R_j S_j)^2}{(\sum_j T_j)^2} + o(1), \tag{1.19}$$

where $R_j = c_j/\sigma_j$, $S_j = (x - c_j a)/\sigma_j$ and

$$T_j = \frac{v_j}{\sigma_j} \exp\left[-\frac{(y - ac_j)^2}{2\sigma_j^2}\right].$$

As explained earlier, the 'correct' asymptotic variance is given by the inverse of the Fisher information, so that the expected change in the inverse of the variance is the Fisher information corresponding to one observation, i.e. the negative of the expected second derivative of

$$\log p(y|\mu) = \text{const.} + \log \left\{ \sum_{j=1}^{J} \frac{v_j}{\sigma_j} \exp\left(-\frac{1}{2\sigma_j^2}(y - c_j\mu)^2 \right) \right\}$$
$$= \text{const.} + \log \sum_j T'_j,$$

where T'_j is the same as T_j except that a is replaced by μ. In what follows, R'_j and S'_j are similarly related to R_j and S_j. Then it is straightforward to show that the observed information is

$$-\frac{\partial^2}{\partial \mu^2} \log p(y|\mu) = \frac{\sum_j R_j'^2 T'_j}{\sum_j T'_j} - \frac{\sum_j T'_j R_j'^2 S_j'^2}{\sum_j T'_j} + \frac{(\sum_j T'_j R'_j S'_j)^2}{(\sum_j T'_j)^2}. \quad (1.20)$$

The right-hand sides of (1.19) and (1.20) differ only in that (1.19) involves a whereas (1.20) involves μ, and (1.19) is correct just to $O(b)$ whereas (1.20) is exact. However, because of the nature of (1.18), stochastic approximation theory as applied by Smith and Makov (1981) will confirm that asymptotically a will converge to μ and terms of $O(b)$ will be negligible.

Thus, the approximate posterior distribution derived in this section behaves as we would wish, in terms of its mean and variance; by construction it is also Gaussian.

Hjort and Titterington (2010) show that, for this problem, the change in precision achieved by the variational approximation in incorporating an extra observation is unrealistically large and in fact equal, to first order, to the change corresponding to the complete-data scenario. They also report details for particular normal mixtures, including Minka's (2001b) 'clutter' problem.

1.4.4 Mixture of two known distributions

Recall that in this case

$$t_i(\lambda) = p(y_i|\lambda) = \lambda f_1(y_i) + (1 - \lambda) f_2(y_i),$$

in which λ is an unknown mixing weight between zero and one and f_1 and f_2 are known densities. The prior density for λ is assumed to be that of $\text{Be}(a^{(0)}, b^{(0)})$, and the beta approximation for the posterior based on D is assumed to be the $\text{Be}(a^{(n)}, b^{(n)})$ distribution, for hyperparameters $a^{(n)}$ and $b^{(n)}$. The expectation and variance of the beta approximation are respectively

$$E_n = a^{(n)}/(a^{(n)} + b^{(n)}),$$
$$V_n = (a^{(n)} b^{(n)})/[(a^{(n)} + b^{(n)})^2 (a^{(n)} + b^{(n)} + 1)]$$
$$= E_n(1 - E_n)/(L_n + 1),$$

where $L_n = a^{(n)} + b^{(n)}$. The limiting behaviour of E_n and V_n is of key interest. We would want E_n to tend to the true λ in some sense and V_n to tend to the variance of the correct posterior distribution of λ. Asymptotic normality of the approximating distribution is also desired. If E_n behaves as desired then $E_n(1 - E_n)$ tends to $\lambda(1 - \lambda)$ and, for V_n, the behaviour of $a^{(n)} + b^{(n)} + 1 = L_n + 1$, and therefore of L_n, is then crucial.

Hjort and Titterington (2010) compile the following results.

- For the case of *complete data*, E_n is, to first order, $\hat{\lambda}_{\text{CO}}$, the proportion of the n observations that belong to the first component, it therefore does tend to λ, by the law of large numbers, and the limiting version of V_n is

$$V_{\text{CO}} = \lambda(1 - \lambda)/n,$$

  for large n. Asymptotic normality of the posterior distribution of λ follows from the Bernstein–von Mises mirror result corresponding to the central limit result for $\hat{\lambda}_{\text{CO}}$.
- The behaviour of the *correct* posterior distribution will be dictated by the behaviour of the maximum likelihood estimator $\hat{\lambda}_{\text{ML}}$; for large n, approximately,

$$\begin{aligned} E(\hat{\lambda}_{\text{ML}}) &= \lambda, \\ \text{var}(\hat{\lambda}_{\text{ML}}) &= \frac{1}{n \int \frac{[f_1(x) - f_2(x)]^2}{f(x)} \text{d}x} \\ &= V_{\text{ML}}, \end{aligned}$$

  say. Again, these properties can be transferred to the posterior distribution of λ, by a Bernstein–von Mises argument. This transference will apply as a general rule.
- For the *variational approximation*, which is of course a beta distribution, the limiting version of V_n is

$$V_{\text{VA}} = \lambda(1 - \lambda)/n,$$

  as already mentioned in Section 1.3.6. This is the same as V_{CO}. It is therefore 'smaller than it should be' and would lead to unrealistically narrow interval estimates for λ.
- For the recursive *quasi-Bayes* approach of Smith and Makov (1978), the posterior mean is consistent for large n, and

$$V_n \simeq V_{\text{QB}} = \lambda(1 - \lambda)/n;$$

  this is the same as for the confirmed-data case and the variational approximation, thereby falling foul of the same criticism as the latter as being 'too small'. Similar remarks apply to a recursive version of the variational approximation, implemented in Humphreys and Titterington (2000).
- For the *probabilistic editor*, the sequence $\{E_n\}$ of posterior means will converge (to the true λ value), and the posterior variance of the PE-based beta

approximation based on n observations is approximately, for large n,

$$V_{\mathrm{PE}} = n^{-1}\lambda(1-\lambda)\left[1 - \int \frac{f_1 f_2}{\lambda f_1 + (1-\lambda) f_2}\right]^{-1},$$

which Hjort and Titterington (2010) then show is equal to V_{ML}. Thus, asymptotically, the probabilistic editor, and by implication the moment-matching version of expectation propagation, get the variance right. The same is shown to be true for the standard EP and ADF approaches based on minimisation of the Kullback–Leibler divergence rather than on moment-matching.

1.4.5 Discussion

The simulation exercise of Humphreys and Titterington (2000), mentioned in Section 1.3.5, compared empirically the nonrecursive variational approximation, its recursive variant, the recursive quasi-Bayes and probabilistic editor, and the Gibbs sampler, the last of which can be regarded as providing a reliable estimate of the true posterior. As can be expected on the basis of the above analysis, the approximation to the posterior density provided by the probabilistic editor is very similar to that obtained from the Gibbs sampler, whereas the other approximations are 'too narrow'. Furthermore, the variances associated with the various approximations are numerically very close to the 'asymptotic' values derived above. The implication is that this is also the case for the corresponding version of the EP algorithm based on moment-matching, mentioned in Section 3.3.3 of Minka (2001a), of which the PE represents an online version. Of course EP updates using KL divergence rather than (always) matching moments, but the two versions perform very similarly in Minka's empirical experiments, and Section 1.4.4 reflects this asymptotically. Recursive versions of the algorithm with KL update, i.e. versions of ADF, are outlined and illustrated by simulation in Chapter 5 of Stephens (1997) for mixtures of known densities, extending earlier work by Bernardo and Girón (1988), and for mixtures of Gaussian densities with all parameters unknown, including the mixing weights. For mixtures of two known densities, Stephens notes that, empirically, the KL update appears to produce an estimate of the posterior density that is indistinguishable from the MCMC estimate, and is much superior to the quasi-Bayes estimate, which is too narrow. For a mixture of four known densities, for which the conjugate prior distributions are four-cell Dirichlet distributions, Stephens shows the KL update clearly to be better than the quasi-Bayes update, but somewhat more 'peaked' than it should be. This is because, in terms of the approach of the present paper, there are insufficient hyperparameters in the conjugate family to match all first and second moments. For a J-cell Dirichlet, with J hyperparameters, there are $J - 1$ independent first moments and $J(J-1)/2$ second moments, so that full moment-matching is not possible for $J > 2$, that is for any case but mixtures of $J = 2$ known densities.

This is implicit in the work of Cowell *et al.* (1996) on recursive updating, following on from Spiegelhalter and Lauritzen (1990) and Spiegelhalter and Cowell (1992), and referred to in Section 3.3.3 of Minka (2001a) and Section 9.7.4 of

Cowell *et al.* (1999). They chose Dirichlet hyperparameters to match first moments and the average variance of the parameters. Alternatively, one could match the average of the variances and covariances. However, the upshot is that there is no hope that a pure Dirichlet approximation will produce a totally satisfactory approximation to the 'correct' posterior for $J > 2$, whether through EP or the recursive alternatives, based on KL updating or moment-matching. Nevertheless, these versions should be a distinct improvement on the quasi-Bayes and variational approximations in terms of variance. A possible way forward for small J is to approximate the posterior by a mixture of a small number of Dirichlets. To match all first- and second-order moments of a posterior distribution of a set of J-cell multinomial probabilities one would need a mixture of K pure J-cell Dirichlets, where

$$KJ + (K - 1) = (J - 1) + (J - 1) + (J - 1)(J - 2)/2,$$

i.e. Dirichlet hyperparameters + mixing weights = first moments + variances + covariances. This gives $K = J/2$. Thus, for even J, the match can be exact, but for odd J there would be some redundancy. In fact, even for J as small as 4, the algebraic details of the moment-matching become formidable.

Acknowledgements

This work has benefited from contact with Philip Dawid, Steffen Lauritzen, Yee Whye Teh, Jinghao Xue and especially Nils Lid Hjort. The chapter was written, in part, while the author was in residence at the Isaac Newton Institute in Cambridge, taking part in the Research Programme on Statistical Theory and Methods for Complex High-Dimensional Data.

References

Athans, M., Whiting, R. and Gruber, M. (1977) A suboptimal estimation algorithm with probabilistic editing for false measurements with application to target tracking with wake phenomena. *IEEE Transactions on Automatic Control*, **AC-22**, 273–384.

Attias, H. (1999) Inferring parameters and structure of latent variable models by variational Bayes. In *Proceedings of the 15th Conference on Uncertainty in Artificial Intelligence*, pp. 21–30. Morgan Kaufman.

Attias, H. (2000) A variational Bayesian framework for graphical models. In *Advances in Neural Information Processes and Systems* **12** (eds S. A. Solla, T. K. Leen, and K. L. Müller), pp. 209–215. MIT Press.

Bernardo, J. M. and Girón, F. J. (1988) A Bayesian analysis of simple mixture problems (with discussion). In *Bayesian Statistics* (eds J. M. Bernardo, M. H. DeGroot, D. V. Lindley and A. F. M. Smith), pp. 67–78.

Bishop, C. M. (2006) *Pattern Recognition and Machine Learning.* Springer.

Corduneanu, A. and Bishop, C. M. (2001) Variational Bayesian model selection for mixture distributions. In *Proceedings of the 8th Conference on Artificial Intelligence and Statistics* (eds T. Richardson and T. Jaakkola), pp. 27–34. Morgan Kaufman.

Cowell, R. G., Dawid, A. P. and Sebastiani, P. (1996) A comparison of sequential learning methods for incomplete data. In *Bayesian Statistics 5* (eds J. M. Bernardo, J. O. Berger, A. P. Dawid and A. F. M. Smith), pp. 533–541. Clarendon Press.

Cowell, R. G., Dawid, A. P., Lauritzen, S. L. and Spiegelhalter, D. J. (1999) *Probabilistic Networks and Expert Systems.* Springer.

Csiszár, I. and Tusnády, G. (1984) Information geometry and alternating minimization procedures. *Statistics & Decisions, Supplement*, **1**, 205–237.

Dempster, A. P., Laird, N. M. and Rubin, D. B. (1977) Maximum likelihood from incomplete data via the EM algorithm (with discussion). *Journal of the Royal Statistical Society, Series B*, **39**, 1–38.

Hjort, N. L. and Titterington, D. M. (2010) On Expectation Propagation and the Probabilistic Editor in some simple mixture problems. Preprint.

Humphreys, K. and Titterington, D. M. (2000) Approximate Bayesian inference for simple mixtures. In *COMPSTAT 2000* (eds J. G. Bethlehem and P. G. M. van der Heijden), pp. 331–336. Physica-Verlag.

Humphreys, K. and Titterington, D. M. (2001) Some examples of recursive variational approximations. In *Advanced Mean Field Methods: Theory and Practice* (eds M. Opper and D. Saad), pp. 179–195. MIT Press.

Jordan, M. I. and Jacobs, R. A. (1994) Hierarchical mixtures of experts and the EM algorithm. *Neural Computation*, **6**, 181–214.

McGrory, C. A. (2005) Variational approximations in Bayesian model selection. PhD Thesis, University of Glasgow.

McGrory, C. A. and Titterington, D. M. (2007) Variational approximations in Bayesian model selection for finite mixture distributions. *Computational Statististics and Data Analysis*, **51**, 5352–5367.

MacKay, D. J. C. (1992) Bayesian interpolation. *Neural Computation*, **4**, 415–447.

MacKay, D. J. C. (1997) Ensemble learning for hidden Markov models. Technical Report, Cavendish Laboratory, University of Cambridge.

McLachlan, G. J. and Krishnan, T. (1997) *The EM Algorithm and Extensions.* John Wiley & Sons, Ltd.

McLachlan, G. J. and Peel, D. (2000) *Finite Mixture Models*. John Wiley & Sons, Ltd.

Makov, U. E. (1983) Approximate Bayesian procedures for dynamic linear models in the presence of jumps. *The Statistician*, **32**, 207–213.

Makov, U. E. and Smith, A. F. M. (1977) A quasi-Bayes unsupervised learning procedure for priors. *IEEE Transactions on Information Theory*, **IT-23**, 761–764.

Maybeck, P. S. (1982) *Stochastic Models, Estimation and Control.* Academic Press.

Minka, T. (2001a) A family of algorithms for approximate Bayesian inference. PhD Dissertation, Massachusetts Institute of Technology.

Minka, T. (2001b) Expectation Propagation for approximate Bayesian inference. In *Proceedings of the 17th Conference on Uncertainty in Artifical Intelligence*.

Neal, R. M. and Hinton, G. E. (1999) A view of the EM algorithm that justifies incremental, sparse, and other variants. In *Learning in Graphical Models* (ed. M. Jordan), pp. 355–368. MIT Press.

Penny, W. D. and Roberts, S. J. (2000) Variational Bayes for 1-dimensional mixture models. Technical Report PARG-2000-01, Oxford University.

Peters, B. C. and Walker, H. F. (1978) An iterative procedure for obtaining maximum-likelihood estimates of the parameters for a mixture of normal distributions. *SIAM Journal on Applied Mathematics*, **35**, 362–378.

Rabiner, L. R. (1989) A tutorial on hidden Markov models and selected applications in speech recogntion. *Proceedings of the IEEE*, **77**, 257–286.

Robert, C. P., Celeux, G. and Diebolt, J. (1993) Bayesian estimation of hidden Markov chains: a stochastic implementation. *Statistics and Probability Letters*, **16**, 77–83.

Smith, A. F. M. and Makov, U. E. (1978) A quasi-Bayes sequential procedure for mixtures. *Journal of the Royal Statistical Society, Series B*, **40**, 106–112.

Smith, A. F. M. and Makov, U. E. (1981) Unsupervised learning for signal versus noise. *IEEE Transactions on Information Theory*, **IT-27**, 498–500.

Spiegelhalter, D. J. and Cowell, R. G. (1992) Learning in probabilistic expert systems. In *Bayesian Statistics 4* (eds J. M. Bernardo, J. O. Berger, A. P. Dawid and A. F. M. Smith), pp. 447–465. Clarendon Press.

Spiegelhalter, D. J. and Lauritzen, S. L. (1990) Sequential updating of conditional probabilities on directed graphical structures. *Networks*, **20**, 579–605.

Stephens, M. (1997) Bayesian methods for mixtures of normal distributions. DPhil Dissertation, University of Oxford.

Titterington, D. M. (1976) Updating a diagnostic system using unconfirmed cases. *Applied Statistics*, **25**, 238–247.

Titterington, D. M. (2004) Bayesian methods for neural networks and related models. *Statistical Science*, **19**, 128–139.

Titterington, D. M., Smith, A. F. M. and Makov, U. E. (1985) *Statistical Analysis of Finite Mixture Distributions*. John Wiley & Sons, Ltd.

Ueda, N. and Ghahramani, Z. (2002) Bayesian model search for mixture models based on optimizing variational bounds. *Neural Networks*, **15**, 1223–1241.

Wang, B. and Titterington, D. M. (2004a) Local convergence of variational Bayes estimators for mixing coefficients. In *2004 JSM Proceedings*, published on CD Rom. American Statistical Association.

Wang, B. and Titterington, D. M. (2004b) Convergence and asymptotic normality of variational Bayesian approximations for exponential family models with missing values. In *Proceedings of the 20th Conference on Uncertainty in Artificial Intelligence* (eds M. Chickering and J. Halperin), pp. 577–584. AUAI Press.

Wang, B. and Titterington, D. M. (2004c) Lack of consistency of mean field and variational Bayes approximations for state space models. *Neural Processing Letters*, **20**, 151–170.

Wang, B. and Titterington, D. M. (2005a) Variational Bayes estimation of mixing coefficients. In *Deterministic and Statistical Methods in Machine Learning*, Lecture Notes in Artificial Intelligence, Vol. 3635 (eds J. Winkler, M. Niranjan and N. Lawrence), pp. 281–295. Springer-Verlag.

Wang, B. and Titterington, D. M. (2005b) Inadequacy of interval estimates corresponding to variational Bayesian approximations. In *Proceedings of the 10th International Workshop on Artificial Intelligence and Statistics*, pp. 373–380.

Wang, B. and Titterington, D. M. (2006) Convergence properties of a general algorithm for calculating variational Bayesian estimates for a normal mixture model. *Bayesian Analysis*, **1**, 625–650.

2

Online expectation maximisation

Olivier Cappé

2.1 Introduction

Before entering into any more details about the methodological aspects, we discuss the motivation behind the association of the two groups of words 'online (estimation)' and 'expectation maximisation (algorithm)' as well as their pertinence in the context of mixtures and more general models involving latent variables.

The adjective *online* refers to the idea of computing estimates of model parameters on-the-fly, without storing the data and by continuously updating the estimates as more observations become available. In the machine learning literature, the phrase *online learning* has been mostly used recently to refer to a specific way of analysing the performance of algorithms that incorporate observations progressively (Césa-Bianchi and Lugosi, 2006). We do not refer here to this approach and will only consider the more traditional set-up in which the objective is to estimate fixed parameters of a statistical model and the performance is quantified by the proximity between the estimates and the parameter to be estimated. In signal processing and control, the sort of algorithm considered in the following is often referred to as *adaptive* or *recursive* (Benveniste *et al.*, 1990; Ljung and Söderström, 1983). The word *recursive* is so ubiquitous in computer science that its use may be somewhat ambiguous and is not recommended. The term *adaptive* may refer to the type of algorithm considered in this chapter but is also often used in contexts

Mixtures: Estimation and Applications, First Edition. Edited by Kerrie L. Mengersen, Christian P. Robert and D. Michael Titterington.

where the focus is on the ability to track slow drifts or abrupt changes in the model parameters, which will not be our primary concern.

Traditional applications of online algorithms involve situations in which the data cannot be stored, due to their volume and rate of sampling, as in real-time signal processing or stream mining. The wide availability of very large datasets involving thousands or millions of examples is also at the origin of the current renewed interest in online algorithms. In this context, online algorithms are often more efficient, i.e. converging faster towards the target parameter value, and need fewer computer resource in terms of memory or disk access than their batch counterparts (Neal and Hinton, 1999). In this chapter, we are interested in both contexts: when the online algorithm is used to process on-the-fly a potentially unlimited amount of data or when it is applied to a fixed but large dataset. We will refer to the latter context as the *batch estimation* mode.

Our main interest is in maximum likelihood estimation and, although we may consider adding a penalty term (i.e. maximum a posteriori estimation), we will not consider 'fully Bayesian' methods, which aim at sequentially simulating from the parameter's posterior. The main motivation for this restriction is to stick to computationally simple iterations, which is an essential requirement of successful online methods. In particular, when online algorithms are used for batch estimation, it is required that each parameter update can be carried out very efficiently for the method to be computationally competitive with traditional batch estimation algorithms. Fully Bayesian approaches (see, for example, Chopin, 2002) typically require Monte Carlo simulations even in simple models and raise some challenging stability issues when used on very long data records (Kantas *et al.*, 2009).

This quest for simplicity of each parameter update is also the reason for focusing on the EM (expectation maximisation) algorithm. Ever since its introduction by Dempster *et al.* (1977), the EM algorithm has been criticised for its often suboptimal convergence behaviour and many variants have been proposed by, among others, Lange (1995) and Meng and Van Dyk (1997). This being said, thirty years after the seminal paper by Dempster and his co-authors, the EM algorithm still is, by far, the most widely used inference tool for latent variable models due to its numerical stability and ease of implementation. Our main point here is not to argue that the EM algorithm is always preferable to other options. However, the EM algorithm, which does not rely on fine numerical tunings involving, for instance, line searches, re-projections or preconditioning, is a perfect candidate for developing online versions with very simple updates. We hope to convince the reader in the rest of this chapter that the online version of EM that is described here shares many of the attractive properties of the original proposal of Dempster *et al.* (1977) and provides an easy to implement and robust solution for online estimation in latent variable models.

Quite obviously, guaranteeing the strict likelihood ascent property of the original EM algorithm is hardly feasible in an online context. On the other hand, a remarkable property of the online EM algorithm is that it can reach asymptotic Fisher efficiency by converging towards the actual parameter value at a rate that is equivalent to that of the maximum likelihood estimator (MLE). Hence, when the number of observations is sufficiently large, the online EM algorithm does become highly competitive and it is not necessary to consider potentially faster-converging alternatives. When used for batch estimation, i.e. on a fixed large dataset, the situation is different but we will nonetheless show that the online algorithm does converge towards the maximum likelihood parameter estimate corresponding to the whole dataset. To achieve this result, one typically needs to scan the dataset repeatedly. In terms of computing time, the advantages of using an online algorithm in this situation will typically depend on the size of the problem. For long data records, however, this approach is certainly to be more recommended than the use of the traditional batch EM algorithm and is preferable to alternatives considered in the literature.

The rest of this chapter is organised as follows. The next section is devoted to the modelling assumptions that are adopted throughout the chapter. In Section 2.3, we consider the large-sample behaviour of the traditional EM algorithm, insisting on the concept of the limiting EM recursion, which is instrumental in the design of the online algorithm. The various aspects of the online EM algorithm are then examined in Sections 2.4 and 2.5.

Although the chapter is intended to be self-contained, we will nonetheless assume that the reader is familiar with the fundamental concepts of classical statistical inference and, in particular, with Fisher information, exponential families of distributions and maximum likelihood estimation, at the level of Lehmann and Casella (1998), Bickel and Doksum (2000), or equivalent texts.

2.2 Model and assumptions

We assume that we are able to observe a sequence of independent identically distributed data $(Y_t)_{t\geq 1}$, with marginal distribution π. An important remark is that we do not necessarily assume that π corresponds to a distribution that is reachable by the statistical model that is used to fit the data. As discussed in Section 2.4.4 below, this distinction is important to analyse the use of the online algorithm for batch maximum-likelihood estimation.

The statistical models that we consider are of the missing-data type, with an unobservable random variable X_t associated with each observation Y_t. The latent variable X_t may be continuous or vector-valued and we will not be restricting ourselves to finite mixture models. Following the terminology introduced by Dempster *et al.* (1977), we will refer to the pair (X_t, Y_t) as the *complete data*. The likelihood function $f_{\theta_\star}$ is thus defined as the marginal

$$f_\theta(y_t) = \int p_\theta(x_t, y_t) \mathrm{d}x_t \,,$$

where $\theta \in \Theta$ is the parameter of interest to be estimated. If the actual marginal distribution of the data belongs to the family of the model distributions, i.e. if $\pi = f_{\theta_\star}$ for some parameter value $\theta_\star$, the model is said to be *well-specified*; however, as mentioned above, we do not restrict ourselves to this case. In the following, the statistical model $(p_\theta)_{\theta\in\Theta}$ is assumed to satisfy the following key requirements.

Assumption 1: Modelling assumptions

(i) The model belongs to a (curved) exponential family

$$p_\theta(x_t, y_t) = h(x_t, y_t) \exp\left[\langle s(x_t, y_t), \psi(\theta)\rangle - A(\theta)\right] , \qquad (2.1)$$

where $s(x_t, y_t)$ is a vector of complete-data sufficient statistics belonging to a convex set $\mathcal{S}$, $\langle\cdot, \cdot\rangle$ denotes the dot product and A is the log-partition function.

(ii) The complete-data maximum-likelihood estimation is explicit, in the sense that the function $\bar{\theta}$ defined by

$$\bar{\theta} : \mathcal{S} \to \Theta,$$

$$S \mapsto \bar{\theta}(S) = \arg\max_{\theta \in \Theta} \langle S, \psi(\theta)\rangle - A(\theta)$$

is available in closed form.

Assumption 1 defines the context where the EM algorithm may be used directly (see, in particular, the discussion of Dempster *et al.*, 1977). Note that (2.1) is not restricted to the specific case of exponential family distributions in canonical (or natural) parameterisation. The latter correspond to the situation where ψ is the identity function, which is special in that p_θ is then log-concave with the complete-data Fisher information matrix $I_p(\theta)$ being given by the Hessian $\nabla_\theta^2 A(\theta)$ of the log-partition function. Of course, if ψ is an invertible function, one could use the reparameterisation $\eta = \psi(\theta)$ to recover the natural parameterisation but it is important to recognise that for many models of interest, ψ is a function that maps low-dimensional parameters to higher-dimensional statistics. To illustrate this situation, we will use the following simple running example.

Example 1: Probabilistic PCA Model

Consider the *probabilistic principal component analysis* (*PCA*) model of Tipping and Bishop (1999). The model postulates that a d-dimensional observation vector Y_t can be represented as

$$Y_t = uX_t + \lambda^{1/2} N_t , \qquad (2.2)$$

where N_t is a centred unit-covariance d-dimensional multivariate Gaussian vector, while the latent variable X_t is also such a vector but of much lower dimension. Hence, u is a $d \times r$ matrix with $r \ll d$.

Equation (2.2) is thus fully equivalent to assuming that Y_t is a centred d-dimensional Gaussian variable with a structured covariance matrix given by $\Sigma(\theta) = uu' + \lambda I_d$, where the prime denotes transposition and I_d is the d-dimensional identity matrix. Clearly, there are in this model many ways to estimate u and λ that do not rely on the probabilistic model of (2.2), the standard PCA being probably the most well-known. Tipping and Bishop (1999) discuss several reasons for using the probabilistic approach that include the use of priors on the parameters, the ability to deal with missing or censored coordinates of the observations but also the access to quantitative diagnostics based on the likelihood that are helpful, in particular, for determining the number of relevant factors.

To cast the model of (2.2) in the form given by (2.1), the complete-data model

$$\begin{pmatrix} X_t \\ \cdots \\ Y_t \end{pmatrix} \sim \mathcal{N}\left(\begin{pmatrix} 0 \\ \cdots \\ 0 \end{pmatrix}, \begin{pmatrix} I_r \vdots & u' \\ \cdots\cdots & \cdots\cdots \\ u \vdots & uu' + \lambda I_d \end{pmatrix} \right)$$

must be reparameterised by the precision matrix $\Sigma^{-1}(\theta)$, yielding

$$p_\theta(x_t, y_t) = (2\pi)^{-d/2} \exp\left(\operatorname{trace}\left[\Sigma^{-1}(\theta) s(x_t, y_t)\right] - \frac{1}{2} \log|\Sigma(\theta)| \right) ,$$

where $s(x_t, y_t)$ is the rank one matrix

$$s(x_t, y_t) = -\frac{1}{2} \begin{pmatrix} X_t \\ \cdots \\ Y_t \end{pmatrix} \begin{pmatrix} X_t \\ \cdots \\ Y_t \end{pmatrix}' .$$

Hence, in the probabilistic PCA model, the ψ function in this case maps the pair (u, λ) to the $(d + 1)$-dimensional symmetric positive definite matrix $\Sigma^{-1}(\theta)$. Assumption 1 (ii) holds in this case and the EM algorithm can be used; see the general formulae given in Tipping and Bishop (1999) as well as the details of the particular case where $r = 1$ below.

In the following, we will more specifically look at the particular case where $r = 1$ and u is thus a d-dimensional vector. This very simple case also provides a nice and concise illustration of more complex situations, such as the *direction of arrival (DOA)* model considered by Cappé *et al.* (2006). The case of a single-factor PCA is also interesting as most of the required calculations can be done explicitly. In particular, it is possible to provide the following expression for $\Sigma^{-1}(\theta)$ using the block matrix inversion and Sherman–Morrison formulae:

$$\Sigma^{-1}(\theta) = \begin{pmatrix} 1 + \|u\|^2/\lambda \vdots & -u'/\lambda \\ \cdots\cdots & \cdots\cdots \\ -u/\lambda \vdots & \lambda^{-1} I_d \end{pmatrix} .$$

The above expression shows that one may redefine the sufficient statistics $s(x_t, y_t)$ as consisting solely of the scalars $s^{(0)}(y_t) = \|y_t\|^2$ and $s^{(2)}(x_t) = x_t^2$ and of the

d-dimensional vector $s^{(1)}(x_t, y_t) = x_t y_t$. The complete-data log-likelihood is given by

$$\log p_\theta(x_t, y_t) = C^{st} - \frac{1}{2}\left[nd \log \lambda + \frac{s^{(0)}(y_t) - 2u' s^{(1)}(x_t, y_t) + s^{(2)}(x_t)\|u\|^2}{\lambda}\right], \tag{2.3}$$

ignoring constant terms that do not depend on the parameters u and λ.

Note that developing an online algorithm for (2.2) in this particular case is equivalent to recursively estimating the largest eigenvalue of a covariance matrix from a series of multivariate Gaussian observations. We return to this example shortly below.

2.3 The EM algorithm and the limiting EM recursion

In this section, we first review core elements regarding the EM algorithm that will be needed in the following. Next, we introduce the key concept of the limiting EM recursion, which corresponds to the limiting deterministic iterative algorithm obtained when the number of observations grows to infinity. This limiting EM recursion is important both for understanding the behaviour of classic batch EM when used with many observations and for motivating the form of the online EM algorithm.

2.3.1 The batch EM algorithm

In the light of Assumption 1 and in particular of the assumed form of the likelihood in (2.1), the classic EM algorithm of Dempster *et al.* (1977) takes the following form.

Given n observations, $Y_1, \ldots, Y_n$ and an initial parameter guess θ_0, do, for $k \geq 1$,

E-step

$$S_{n,k} = \frac{1}{n}\sum_{t=1}^{n} \mathrm{E}_{\theta_{k-1}}\left[s(X_t, Y_t) \middle| Y_t\right], \tag{2.4}$$

M-step

$$\theta_k = \bar{\theta}\left(S_{n,k}\right). \tag{2.5}$$

For the single-factor PCA model of Example 1, it is easy to check from the expression of the complete-data log-likelihood in (2.3) and the definition of the sufficient statistics that the E-step reduces to the computation of

$$\begin{aligned} s^{(0)}(Y_t) &= \|Y_t\|^2 , \\ \mathrm{E}_\theta \left[s^{(1)}(X_t, Y_t) \middle| Y_t \right] &= \frac{Y_t Y_t' u}{\lambda + \|u\|^2} , \\ \mathrm{E}_\theta \left[s^{(2)}(X_t) \middle| Y_t \right] &= \frac{\lambda}{\lambda + \|u\|^2} + \frac{(Y_t' u)^2}{(\lambda + \|u\|^2)^2} , \end{aligned} \tag{2.6}$$

with the corresponding empirical averages

$$\begin{aligned} S_{n,k}^{(0)} &= \frac{1}{n} \sum_{t=1}^{n} s^{(0)}(Y_t) , \\ S_{n,k}^{(1)} &= \frac{1}{n} \sum_{t=1}^{n} \mathrm{E}_{\theta_{k-1}} \left[s^{(1)}(X_t, Y_t) \middle| Y_t \right] , \\ S_{n,k}^{(2)} &= \frac{1}{n} \sum_{t=1}^{n} \mathrm{E}_{\theta_{k-1}} \left[s^{(2)}(X_t) \middle| Y_t \right] . \end{aligned}$$

The M-step equations that define the function $\bar{\theta}$ are given by

$$\begin{aligned} u_k &= \bar{\theta}^{(u)}(S_{n,k}) = S_{n,k}^{(1)} / S_{n,k}^{(2)} , \\ \lambda_k &= \bar{\theta}^{(\lambda)}(S_{n,k}) = \frac{1}{d} \left(S_{n,k}^{(0)} - \|S_{n,k}^{(1)}\|^2 / S_{n,k}^{(2)} \right) . \end{aligned} \tag{2.7}$$

2.3.2 The limiting EM recursion

If we return to the general case, a very important remark is that the algorithm in Section 2.3.1 can be fully reparameterised in the domain of sufficient statistics, reducing to the recursion

$$S_{n,k} = \frac{1}{n} \sum_{t=1}^{n} \mathrm{E}_{\bar{\theta}(S_{n,k-1})} \left[s(X_t, Y_t) \middle| Y_t \right]$$

with the convention that $S_{n,0}$ is such that $\theta_0 = \bar{\theta}(S_{n,0})$. Clearly, if a uniform (with respect to $S \in \mathcal{S}$) law of large numbers holds for the empirical averages of $\mathrm{E}_{\bar{\theta}(S)} \left[s(X_t, Y_t) \middle| Y_t \right]$, the EM update tends, as the number of observations n tends to infinity, to the deterministic mapping $T_{\mathcal{S}}$ on $\mathcal{S}$ defined by

$$T_{\mathcal{S}}(S) = \mathrm{E}_\pi \left(\mathrm{E}_{\bar{\theta}(S)} \left[s(X_1, Y_1) \middle| Y_1 \right] \right) . \tag{2.8}$$

Hence, the sequence of EM iterates $(S_{n,k})_{k \geq 1}$ converges to the sequence $(T_{\mathcal{S}}^k(S_0))_{k \geq 1}$, which is deterministic except for the choice of S_0. We refer to the limiting mapping $T_{\mathcal{S}}$ defined in (2.8) as the limiting EM recursion. Of course, this mapping on $\mathcal{S}$ also induces a mapping T_Θ on Θ by considering the values of θ associated with values of S by the function $\bar{\theta}$. This second mapping is defined as

$$T_\Theta(\theta) = \bar{\theta} \left(\mathrm{E}_\pi \left\{ \mathrm{E}_\theta \left[s(X_1, Y_1) \middle| Y_1 \right] \right\} \right) . \tag{2.9}$$

By exactly the same arguments as those of Dempster *et al.* (1977) for the classic EM algorithm, it is straightforward to check that, under suitable regularity assumptions, T_Θ is such that the following hold:

1. The Kullback–Leibler divergence $D(\pi|f_\theta) = \int \log[\pi(x)/f_\theta(x)]\pi(x)\mathrm{d}x$ is a Lyapunov function for the mapping T_Θ; that is

$$D\left(\pi|f_{T_\Theta(\theta)}\right) \leq D\left(\pi|f_\theta\right) .$$

2. The set of fixed points of T_Θ, i.e. such that $T_\Theta(\theta) = \theta$, is given by

$$\{\theta : \nabla_\theta D(\pi|f_\theta) = 0\} ,$$

where ∇_θ denotes the gradient.

Obviously, (2.9) is not directly exploitable in a statistical context as it involves integrating under the unknown distribution of the observations. This limiting EM recursion can, however, be used in the context of adaptive Monte Carlo methods (Cappé *et al.*, 2008) and is known in machine learning as part of the information bottleneck framework (Slonim and Weiss, 2003).

2.3.3 Limitations of batch EM for long data records

The main interest in (2.9) is to provide a clear understanding of the behaviour of the EM algorithm when used with very long data records, justifying much of the intuition of Neal and Hinton (1999). Note that a large part of the post 1980s literature on the EM algorithm focuses on accelerating convergence towards the MLE for a fixed data size. Here, we consider the related but very different issue of understanding the behaviour of the EM algorithm when the data size increases.

Figure 2.1 displays the results obtained with the batch EM algorithm for the single component PCA model estimated from data simulated under the model with u being a 20-dimensional vector of unit norm and $\lambda = 5$. Both u and λ are treated as unknown parameters but only the estimated squared norm of u is displayed on Figure 2.1, as a function of the number of EM iterations and for two different data sizes. Note that in this very simple model, with the observation likelihood being given the $\mathcal{N}(0, uu' + \lambda I_d)$ density, it is straightforward to check that the Fisher information for the parameter $\|u\|^2$ equals $[2(\lambda + \|u\|^2)^2]^{-1}$, which has been used to represent the asymptotic confidence interval in grey. The width of this interval is meant to be directly comparable to that of the whiskers in the box-and-whisker plots. Boxplots are used to summarise the results obtained from one thousand independent runs of the method.

Comparing the top and bottom plots clearly shows that, when the number of iterations increases, the trajectories of the EM algorithm converge to a fixed deterministic trajectory, which is that of the limiting EM recursion $(T_\Theta^k(\theta_0))_{k\geq 1}$. Of course, this trajectory depends on the choice of the initial point θ_0, which was fixed

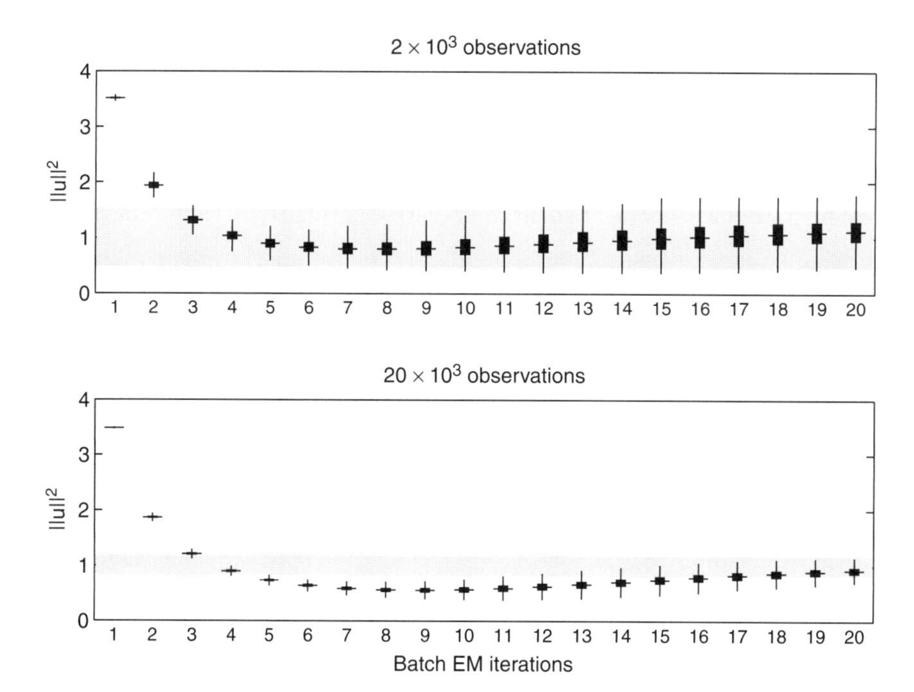

Figure 2.1 Convergence of batch EM estimates of $\|u\|^2$ as a function of the number of EM iterations for 2000 (top) and 20 000 (bottom) observations. The box-and-whisker plots (outlier-plotting suppressed) are computed from 1000 independent replications of the simulated data. The grey region corresponds to ± 2 interquartile ranges (approx. 99.3 % coverage) under the asymptotic Gaussian approximation of the MLE.

throughout this experiment. It is also observed that the monotone increase in the likelihood guaranteed by the EM algorithm does not necessarily imply a monotone convergence of all parameters towards the MLE. Hence, if the number of EM iterations is kept fixed, the estimates returned by the batch EM algorithm with the larger number of observations (bottom plot) are not significantly improved despite the tenfold increase in computation time; the E step computations must be done for all observations and hence the complexity of the E-step scales proportionally to the number n of observations. From a statistical perspective, the situation is even less satisfying as estimation errors that were not statistically significant for 2000 observations can become significant when the number of observations increases. In the upper plot, interrupting the EM algorithm after three or four iterations does produce estimates that are statistically acceptable (comparable to the exact MLE) but, in the lower plot, about 20 iterations are needed to achieve the desired precision. As suggested by Neal and Hinton (1999), this paradoxical situation can to some extent be avoided by updating the parameter (that is applying the M-step) more often, without waiting for a complete scan of the data record.

2.4 Online expectation maximisation

2.4.1 The algorithm

The limiting EM argument developed in the following section shows that, when the number of observations tends to infinity, the EM algorithm is trying to locate the fixed points of the mapping T_S defined in (2.8), that is the roots of the equation

$$\mathrm{E}_{\pi}\left(\mathrm{E}_{\bar{\theta}(S)}\left[s(X_1, Y_1) | Y_1\right]\right) - S = 0 . \tag{2.10}$$

Although we cannot compute the required expectation with respect to the unknown distribution π of the data, each new observation provides us with an unbiased noisy observation of this quantity through $\mathrm{E}_{\bar{\theta}(S)}\left[s(X_n, Y_n) | Y_n\right]$. Solving (2.10) is thus an instance of the most basic case where the stochastic approximation (or Robbins–Monro) method can be used. The literature on stochastic approximation is huge but we recommend the textbooks by Benveniste *et al.* (1990) and Kushner and Yin (2003) for more details and examples as well as the review paper by Lai (2003) for a historical perspective. The standard stochastic approximation approach in approximating the solution of (2.10) is to compute

$$S_n = S_{n-1} + \gamma_n \left(\mathrm{E}_{\bar{\theta}(S_{n-1})}\left[s(X_n, Y_n) | Y_n\right] - S_{n-1}\right) , \tag{2.11}$$

for $n \geq 1$, S_0 being arbitrary and $(\gamma_n)_{n \geq 1}$ denoting a sequence of positive step-sizes that decrease to zero. This equation, rewritten in an equivalent form below, is the main ingredient of the online EM algorithm.

Given S_0, θ_0 and a sequence of step-sizes $(\gamma_n)_{n \geq 1}$, do, for $n \geq 1$,

Stochastic E-step

$$S_n = (1 - \gamma_n) S_{n-1} + \gamma_n \mathrm{E}_{\theta_{n-1}}\left[s(X_n, Y_n) | Y_n\right] , \tag{2.12}$$

M-step

$$\theta_n = \bar{\theta}(S_n) . \tag{2.13}$$

Rewriting (2.10) in the form displayed in (2.12) is very enlightening as it shows that the new statistic S_n is obtained as a convex combination of the previous statistic S_{n-1} and of an update that depends on the new observation Y_n. In particular it shows that the step-sizes γ_n have a natural scale as their highest admissible value is equal to one. This means that one can safely take $\gamma_1 = 1$ and that only the rate at which the step-size decreases needs to be selected carefully (see below). It is also observed that if γ_1 is set to one, the initial value of S_0 is never used and it suffices to select the initial parameter guess θ_0; this is the approach used in the following simulations.

The only adjustment to this algorithm that is necessary in practice is to omit the M-step of (2.13) for the first observations. It typically takes a few observations before the complete-data maximum-likelihood solution is well-defined and the parameter update should be inhibited during this early phase of training (Cappé and Moulines, 2009). In the simulations presented below in Sections 2.4.2 and 2.4.4, the M-step was omitted for the first five observations only, but in most complex scenarios a longer initial parameter freezing phase may be necessary.

When $\gamma_1 < 1$, it is tempting to interpret the value of S_0 as being associated with a prior distribution on the parameters. Indeed, the choice of a conjugate prior for θ in the exponential family defined by (2.1) does result in a complete-data maximum a posteriori (MAP) estimate of $\bar{\theta}$ given by $\bar{\theta}(S_{n,k} + S_0/n)$ (instead of $\bar{\theta}(S_{n,k})$ for the MLE), where S_0 is the hyper-parameter of the prior (Robert, 2001). However, it is easily checked that the influence of S_0 in S_n decreases as $\prod_{k=1}^{n}(1 - \gamma_k)$, which for the suitable step-size decrease schemes (see the beginning of Section 2.4.2 below) decreases faster than $1/n$. Hence, the value of S_0 has a rather limited impact on the convergence of the online EM algorithm. To achieve MAP estimation (assuming a conjugate prior on θ) it is thus recommended to replace (2.13) by $\theta_n = \bar{\theta}(S_n + S_0/n)$, where S_0 is the hyperparameter of the prior on θ.

A last remark of importance is that this algorithm can most naturally be interpreted as a stochastic approximation recursion on the sufficient statistics rather than on the parameters. There does not exist a similar algorithm that operates directly on the parameters because unbiased approximations of the right-hand side of (2.9) based on the observations Y_t are not easily available. As we will see below, the algorithm is asymptotically equivalent to a gradient recursion on θ_n, which involves an additional matrix weighting that is not necessary in (2.12).

2.4.2 Convergence properties

Under the assumption that the step-size sequence satisfies $\sum_n \gamma_n = \infty, \sum_n \gamma_n^2 < \infty$ and other regularity hypotheses that are omitted here (see Cappé and Moulines, 2009, for details), the following properties characterise the asymptotic behaviour of the online EM algorithm.

(i) The estimates θ_n converge to the set of roots of the equation $\nabla_\theta D(\pi | f_\theta) = 0$.

(ii) The algorithm is asymptotically equivalent to a gradient algorithm

$$\theta_n = \theta_{n-1} + \gamma_n J^{-1}(\theta_{n-1}) \nabla_\theta \log f_{\theta_{n-1}}(Y_n) , \tag{2.14}$$

where $J(\theta) = -\mathrm{E}_\pi \left\{ \mathrm{E}_\theta \left[\nabla_\theta^2 \log p_\theta(X_1, Y_1) \middle| Y_1 \right] \right\}$.

(iii) For a well-specified model (i.e. if $\pi = f_{\theta_\star}$) and under Polyak–Ruppert averaging, θ_n is Fisher efficient: sequences that do converge to $\theta_\star$ are such that $\sqrt{n}(\theta_n - \theta_\star)$ converges in distribution to a centred multivariate Gaussian variable with covariance matrix $I_\pi(\theta_\star)$, where $I_\pi(\theta) = -\mathrm{E}_\pi[\nabla_\theta^2 \log f_\theta(Y_1)]$ is the Fisher information matrix corresponding to the observed data.

Polyak–Ruppert averaging refers to a postprocessing step, which simply consists of replacing the estimated parameter values θ_n produced by the algorithm by their average

$$\tilde{\theta}_n = \frac{1}{n - n_0} \sum_{t=n_0+1}^{n} \theta_n ,$$

where n_0 is a positive index at which averaging is started (Polyak and Juditsky, 1992; Ruppert, 1988). Regarding the statements (i) and (ii) above, it is important to understand that the limiting estimating equation $\nabla_\theta D(\pi | f_\theta) = 0$ may have multiple solutions, even in well-specified models. In practice, the most important factor that influences the convergence to one of the stationary points of the Kullback–Leibler divergence $D(\pi | f_\theta)$ rather than the other is the choice of the initial value θ_0. An additional important remark about (i) to (iii) is the fact that the asymptotic equivalent gradient algorithm in (2.14) is not a practical algorithm as the matrix $J(\theta)$ depends on π and hence cannot be computed. Note also that $J(\theta)$ is (in general) equal neither to the complete-data information matrix $I_p(\theta)$ nor to the actual Fisher information in the observed model $I_\pi(\theta)$. The form of $J(\theta)$ as well as its role to approximate the convergence behaviour of the EM algorithm follows the idea of Lange (1995).

From our experience, it is generally sufficient to consider step-size sequences of the form $\gamma_n = 1/n^\alpha$, where the useful range of values for α is from 0.6 to 0.9. The most robust setting is obtained when taking α close to 0.5 and using Polyak–Ruppert averaging. However, the latter requires us to chose an index n_0 that is sufficiently large and, hence, some idea of the convergence time is necessary. To illustrate these observations, Figure 2.2 displays the results of online EM estimation for the single component PCA model, exactly in the same conditions as those considered previously for batch EM estimation in Section 2.3.3.

From a computational point of view, the main difference between the online EM algorithm and the batch EM algorithm of Section 2.3.1 is that the online algorithm performs the M-step update in (2.7) after each observation, according to (2.13), while the batch EM algorithm only applies the M-step update after a complete scan of all available observations. However, both algorithms require the computation of the E-step statistics following (2.7) for each observation. In batch EM, these local E-step computations are accumulated, following (2.4), while the online algorithm recursively averages these according to (2.12). Hence, as the computational complexity of the E- and M-steps are, in this case, comparable, the computational complexity of the online estimation is equivalent to that of one or two batch EM iterations. With this in mind, it is obvious that the results of Figure 2.2 compare very favourably to those of Figure 2.1 with an estimation performance that is compatible with the statistical uncertainty for observation lengths of 2000 and larger (see the last two boxplots on the right in each display).

Regarding the choice of the step-size decrease exponent α, it is observed that, while the choice of $\alpha = 0.6$ (middle plot) does result in more variability than the choice of $\alpha = 0.9$ (top plot), especially for smaller observation sizes, the long-run performance is somewhat better with the former. In particular, when Polyak–Ruppert averaging is used (bottom plot), the performance for the longest data sequence (20 000 observations) is clearly compatible with the claim that online

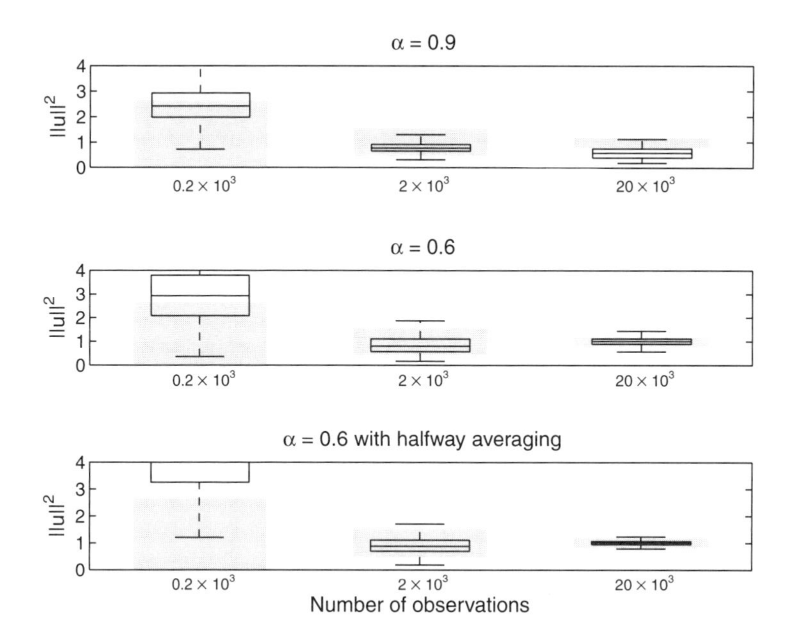

Figure 2.2 Online EM estimates of $\|u\|^2$ for various data sequences (200, 2000 and 20 000 observations, from left to right) and algorithm settings ($\alpha = 0.9$, $\alpha = 0.6$ and $\alpha = 0.6$ with Polyak–Ruppert averaging, from top to bottom). The box-and-whisker plots (outliers plotting suppressed) are computed from 1000 independent replications of the simulated data. The grey regions correspond to ± 2 interquartile range (approx. 99.3 % coverage) under the asymptotic Gaussian approximation of the MLE.

EM is Fisher-efficient in this case. A practical concern associated with averaging is the choice of the initial instant n_0 where averaging starts. In the case of Figure 2.2, we choose n_0 to be equal to half the length of each data record, and hence averaging is used only on the second half of the data. While it produces refined estimates for the longer data sequences, one can observe that the performance is rather degraded for the smallest observation size (200 observations) because the algorithm is still very far from having converged after just 100 observations. Hence, averaging is efficient but does require a value of n_0 to be chosen that is sufficiently large so as to avoid introducing a bias due to the lack of convergence.

In our experience, the fact that choices of α close to 0.5 are more reliable than values closer to the upper limit of 1 is a very constant observation. It may come as some surprise for readers familiar with the gradient descent algorithm used in numerical optimisation, which shares some similarity with (2.12). In this case, it is known that the optimal step-size choice is of the form $\gamma_n = a(n + b)^{-1}$ for a broad class of functions (Nesterov, 2003). However, the situation here is very different as we do not observe exact gradients or expectations but only noisy versions of them. We note that, for a complete-data exponential family model in natural parameterisation, it is easily checked that $\mathrm{E}_\theta\left[s(X_n, Y_n) | Y_n\right] = \mathrm{E}_\theta[s(X_n, Y_n)] + \nabla_\theta \log f_\theta(Y_n)$

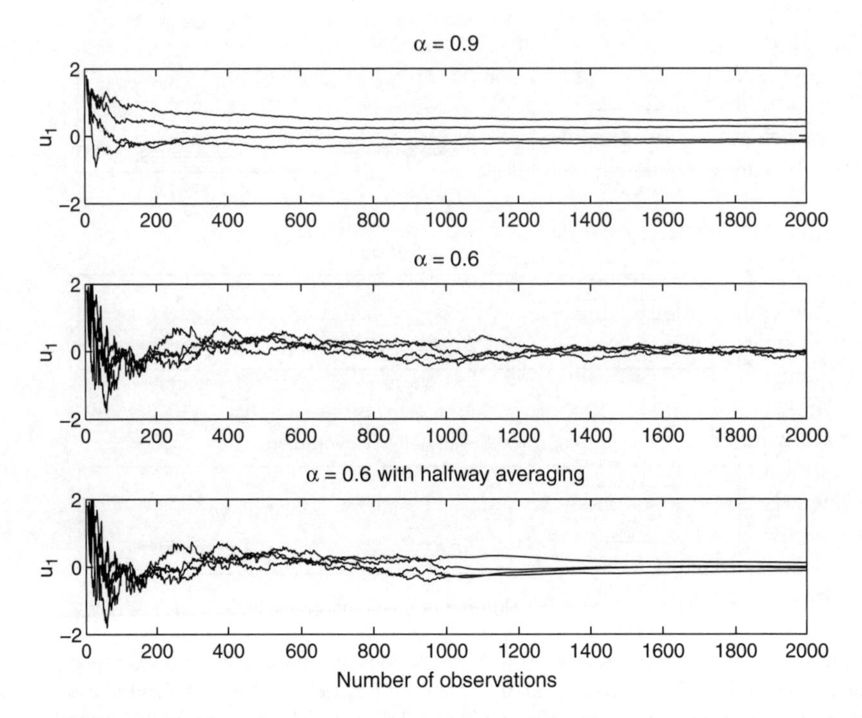

Figure 2.3 Four superimposed trajectories of the estimate of u_1 (first component of u) for various algorithm settings ($\alpha = 0.9$, $\alpha = 0.6$ and $\alpha = 0.6$ with Polyak–Ruppert averaging, from top to bottom). The actual value of u_1 is equal to zero.

and hence that the recursion in the space of sufficient statistics is indeed very close to a gradient ascent algorithm. However, we only have access to $\nabla_\theta \log f_\theta(Y_n)$, which is a noisy version of the gradient of the actual limiting objective function $D(\pi | f_\theta)$ that is minimised. Figure 2.3 shows that, while the trajectory of the parameter estimates appears to be much rougher and variable with $\alpha = 0.6$ than with $\alpha = 0.9$, the bias caused by the initialisation is also forgotten much more rapidly. It is also observed that the use of averaging (bottom display) makes it possible to achieve the best of both worlds (rapid forgetting of the initial condition and smooth trajectories).

Liang and Klein (2009) considered the performance of the online EM algorithm for large-scale natural language processing applications. This domain of application is characterised by the use of very large-dimensional models, most often related to the multinomial distribution, involving tens of thousands of different words and tens to hundreds of semantic tags. As a consequence, each observation, be it a sentence or a whole document, is poorly informative about the model parameters (typically a given text contains only a very limited portion of the whole available vocabulary). In this context, Liang and Klein (2009) found that the algorithm was highly competitive with other approaches, but only when combined with mini-batch blocking; rather than applying the algorithm in Section 2.4.1 at the observation scale, the algorithm is used on mini-batches consisting of m consecutive observations $(Y_{m(k-1)+1}, Y_{mk+2} \dots, Y_{mk})_{k\geq 1}$. For the models and data considered by Liang and Klein (2009), values of m of up to a few thousands yielded optimal performance. More generally, mini-batch blocking can be useful in dealing with mixture-like models with rarely active components.

2.4.3 Application to finite mixtures

Although we have considered so far only the simple case of Example 1, which allows for the computation of the Fisher information matrix and hence for quantitative assessment of the asymptotic performance, the online EM algorithm is easy to implement in models involving finite mixtures of distributions.

Figure 2.4 displays the Bayesian network representation corresponding to a mixture model: for each observation Y_t there is an unobservable mixture indicator or allocation variable X_t that takes its value in the set $\{1, \dots, m\}$. A typical parameterisation for this model is to have separate sets of parameters ω and β for, respectively, the parameters of the prior on X_t and of the conditional distribution of Y_t given X_t. Usually, $\omega = (\omega^{(1)}, \dots, \omega^{(m)})$ is chosen to be the collection of component proportions, $\omega^{(i)} = \mathrm{P}_\theta(X_t = i)$, and hence ω is constrained to the probability simplex ($\omega^{(i)} \geq 0$ and $\sum_{i=1}^{m} \omega^{(i)} = 1$). The observation probability density function is parameterised as

$$f_\theta(y_t) = \sum_{i=1}^{m} \omega^{(i)} g_{\beta^{(i)}}(y_t) ,$$

where $(g_\lambda)_{\lambda\in\Lambda}$ is a parametric family of probability densities and $\beta = (\beta^{(1)}, \dots, \beta^{(m)})$ are the component-specific parameters. We assume that

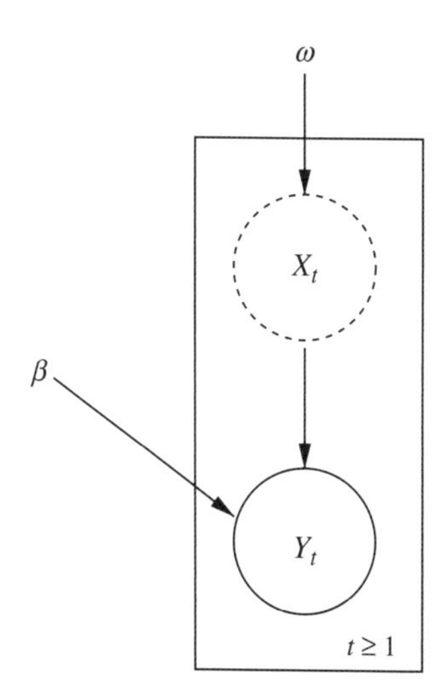

Figure 2.4 Bayesian network representation of a mixture model.

$g_\lambda(y_t)$ has an exponential family representation similar to that of (2.1) with sufficient statistic $s(y_t)$ and maximum-likelihood function $\bar{\lambda} : \mathcal{S} \mapsto \Lambda$, which is such that $\bar{\lambda}[\frac{1}{n}\sum_{t=1}^{n} s(Y_t)] = \arg\max_{\lambda \in \Lambda} \frac{1}{n}\sum_{t=1}^{n} \log g_\lambda(Y_t)$. It is then easily checked that the complete-data likelihood p_θ belongs to an exponential family with sufficient statistics

$$
\begin{aligned}
s^{(\omega,i)}(X_t) &= \mathbb{1}\{X_t = i\}\,,\\
s^{(\beta,i)}(X_t, Y_t) &= \mathbb{1}\{X_t = i\} s(Y_t)\,, \quad \text{for } i = 1, \ldots, m.
\end{aligned}
$$

The function $\bar{\theta}(S)$ can then be decomposed as

$$
\begin{aligned}
\omega^{(i)} &= S^{(\omega,i)}\,,\\
\beta^{(i)} &= \bar{\lambda}\left(S^{(\lambda,i)}/S^{(\omega,i)}\right)\,, \quad \text{for } i = 1, \ldots, m.
\end{aligned}
$$

Hence the online EM algorithm takes the following specific form.

Given S_0, θ_0 and a sequence of step-sizes $(\gamma_n)_{n\geq 1}$, do, for $n \geq 1$,

Stochastic E-step

$$
P_{\theta_{n-1}}(X_t = i|Y_t) = \frac{\omega_{n-1}^{(i)}\, g_{\beta_{n-1}^{(i)}}(Y_t)}{\sum_{j=1}^{m} \omega_{n-1}^{(j)}\, g_{\beta_{n-1}^{(j)}}(Y_t)}\,,
$$

and (2.15)

$$
\begin{aligned}
S_n^{(\omega,i)} &= (1-\gamma_n) S_{n-1}^{(\omega,i)} + \gamma_n P_{\theta_{n-1}}(X_t = i|Y_t)\,,\\
S_n^{(\beta,i)} &= (1-\gamma_n) S_{n-1}^{(\beta,i)} + \gamma_n s(Y_t) P_{\theta_{n-1}}(X_t = i|Y_t)\,,
\end{aligned}
$$

for $i = 1, \ldots, m$.

M-step

$$
\begin{aligned}
\omega_n^{(i)} &= S_n^{(\omega,i)}\,,\\
\beta_n^{(i)} &= \bar{\lambda}\left(S_n^{(\lambda,i)}/S_n^{(\omega,i)}\right)
\end{aligned}
\tag{2.16}
$$

Example 2: Online EM for mixtures of Poisson distributions

We consider a simplistic instance of the algorithm corresponding to the mixture of Poisson distributions (see also Section 2.4 of Cappé and Moulines, 2009). In the case of the Poisson distribution, $g_\lambda(y_t) = (1/y_t!)\,\mathrm{e}^{-\lambda}\lambda^{y_t}$, the sufficient statistic reduces to $s(y_t) = y_t$ and the MLE function $\bar{\lambda}$ is the identity $\bar{\lambda}(S) = S$.

Hence, the online EM recursion for this case simply consists of the following versions of (2.15) and (2.16):

$$S_n^{(\omega,i)} = (1-\gamma_n)S_{n-1}^{(\omega,i)} + \gamma_n P_{\theta_{n-1}}(X_t = i|Y_t)\,,$$
$$S_n^{(\beta,i)} = (1-\gamma_n)S_{n-1}^{(\beta,i)} + \gamma_n Y_t P_{\theta_{n-1}}(X_t = i|Y_t)\,.$$

and

$$\omega_n^{(i)} = S_n^{(\omega,i)}\,,$$
$$\beta_n^{(i)} = S_n^{(\lambda,i)}/S_n^{(\omega,i)}\,.$$

2.4.4 Use for batch maximum-likelihood estimation

An interesting issue that deserves some more comments is the use of the online EM algorithm for batch estimation from a fixed data record, $Y_1, \ldots, Y_N$. In this case, the objective is to save on computational effort compared to the use of the batch EM algorithm.

The analysis of the convergence behaviour of online EM in this context is made easy by the following observation: properties (i) and (ii) stated at the beginning of Section 2.4.2 do not rely on the assumption that the model is well-specified (i.e. that $\pi = f_{\theta_\star}$) and can thus be applied with π being the empirical distribution $\hat{\pi}_N(\mathrm{d}y) = (1/N)\sum_{t=1}^N \delta_{Y_t}(\mathrm{d}y)$ associated with the observed sample, where $\delta_u(dy)$ denotes the Dirac measure localised in u. Hence, if the online EM algorithm is applied by randomly drawing (with replacement) subsequent 'pseudo-observations' from the finite set $\{Y_1, \ldots, Y_N\}$, it converges to points such that

$$\nabla_\theta D(\hat{\pi}_N|f_\theta) = -\frac{1}{N}\nabla_\theta \sum_{t=1}^{N} \log f_\theta(Y_t) = 0\,,$$

that is stationary points of the log-likelihood of the observations $Y_1, \ldots, Y_N$. Property (ii) also provides an asymptotic equivalent of the online EM update, where the index n in (2.14) should be understood as the number of online EM steps rather than the number of actual observations, which is here fixed at N.

In practice, it does not appear that drawing the pseudo-observations randomly makes any real difference when the observations $Y_1, \ldots, Y_N$ are themselves already independent, except for very short data records. Hence, it is more convenient to scan the observations systematically in tours of length N in which each observation is visited in a predetermined order. At the end of each tour, $k = n/N$ is equal to the number of batch tours completed since the start of the online EM algorithm.

To compare the numerical efficiency of this approach with that of the usual batch EM it is interesting to bring together two results. For online EM, based on (2.14), it is possible to show that $\sqrt{n}(\theta_n - \theta_\infty)$ converges in distribution to a multivariate Gaussian distribution, where θ_∞ denotes the limit of θ_n. (Strictly speaking, this has only been shown for random batch scans.) In contrast, the batch EM algorithm achieves so-called linear convergence, which means that, for sufficiently large ks, there exists $\rho < 1$ such that $\|\theta_k - \theta_\infty\| \leq \rho^k$, where θ_k denotes the parameter estimated after k batch EM iterations (Dempster *et al.*, 1977; Lange, 1995). In terms of computing effort, the number k of batch EM iterations is comparable to the number $k = n/N$ of batch tours in the online EM algorithm. Hence the previous theoretical results suggest the following conclusions.

- If the number of available observations N is small, batch EM can be much faster than the online EM algorithm, especially if one wants to obtain a very accurate numerical approximation of the MLE. From a statistical viewpoint, this may be unnecessary as the MLE itself is only a proxy for the actual parameter value, with an error that is of order $1/\sqrt{N}$ in regular statistical models.
- When N increases, the online EM algorithm becomes preferable and, indeed, arbitrarily so if N is sufficiently large. Recall in particular from Section 2.4.2 that when N increases, the online EM estimate obtained after a single batch tour is asymptotically equivalent to the MLE whereas the estimate obtained after a single batch EM iteration converges to the deterministic limit $T_\Theta(\theta_0)$.

In their pioneering work on this topic, Neal and Hinton (1999) suggested an algorithm called incremental EM as an alternative to the batch EM algorithm. The incremental EM algorithm turns out to be exactly equivalent to the algorithm in Section 2.4.1 used with $\gamma_n = 1/n$ up to the end of the batch tour only. After this initial batch scan, the incremental EM proceeds somewhat differently by replacing one by one the previously computed values of $\mathrm{E}_{\theta_{t-1}}[s(X_t, Y_t)|\, Y_t]$ with $\mathrm{E}_{\theta_{t-1+(k-1)N}}[s(X_t, Y_t)|\, Y_t]$ when processing the observation at position t for the kth time. This incremental EM algorithm is indeed more efficient than batch EM, although they do not necessarily have the same complexity, a point that will be further discussed in the next section. For large values of N, however, incremental EM becomes impractical (due to the use of storage space that increases proportionally to N) and is less to be recommended than the online EM algorithm, as shown by the following example; see also the experiments of Liang and Klein (2009) for similar conclusions.

Figure 2.5 displays the normalised log-likelihood values corresponding to the three estimation algorithms (batch EM, incremental EM and online EM) used to estimate the parameters of a mixture of two Poisson distributions (see the example in Section 2.4.1 for implementation details. (Obviously, the fact that only log-likelihoods normalised by the length of the data are plotted hides some important aspects of the problem, in particular the lack of identifiability caused by the unknown labelling of the mixture components.) All data are simulated from a mixture model with parameters $\omega^{(1)} = 0.8$, $\beta^{(1)} = 1$ and $\beta^{(2)} = 3$. In this setting, where the sample size is fixed, it is more difficult to come up with a faithful illustration of the merits of each approach as the convergence behaviour of the algorithms depends very much on the data record and on the initialisation. In Figures 2.5 and 2.6, the two datasets were kept fixed throughout the simulations but the initialisation of each Poisson mean was randomly chosen from the interval [0.5,5]. This randomisation avoids focusing on particular algorithm trajectories and gives a good idea of the general

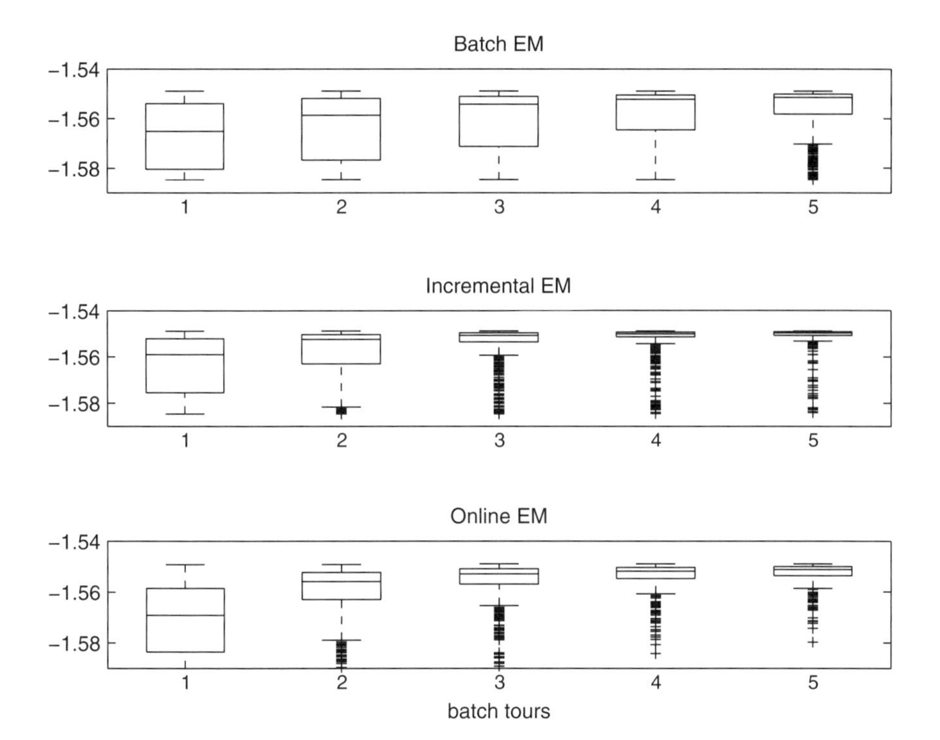

Figure 2.5 Normalised log-likelihood of the estimates obtained with, from top to bottom, batch EM, incremental EM and online EM as a function of the number of batch tours (or iterations, for batch EM). The data sequence is of length $N = 100$ and the box-and-whiskers plots summarise the results of 500 independent runs of the algorithms started from randomly generated starting points θ_0.

situation, although some variations can still be observed when the observation records are varied.

Figure 2.5 corresponds to the case of an observation sequence of length $N = 100$. In this case it is observed that the performance of incremental EM dominates that of the other two algorithms, while the online EM algorithm only becomes preferable to batch EM after the second batch tour. For the data record of length $N = 1000$ (Figure 2.6), online EM now dominates the other two algorithms, with incremental EM still being preferable to batch EM. This is also the case after the first batch tour, illustrating our claim that the choice of $\gamma_n = n^{-0.6}$ used here for the online EM algorithm is indeed preferable to $\gamma_n = n^{-1}$, which coincides with the update used by the incremental EM algorithm during the first batch tour. Finally, one can observe on Figure 2.6 that even after five EM iterations there are a few starting values for which batch EM or incremental EM becomes stuck in regions of low normalised log-likelihood (around -1.6). These indeed correspond to local maxima of the log-likelihood and some trajectories of batch EM converge to these

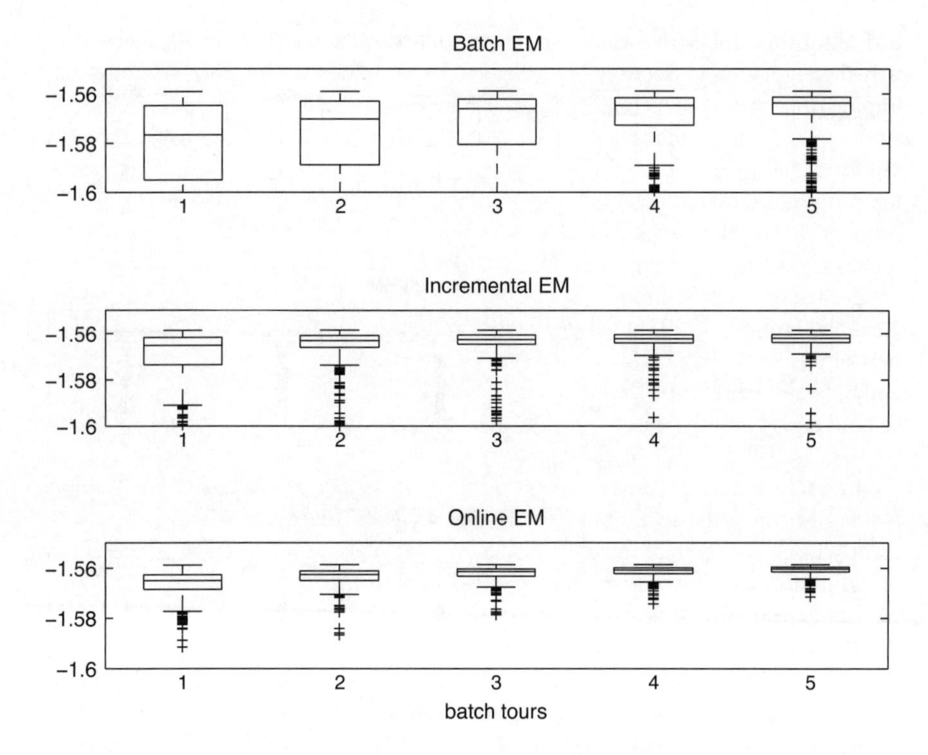

Figure 2.6 The same display as in Figure 2.5 for a data record of length $N = 1000$.

regions, depending on the value of initialisation. Online EM with the above choice of step-size appears to be less affected by this issue, although we have seen that in general only convergence to a stationary point of the log-likelihood is guaranteed.

2.5 Discussion

We conclude this chapter by a discussion of the online EM algorithm.

First, the approach presented here is not the only option for online parameter estimation in latent variable models. One of the earliest and most widely used (see, for example, Liu *et al.*, 2006) algorithm is that of Titterington (1984) consisting of the following gradient update:

$$\theta_n = \theta_{n-1} + \gamma_n I_p^{-1}(\theta_{n-1}) \nabla_\theta \log f_{\theta_{n-1}}(Y_n) , \tag{2.17}$$

where the matrix I_p is the complete-data Fisher information matrix. For complete-data models in natural parameterisation, i.e. with $\psi \equiv 1$ in (2.1), $I_p(\theta)$ coincides with $\nabla_\theta^2 A(\theta)$ and, as it does not depend on X_t or Y_t, is also equal to the matrix $J(\theta)$ that appears in (2.14). Thus, Titterington's algorithm is in this case asymptotically equivalent to online EM. In other cases, I_p and J usually differ, which implies that the recursion of (2.17) converges at the same rate as the online

EM algorithm but with a different asymptotic variance. Another difference is the way the algorithm in Section 2.4.1 deals with parameter constraints. Assumption 1 implies that the M-step update, taking into account the possible parameter constraints, is explicit. Hence, $\theta_n = \bar{\theta}(S_n)$ does satisfy the parameter constraint by definition of the function $\bar{\theta}$. In the case of the example in Section 2.4.1, for instance, the proposed update does guarantee that the mixture weight vector ω stays in the probability simplex. This is not the case with the update of (2.17), which requires reparameterisation or reprojection to handle possible parameter constraints.

As discussed in Section 2.4.4, the online EM algorithm is inspired by the work of Neal and Hinton (1999) but is distinct from their incremental EM approach. To the best of our knowledge, the online EM algorithm was first proposed by Sato (2000) and Sato and Ishii (2000), who described the algorithm and provided some analysis of convergence in the case of exponential families in natural parameterisation and for the case of mixtures of Gaussians.

In Section 2.4.4, we have seen that the online EM algorithm is preferable to batch EM, in terms of computational effort, when the observation size is sufficiently large. To operate in batch mode, online EM needs to be used repeatedly by scanning the data record several times so as to converge towards the maximum-likelihood estimate. It is important to note, however, that one iteration of batch EM operating on N observations requires N individual E-step computations but only one application of the M-step update, whereas online EM applies the M-step update at each observation and, hence, requires N M-step updates per complete batch scan. Thus, the comparison between the two approaches also depends on the respective numerical complexities of the E and M steps. The online approach is more attractive for models in which the M-step update is relatively simple. The availability of parallel computing resources would be most favourable for batch EM for which the E-step computations pertaining to each observation may be done independently. In contrast, in the online approach the computations are necessarily sequential as the parameter is updated when each observation is processed.

As indicated in the beginning of this chapter, the main strength of the online EM algorithm is its simplicity and ease of implementation. As discussed above, this is particularly true in the presence of constraints on the parameter values. We have also seen in Section 2.4.2 that the algorithm of Section 2.4.1 is very robust with respect to the choice of the stepsize γ_n. In particular, the absolute scale of the step-size is fixed due to the use of a convex combination in (2.12) and it is generally sufficient to consider step-sizes of the form $\gamma_n = n^{-\alpha}$. We have shown that values of α of the order of 0.6 (i.e. closer to the lower limit of 0.5 than to the upper limit of 1) yield more robust convergence. In addition, Polyak–Ruppert averaging can be used to smooth the parameter estimates and reach the optimal asymptotic rate of convergence that makes the online estimate equivalent to the actual MLE.

As illustrated by Figure 2.2, the online EM algorithm is not optimal for short data records of, say, less than 100 to 1000 observations and in this case performing batch-mode estimation by repeatedly scanning the data is recommended (see Figure 2.5 for the typical improvement to be expected from this procedure). The main limitation of the online EM algorithm is the requirement that the M-step update $\bar{\theta}$ be available in closed form. In particular, it would be interesting to extend the approach to cases where $\bar{\theta}$ needs to be determined numerically, thus making it possible to handle mixtures of generalised linear models, for instance, and not only mixtures of linear regressions as in Cappé and Moulines (2009).

We finally mention two more directions in which recent papers have proposed to extend the online EM framework. The first one concerns the case of nonindependent observations and, in particular, of observations that follows a hidden Markov model (HMM) with Markov dependencies between successive states. Mongillo and Denève (2008) and Cappé (2009a) have proposed an algorithm for HMMs that appears to be very reminiscent of the algorithm in Section 2.4.1, although not directly interpretable as a stochastic approximation recursion on the expected sufficient statistics. The other topic of importance is motivated by the many cases where the E-step computation of $E_{\theta_{n-1}}[s(X_n, Y_n)| Y_n]$ is not feasible. This typically occurs when the hidden variable X_n is a continuous variable. For such cases, a promising solution consists in approximating the E-step using some form of Monte Carlo simulations; see, for example, Cappé (2009b) and Del Moral *et al.* (2009) for methods that use sequential Monte Carlo approaches. However, much conceptual and theoretical work remains to be done, as the consistency result summarised in Section 2.4.2 only extends straightforwardly to the rather limited case of independent Monte Carlo draws $X_n^{(\theta_{n-1})}$ from the conditionals $p_{\theta_{n-1}}(x_n|Y_n)$. In that case, $s(X_n^{(\theta_{n-1})}, Y_n)$ still provides an unbiased estimate of the limiting EM mapping in θ_{n-1} and the theory is very similar. In the more general case where Markov chain or sequential Monte Carlo simulations are used to produce the draws $X_n^{(\theta_{n-1})}$, the convergence of the online estimation procedure needs to be investigated with care.

References

Benveniste, A., Métivier, M. and Priouret, P. (1990) *Adaptive Algorithms and Stochastic Approximations*, Vol. 22. Springer, Berlin. Translated from the French by Stephen S. S. Wilson.

Bickel, P. J. and Doksum, K. A. (2000) *Mathematical Statistics. Basic Ideas and Selected Topics*, 2nd edn. Prentice-Hall.

Cappé, O. (2009a) Online EM algorithm for hidden Markov models. Preprint.

Cappé, O. (2009b) Online sequential Monte Carlo EM algorithm. In *IEEE Workshop on Statistical Signal Process (SSP)*, Cardiff, Wales, UK.

Cappé, O. and Moulines, E. (2009) On-line expectation-maximization algorithm for latent data models. *Journal of the Royal Statistical Society, Series B*, **71**, 593–613.

Cappé, O., Charbit, M. and Moulines, E. (2006) Recursive EM algorithm with applications to DOA estimation. In *Proceedings of the IEEE International Conference on Acoustics Speech, and Signal Processing*, Toulouse, France.

Cappé, O., Douc, R., Guillin, A., Marin, J. M. and Robert, C. P. (2008) Adaptive importance sampling in general mixture classes. *Statistics and Computing*, **18**, 447–459.

Césa-Bianchi, N. and Lugosi, G. (2006) *Prediction, Learning, and Games*. Cambridge University Press, Cambridge.

Chopin, N. (2002) A sequential particle filter method for static models. *Biometrika*, **89**, 539–552.

Del Moral, P., Doucet, A. and Singh, S. S. (2009) Forward smoothing using sequential Monte Carlo. Technical Report CUED/F-INFENG/TR 638, Cambridge University Engineering Department.

Dempster, A. P., Laird, N. M. and Rubin, D. B. (1977) Maximum likelihood from incomplete data via the EM algorithm. *Journal of the Royal Statistical Society, Series B*, **39**, 1–38 (with discussion).

Kantas, N., Doucet, A., Singh, S. and Maciejowski, J. (2009) An overview of sequential Monte Carlo methods for parameter estimation in general state-space models. In *Proceedings of the IFAC Symposium on System Identification (SYSID)*.

Kushner, H. J. and Yin, G. G. (2003) *Stochastic Approximation and Recursive Algorithms and Applications*, Vol. 35, 2nd edn. Springer, New York.

Lai, T. L. (2003) Stochastic approximation. *Annals of Statistics*, **31**, 391–406.

Lange, K. (1995) A gradient algorithm locally equivalent to the EM algorithm. *Journal of the Royal Statistical Society, Series B*, **57**, 425–437.

Lehmann, E. L. and Casella, G. (1998) *Theory of Point Estimation*, 2nd edn. Springer, New York.

Liang, P. and Klein, D. (2009) Online EM for unsupervised models. In *Conference of the North American Chapter of the Association for Computational Linguistics (NAACL)*, pp. 611–619. Association for Computational Linguistics, Boulder, Colorado.

Liu, Z., Almhana, J., Choulakian, V. and McGorman, R. (2006) Online EM algorithm for mixture with application to internet traffic modeling. *Computational Statistics and Data Analysis*, **50**, 1052–1071.

Ljung, L. and Söderström, T. (1983) *Theory and Practice of Recursive Identification*. MIT Press, Cambridge, Massachusetts.

Meng, X. L. and Van Dyk, D. (1997) The EM algorithm – an old folk song sung to a fast new tune. *Journal of the Royal Statistical Society, Series B*, **59**, 511–567.

Mongillo, G. and Denève, S. (2008) Online learning with hidden Markov models. *Neural Computation*, **20**(7), 1706–1716.

Neal, R. M. and Hinton, G. E. (1999) A view of the EM algorithm that justifies incremental, sparse, and other variants. In *Learning in Graphical Models* (ed. M. I. Jordan) pp. 355–368. MIT Press, Cambridge, Massachusetts.

Nesterov, Y. (2003) *Introductory Lectures on Convex Optimization: A Basic Course*. Kluwer.

Polyak, B. T. and Juditsky, A. B. (1992) Acceleration of stochastic approximation by averaging. *SIAM Journal on Control and Optimization*, **30**, 838–855.

Robert, C. P. (2001) *The Bayesian Choice*, 2nd edn. Springer, New York.

Ruppert, D. (1988) Efficient estimation from a slowly convergent Robbins-Monro process. Technical Report 781, Cornell University, School of Operations Research and Industrial Engineering.

Sato, M. (2000) Convergence of on-line EM algorithm. In *Prooceedings of the International Conference on Neural Information Processing*, vol. 1, pp. 476–481.

Sato, M. and Ishii, S. (2000) On-line EM algorithm for the normalized Gaussian network. *Neural Computation*, **12**, 407–432.

Slonim, N. and Weiss, Y. (2003) Maximum likelihood and the information bottleneck. In *Advances in Neural Information Processing Systems (NIPS)*, Vol. 15, pp. 335–342.

Tipping, M. and Bishop, C. (1999) Probabilistic principal component analysis. *Journal of the Royal Statistical Society, Series B*, **6**, 611–622.

Titterington, D. M. (1984) Recursive parameter estimation using incomplete data. *Journal of the Royal Statistical Society, Series B*, **46**, 257–267.

3

The limiting distribution of the EM test of the order of a finite mixture

Jiahua Chen and Pengfei Li

3.1 Introduction

Let $f(x; \theta)$ be a parametric density function with respect to some σ-finite measure with parameter space Θ and

$$f(x; \Psi) = \sum_{h=1}^{m} \alpha_h f(x; \theta_h) = \int_{\Theta} f(x; \theta) \mathrm{d}\Psi(\theta) \tag{3.1}$$

for some mixing distribution on Θ given by $\Psi(\theta) = \sum_{h=1}^{m} \alpha_h I(\theta_h \leq \theta)$, where $I(\cdot)$ is the indicator function. In this set-up, $(\theta_1, \ldots, \theta_m)$ are the mixing parameters and $(\alpha_1, \ldots, \alpha_m)$ are the mixing proportions. The positive integer m is the order of the finite mixture and its value is the subject of a statistical hypothesis test. For a comprehensive review of finite mixture models, see Titterington *et al.* (1985).

Due to the nonregularity of finite mixture models, statisticians have met serious challenges in developing effective and easy-to-use methods for testing hypotheses on the size of m, based on a random sample from a mixture model. In one of the earliest papers, Hartigan (1985) revealed that the likelihood ratio test (LRT) statistic under a simplified normal mixture model diverges to infinity instead of having a chi-square limiting distribution as for regular models. The ultimate factors behind

---

Mixtures: Estimation and Applications, First Edition. Edited by Kerrie L. Mengersen, Christian P. Robert and D. Michael Titterington.

this phenomenon are the partial nonidentifiability and the unbounded Θ. Most later developments study the likelihood ratio test under the assumption that Θ is naturally or unnaturally bounded. Some representative papers include Chernoff and Lander (1995), Dacunha-Castelle and Gassiat (1999), Chen and Chen (2001) and Liu and Shao (2003). In spite of these advances, obtaining approximate p-values of the likelihood ratio test remains a largely unsolved problem unless we place additional range restrictions on the mixing proportions. The results in Chen and Cheng (1995) and Lemdani and Pons (1995) are of this nature. The modified likelihood ratio test (MLRT) discussed in Chen (1998) and Chen *et al.* (2001, 2004) replaces the hard restriction on the mixing proportions with soft restrictions. This approach results in some easy-to-implement methods under more natural settings.

Most recently, a class of EM tests has been proposed in Chen and Li (2009), Li *et al.* (2009) and Li and Chen (2010). Some of the ideas in EM tests can be traced back to the MLRT, but the EM test has several additional advantages. The seemingly complicated test statistic not only has a convenient limiting distribution for practical implementation, it also makes the theoretical analysis less challenging and the large-sample conclusion applicable to a broader range of finite mixture models. Most noticeably, the asymptotic results for the EM test are derived without requiring Θ to be bounded, and are applicable to finite mixtures of more kernel densities such as the exponential distribution.

Studying the asymptotic behaviour of a new inference method has always been the focus of its development. The EM test has relatively simple asymptotic properties for the homogeneity test, as shown in Chen and Li (2009), Li *et al.* (2009). The EM test for the general order of the finite mixture model has more complex large sample properties. In this paper, we develop a number of innovative techniques to meet the challenges in deriving the limiting distributions of the EM test for the general order. Through the notion of the partial EM iteration and the constrained maximisation of the quadratic form in normally distributed random variables, we show that the limiting distribution is a mixture of chi squares. We present the detailed accounts of mixing proportions and their consistent estimations. The paper is organised as follows. In Section 2, we give a brief but complete description of the EM test. In Section 3, we present detailed and rigorous proofs. We end the paper with a brief discussion section.

3.2 The method and theory of the EM test

Given a set of independent and identically distributed (iid) samples $X_1, \ldots, X_n$ from $f(x; \Psi)$ defined by (3.1), we aim to test

$$H_0 : m = m_0 \text{ versus } H_A : m > m_0$$

for some positive interger m_0 of interest. Denote the log-likelihood function as $\ell_n(\Psi) = \sum_{i=1}^{n} \log\{f(X_i; \Psi)\}$. We first give a brief description of the EM test and then present the asymptotic results. The iid structure will be assumed throughout the paper without further comment.

3.2.1 The definition of the EM test statistic

The EM test statistic is constructed iteratively. Let $\hat{\Psi}_0$ be the maximum-likelihood estimator (MLE) of Ψ under the null hypothesis such that $\hat{\Psi}_0 = \sum_{h=1}^{m_0} \hat{\alpha}_{0h} I(\hat{\theta}_{0h} \le \theta)$ and $\hat{\theta}_{01} \le \hat{\theta}_{02} \le \cdots \le \hat{\theta}_{0m_0}$. We divide the parameter space Θ into m_0 intervals $I_h = (\eta_{h-1}, \eta_h]$ with $\eta_h = (\hat{\theta}_{0h} + \hat{\theta}_{0\,h+1})/2$, $h = 1, 2, \ldots, m_0 - 1$. We interpret η_0 and η_{m_0} as the lower and upper bounds of Θ. For instance, if $\Theta = (-\infty, +\infty)$, we put $\eta_0 = -\infty$ and $\eta_{m_0} = +\infty$.

For each vector $\boldsymbol{\beta} = (\beta_1, \ldots, \beta_{m_0})^\tau$ such that $\beta_h \in (0, 0.5]$, we create a class of mixing distributions of order $2m_0$ with the help of $\hat{\Psi}_0$:

$$\Omega_{2m_0}(\boldsymbol{\beta}) = \Big\{ \sum_{h=1}^{m_0} [\alpha_h \beta_h I(\theta_{1h} \le \theta) + \alpha_h (1 - \beta_h) I(\theta_{2h} \le \theta)] : \theta_{1h}, \theta_{2h} \in I_h \Big\}.$$

For each $\Psi \in \Omega_{2m_0}(\boldsymbol{\beta})$ define the modified log-likelihood function as

$$p\ell_n(\Psi) = \ell_n(\Psi) + \sum_{h=1}^{m_0} p(\beta_h) = \ell_n(\Psi) + p(\boldsymbol{\beta})$$

for some continuous $p(\beta)$ that is maximised at 0.5 and goes to negative infinity as β goes to 0 or 1. When $\boldsymbol{\beta}$ is a vector, $p(\boldsymbol{\beta})$ represents the sum. Without loss of generality, we set $p(0.5) = 0$. Note that the penalty depends on β_h but not on α_h.

We next specify a finite set of numbers from $(0, 0.5]$ and call it B. One example is $\mathsf{B} = \{0.1, 0.3, 0.5\}$. Assume that B contains J elements; then B^{m_0} contains J^{m_0} vectors of $\boldsymbol{\beta}$. For each $\boldsymbol{\beta}_0 \in \mathsf{B}^{m_0}$, we compute

$$\Psi^{(1)}(\boldsymbol{\beta}_0) = \arg\max\{p\ell_n(\Psi) : \Psi \in \Omega_{2m_0}(\boldsymbol{\beta}_0)\}$$

where the maximisation is with respect to $\boldsymbol{\alpha} = (\alpha_1, \ldots, \alpha_{m_0})^\tau, \boldsymbol{\theta}_1 = (\theta_{11}, \ldots, \theta_{1m_0})^\tau$ and $\boldsymbol{\theta}_2 = (\theta_{21}, \ldots, \theta_{2m_0})^\tau$. Note that $\Psi^{(1)}(\boldsymbol{\beta}_0)$ is a member of $\Omega_{2m_0}(\boldsymbol{\beta}_0)$. Let $\boldsymbol{\beta}^{(1)} = \boldsymbol{\beta}_0$. The EM iteration starts from here.

Suppose we have $\Psi^{(k)}(\boldsymbol{\beta}_0)$ already calculated with $\boldsymbol{\alpha}^{(k)}$, $\boldsymbol{\theta}_1^{(k)}$, $\boldsymbol{\theta}_2^{(k)}$ and $\boldsymbol{\beta}^{(k)}$ available. The calculation for $k = 1$ has been illustrated with $\boldsymbol{\alpha}^{(1)}$, $\boldsymbol{\theta}_1^{(1)}$, $\boldsymbol{\theta}_2^{(1)}$, and $\boldsymbol{\beta}^{(1)}$ being constituent entities of $\Psi^{(1)}(\boldsymbol{\beta}_0)$. For each $i = 1, 2, \ldots, n$ and $h = 1, \ldots, m_0$, let

$$w_{i1h}^{(k)} = \frac{\alpha_h^{(k)} \beta_h^{(k)} f\left(X_i; \theta_{1h}^{(k)}\right)}{f(X_i; \Psi^{(k)}(\boldsymbol{\beta}_0))} \quad \text{and} \quad w_{i2h}^{(k)} = \frac{\alpha_h^{(k)} \left(1 - \beta_h^{(k)}\right) f\left(X_i; \theta_{2h}^{(k)}\right)}{f(X_i; \Psi^{(k)}(\boldsymbol{\beta}_0))}.$$

We then proceed to obtain $\Psi^{(k+1)}(\boldsymbol{\beta}_0)$ by letting

$$\alpha_h^{(k+1)} = n^{-1} \sum_{i=1}^{n} \{w_{i1h}^{(k)} + w_{i2h}^{(k)}\},$$

$$\theta_{jh}^{(k+1)} = \arg\max_{\theta} \Big\{ \sum_{i=1}^{n} w_{ijh}^{(k)} \log f(X_i; \theta) \Big\}, \; j = 1, 2,$$

and

$$\beta_h^{(k+1)} = \arg\max_{\beta} \Big\{ \sum_{i=1}^{n} w_{i1h}^{(k)} \log(\beta) + \sum_{i=1}^{n} w_{i2h}^{(k)} \log(1-\beta) + p(\beta) \Big\}.$$

The computation is iterated a prespecified number of times, K.

For each $\boldsymbol{\beta}_0 \in \mathsf{B}^{m_0}$ and k, we define

$$M_n^{(k)}(\boldsymbol{\beta}_0) = 2\{p\ell_n(\Psi^{(k)}(\boldsymbol{\beta}_0)) - \ell_n(\hat{\Psi}_0)\}.$$

The EM test statistic, for a prespecified K, is then defined to be

$$EM_n^{(K)} = \max\{M_n^{(K)}(\boldsymbol{\beta}_0) : \boldsymbol{\beta}_0 \in \mathsf{B}^{m_0}\}.$$

We reject the null hypothesis when $EM_n^{(K)}$ exceeds some critical value to be determined. To determine the critical value or, equivalently, to compute the p-value of the hypothesis test, we need to know the distribution of $EM_n^{(K)}$, at least asymptotically. This task leads us to the next subsection.

3.2.2 The limiting distribution of the EM test statistic

To present the main results, we must first introduce some notation. First, we denote the true mixing distribution Ψ_0 and similarly for its compositions. For $i = 1, \ldots, n$ and $h = 1, \ldots, m_0$, let

$$Y_i(\theta) = \frac{f'(X_i;\theta)}{f(X_i;\Psi_0)}, \quad Z_i(\theta) = \frac{f''(X_i;\theta)}{2f(X_i;\Psi_0)}, \quad \Delta_{ih} = \frac{f(X_i;\theta_{0h}) - f(X_i;\theta_{0m_0})}{f(X_i;\Psi_0)}, \tag{3.2}$$

and

$$\mathbf{b}_{1i} = \Big(\Delta_{i1}, \ldots, \Delta_{i\,m_0-1}, Y_i(\theta_{01}), \ldots, Y_i(\theta_{0m_0})\Big)^{\tau},$$

$$\mathbf{b}_{2i} = \Big(Z_i(\theta_{01}), \ldots, Z_i(\theta_{0m_0})\Big)^{\tau}.$$

For $j, k = 1, 2$, set $\mathbf{B}_{jk} = E\left[\{\mathbf{b}_{ji} - E(\mathbf{b}_{ji})\}\{\mathbf{b}_{ki} - E(\mathbf{b}_{ki})\}^{\tau}\right]$. Furthermore, we orthogonalize $\mathbf{b}_{1i}$ and $\mathbf{b}_{2i}$ by introducing $\tilde{\mathbf{b}}_{2i} = \mathbf{b}_{2i} - \mathbf{B}_{21}\mathbf{B}_{11}^{-1}\mathbf{b}_{1i}$, which has the covariance matrix $\tilde{\mathbf{B}}_{22} = \mathbf{B}_{22} - \mathbf{B}_{21}\mathbf{B}_{11}^{-1}\mathbf{B}_{12}$.

Our asymptotic results are derived under the following regularity conditions on the penalty function $p(\beta)$ and the kernel density function $f(x;\theta)$. All the expectations are under the true distribution specified by $f(x;\Psi_0)$.

C0. The penalty $p(\beta)$ is a continuous function such that it is maximised at $\beta = 0.5$ and goes to negative infinity as β goes to 0 or 1.

C1. The kernel function $f(x;\theta)$ is such that the mixture distribution $f(x;\Psi)$ satisfies Wald's integrability conditions for consistency of the maximum-likelihood estimate. For this, it suffices to require that

(a) $E|\log f(X;\Psi_0)| < \infty$;
(b) for sufficiently small ρ and for sufficiently large r, $E\log[1+f(X;\theta,\rho)] < \infty$ for $\theta \in \Theta$ and $E\log[1+\varphi(X,r)] < \infty$, where $f(x;\theta,\rho) = \sup_{|\theta'-\theta|\le\rho} f(x;\theta')$ and $\varphi(x;r) = \sup_{|\theta|\ge r} f(x;\theta)$;
(c) $\lim_{|\theta|\to\infty} f(x;\theta) = 0$ for all x except on a set with probability zero.

C2. The kernel function $f(x;\theta)$ has common support and is four times continuously differentiable with respect to θ.

C3. For any two mixing distribution functions Ψ_1 and Ψ_2 such that

$$\int f(x;\theta)\mathrm{d}\Psi_1(\theta) = \int f(x;\theta)\mathrm{d}\Psi_2(\theta), \text{ for all } x,$$

we must have $\Psi_1 = \Psi_2$.

C4. Let $N(\theta,\epsilon) = \{\theta' \in \Theta : |\theta' - \theta| \le \epsilon\}$ for some positive ϵ. There exists an integrable $g(\cdot)$ and a small positive ϵ_0 such that

$$|\Delta_{ih}|^3 \le g(X_i),\ |Y_i(\theta)|^3 \le g(X_i),\ |Z_i^{(k)}(\theta)|^3 \le g(X_i),$$

for $\theta \in N(\theta_{0h},\epsilon_0)$, $h = 1,\ldots,m_0$ and $k = 0,1,2$, with $Z_i^{(k)}(\theta)$ being the kth derivative.

C5. The variance–covariance matrix $\mathbf{B}$ of

$$\mathbf{b}_i = (\Delta_{i1},\ldots,\Delta_{i\,m_0-1}, Y_i(\theta_{01}),\ldots,Y_i(\theta_{0m_0}), Z_i(\theta_{01}),\ldots,Z_i(\theta_{0m_0}))^\tau \quad (3.3)$$

with its elements defined in (3.2) is positive definite.

These conditions are comparable to those of Wald (1949). The most commonly used parametric distribution families such as Poisson, binomial, exponential and normal with a location parameter satisfy these conditions. C3 and C5 are identifiability conditions in the context of finite mixture models. Because the inequalities in C4 are required only in a neighbourhood of θ_{0h}, our result is applicable to a finite mixture of exponential distributions or, more generally, a mixture of scale families. This is where the results in Dacunha-Castelle and Gassiat (1999) and Chen *et al.* (2004) are not applicable. The parameter space of kernel distributions such as Poisson and exponential are naturally unbounded. Again, the results in Dacunha-Castelle and Gassiat (1999), Chen *et al.* (2004) and so on are applicable only if the parameter space is bounded. Adding an artificial restriction is not part of the above regularity conditions.

Theorem 1: Under the regularity conditions C0 to C5 and assuming $0.5 \in \mathsf{B}$,

$$EM_n^{(K)} = \sup_{\mathbf{v}\ge 0}\left(2\mathbf{v}^\tau \sum_{i=1}^{n} \tilde{\mathbf{b}}_{2i} - n\mathbf{v}^\tau \tilde{\mathbf{B}}_{22}\mathbf{v}\right) + o_p(1)$$

where the supremum over $\mathbf{v}$ is taken over the range specified by

$$\{\mathbf{v} \ge 0\} = \{\mathbf{v} = (v_1,\ldots,v_{m_0})^\tau : v_j \ge 0,\ j = 1,\ldots,m_0\}$$

under the null hypothesis, as $n \to \infty$.

Furthermore, let $\mathbf{w} = (w_1, \ldots, w_{m_0})^\tau$ be a 0-mean multivariate normal random vector with covariance matrix $\tilde{\mathbf{B}}_{22}$; then we have

$$EM_n^{(K)} \to \sup_{\mathbf{v}\geq 0}(2\mathbf{v}^\tau\mathbf{w} - \mathbf{v}^\tau\tilde{\mathbf{B}}_{22}\mathbf{v}) = \sum_{h=0}^{m_0} a_h\chi_h^2$$

in distribution, for some $a_h \geq 0$ and $\sum_{h=0}^{m_0} a_h = 1$.

Some analytical expressions for a_h are given as follows. Let ω_{ij} be the correlation between w_i and w_j.

1. When $m_0 = 1$, we have $a_0 = a_1 = 0.5$.
2. When $m_0 = 2$, we have $a_0 = (\pi - \arccos\omega_{12})/(2\pi)$, $a_1 = 0.5$ and $a_0 + a_2 = 0.5$.
3. When $m_0 = 3$, we have $a_0 + a_2 = a_1 + a_3 = 0.5$ and

$$\begin{aligned} a_0 &= (2\pi - \arccos\omega_{12} - \arccos\omega_{13} - \arccos\omega_{23})/(4\pi), \\ a_1 &= (3\pi - \arccos\omega_{12:3} - \arccos\omega_{13:2} - \arccos\omega_{23:1})/(4\pi), \end{aligned}$$

where $\omega_{ij:k}$, the conditional correlation between w_i and w_j given w_k, is given by

$$\omega_{ij:k} = \frac{\omega_{ij} - \omega_{ik}\omega_{jk}}{\sqrt{(1-\omega_{ik}^2)(1-\omega_{jk}^2)}}.$$

Exact expressions of a_h for $m_0 \geq 4$ are more difficult to derive. When necessary we can use computer simulation to obtain approximate values of a_h based on an estimated $\tilde{\mathbf{B}}_{22}$.

3.3 Proofs

The proof of our main result involves a few major intermediate conclusions. We present three lemmas together with some explanation of their significance. On the one hand, the first three lemmas are similar, showing how different $\Psi^{(k)}$ is from Ψ_0. On the other hand, the claims of the successive lemmas are stronger. The first merely claims that the difference between $\Psi^{(k)}$ and Ψ_0 is $o_p(1)$ when $k = 1$; the second improves $o_p(1)$ to $O_p(n^{-1/4})$ or so; and the third strengthens the result to be applicable to all finite k.

Recall that we introduced $\Omega_{2m_0}(\boldsymbol{\beta}_0)$ and the modified log-likelihood in the definition of the EM test statistic. With their help, the support points of the mixing distribution $\Psi^{(1)}$ are found to be close to those of $\hat{\Psi}_0$ when the data are from a null model. Because $\hat{\Psi}_0$ is a consistent estimator of Ψ_0, the deliberate way of defining the EM test statistic keeps the support points of $\Psi^{(k)}$ close to those of Ψ_0 when the data are from a null model. As a consequence, the asymptotic property of the EM test is determined by the algebraic structure of $f(x, \theta)$ for θ close to one of the support points of Ψ_0. Because of this, θ values distant from the support of Ψ_0 are not relevant in the limiting process. Thus, whether or not Θ is bounded has no effect on the limiting distribution of the EM test, whereas boundedness is a crucial assumption for other methods.

Lemma 1: Suppose that $f(x;\theta)$ and $p(\beta)$ satisfy conditions C0 to C3. Under the null distribution $f(x;\Psi_0)$, and for each given $\boldsymbol{\beta}_0 \in \mathsf{B}^{m_0}$, we have

$$\boldsymbol{\alpha}^{(1)} - \boldsymbol{\alpha}_0 = o_p(1), \;\; \boldsymbol{\beta}^{(1)} - \boldsymbol{\beta}_0 = o_p(1),$$
$$\boldsymbol{\theta}_1^{(1)} - \boldsymbol{\theta}_0 = o_p(1), \;\; \boldsymbol{\theta}_2^{(1)} - \boldsymbol{\theta}_0 = o_p(1).$$

Proof: By definition, $\Psi^{(1)} \in \Omega_{2m_0}(\boldsymbol{\beta}_0)$ and it maximises $p\ell_n(\Psi)$ in this class. At the same time, for any $\boldsymbol{\beta}_0 \in \mathsf{B}^{m_0}$, we can write

$$\hat{\Psi}_0 = \sum_{h=1}^{m_0} \{\hat{\alpha}_{0h}\beta_{0h} I(\hat{\theta}_{0h} \leq \theta) + \hat{\alpha}_{0h}(1-\beta_{0h}) I(\hat{\theta}_{0h} \leq \theta)\}.$$

Hence, $\hat{\Psi}_0 \in \Omega_{2m_0}(\boldsymbol{\beta}_0)$. This membership claim implies that

$$p\ell_n(\Psi^{(1)}) \geq p\ell_n(\hat{\Psi}_0) \geq \ell_n(\Psi_0) + p(\boldsymbol{\beta}_0). \tag{3.4}$$

The second inequality is true because $\ell_n(\hat{\Psi}_0) > \ell_n(\Psi_0)$ and the penalties are balanced.

The classical result of Wald (1949) on the consistency of the maximum-likelihood estimator can be summarised as follows. Based on a set of independent and identically distributed observations from a distribution family parameterised by Ψ, if $\hat{\Psi}_n$ is an estimator of Ψ such that

$$\ell_n(\hat{\Psi}_n) \geq \ell_n(\Psi_0) - o(n)$$

as $n \to \infty$, then $\hat{\Psi}_n \to \Psi_0$ almost surely. Using this result, and because of (3.4) and the nonpositiveness of the penalty, we must have

$$\int_{\Theta} |\Psi^{(1)}(\boldsymbol{\beta}_0) - \Psi_0| \exp(-|\theta|) d\theta \to 0 \tag{3.5}$$

almost surely. With fixed $\boldsymbol{\beta}_0$ and the confinement of $\Psi^{(1)}$ in $\Omega_{2m_0}(\boldsymbol{\beta}_0)$, the consistency result (3.5) is possible only if $\boldsymbol{\alpha}^{(1)} - \boldsymbol{\alpha}_0 = o_p(1)$ and the other conclusions of this lemma are true. This ends the proof.

If $\beta_{0h} = 0$ for some h, then $\theta_{1h}^{(1)}$ becomes nonidentifiable and the claim about $\boldsymbol{\theta}_1^{(1)}$ in the above proof would be invalid. Confining $\boldsymbol{\beta}_0 \in \mathsf{B}^{m_0}$, which excludes the 0 value, is hence an important condition in the above proof.

In the next lemma, we strengthen Lemma 1 by providing an order assessment. For this purpose, we first provide a technical remark used in the proof. For

$i = 1, 2, \ldots, n$, let

$$U_{ih}(\theta) = \frac{f(X_i;\theta) - f(X_i;\theta_{0h}) - f'(X_i;\theta_{0h})(\theta-\theta_{0h}) - f''(X_i;\theta_{0h})(\theta-\theta_{0h})^2/2}{f(X_i;\Psi_0)(\theta-\theta_{0h})^3}.$$

Condition C4 ensures that for $h = 1, \ldots, m_0$, the processes

$$n^{-1/2}\sum_{i=1}^{n} U_{ih}(\theta)$$

are tight for θ in a small neighbourhood of θ_{0h}. See Billingsley (1968) or Chen *et al.* (2004) for the notion of tightness. In this paper, we use this fact to claim that

$$n^{-1/2}\sum_{i=1}^{n} U_{ih}(\tilde{\theta}) = O_p(1)$$

when $\tilde{\theta}$ is consistent for θ_{0h}; this will be used in the next proof.

We now define some local moments before stating the next lemma:

$$m_{1h}^{(k)} = \beta_h^{(k)}(\theta_{1h}^{(k)} - \theta_{0h}) + (1-\beta_h^{(k)})(\theta_{2h}^{(k)} - \theta_{0h});$$
$$m_{2h}^{(k)} = \beta_h^{(k)}(\theta_{1h}^{(k)} - \theta_{0h})^2 + (1-\beta_h^{(k)})(\theta_{2h}^{(k)} - \theta_{0h})^2;$$

$$\mathbf{t}^{(k)} = \left(\alpha_1^{(k)} - \alpha_{01}, \ldots, \alpha_{m_0-1}^{(k)} - \alpha_{0\,m_0-1}; \alpha_1^{(k)}m_{11}^{(k)}, \ldots, \alpha_{m_0}^{(k)}m_{1m_0}^{(k)}; \alpha_1^{(k)}m_{21}^{(k)}, \ldots, \alpha_{m_0}^{(k)}m_{2m_0}^{(k)}\right)^{\tau}. \tag{3.6}$$

Roughly, $\mathbf{t}^{(k)}$ is a vector representing the difference between $\Psi^{(k)}$ and Ψ_0.

Lemma 2: Suppose that $f(x;\theta)$ and $p(\beta)$ satisfy conditions C0 to C5. Under the null distribution $f(x;\Psi_0)$, and for each given $\boldsymbol{\beta}_0 \in \mathsf{B}^{m_0}$, we have

$$\boldsymbol{\alpha}^{(1)} - \boldsymbol{\alpha}_0 = O_p(n^{-1/2}),\ \ \boldsymbol{\beta}^{(1)} - \boldsymbol{\beta}_0 = O_p(n^{-1/6}),$$
$$\boldsymbol{\theta}_1^{(1)} - \boldsymbol{\theta}_0 = O_p(n^{-1/4}),\ \ \boldsymbol{\theta}_2^{(1)} - \boldsymbol{\theta}_0 = O_p(n^{-1/4})$$

and

$$\boldsymbol{m}_1^{(1)} = O_p(n^{-1/2}).$$

PROOF: Because the details of the proof are overly technical, to avoid anxiety we first give a simplified argument. Let $R_{1n}(\Psi^{(1)}(\boldsymbol{\beta}_0)) = 2\{pl_n[\Psi^{(1)}(\boldsymbol{\beta}_0)] - l_n(\Psi_0)\}$. It has an apparent lower bound

$$R_{1n}[\Psi^{(1)}(\boldsymbol{\beta}_0)] = 2\{\ell_n[\Psi^{(1)}(\boldsymbol{\beta}_0)] - \ell_n(\Psi_0)\} + 2p(\boldsymbol{\beta}_0) \geq 2p(\boldsymbol{\beta}_0).$$

A large chunk of the proof will be used to establish an upper bound:

$$R_{1n}[\Psi^{(1)}(\boldsymbol{\beta}_0)] \leq 2(\mathbf{t}^{(1)})^{\tau}\sum_{i=1}^{n}\mathbf{b}_i - n(\mathbf{t}^{(1)})^{\tau}\mathbf{B}(\mathbf{t}^{(1)})\left[1 + o_p(1)\right] + o_p(1) \tag{3.7}$$

with $\mathbf{b}_i$ defined in C5.

Because of the iid structure in $\mathbf{b}_i$, and since $E\{\mathbf{b}_i\} = 0$ and $\text{Var}(\mathbf{b}_i) = \mathbf{B}$, the upper bound of $R_{1n}[\Psi^{(1)}(\boldsymbol{\beta}_0)]$ itself is bounded by

$$n^{-1}\left\{\sum_{i=1}^{n}\mathbf{b}_i\right\}^{\tau}\mathbf{B}\left\{\sum_{i=1}^{n}\mathbf{b}_i\right\}[1+o_p(1)] = O_p(1).$$

It is seen that only when $\mathbf{t}^{(1)} = O_p(n^{-1/2})$ are the lower and upper bounds possible simultaneously. Because $\boldsymbol{\beta}^{(1)} = \boldsymbol{\beta}_0 \neq 0$ in our setting, the claims of the lemma are then simple interpretations of $\mathbf{t}^{(1)} = O_p(n^{-1/2})$.

With the above simplified argument, we conclude that the real task left in order to prove this lemma is to establish the upper bound (3.7). Here is how we proceed.

Since the penalty is negative,

$$R_{1n}[\Psi^{(1)}(\boldsymbol{\beta}_0)] \leq 2\{\ell_n[\Psi^{(1)}(\boldsymbol{\beta}_0)] - \ell_n(\Psi_0)\} = 2\sum_{i=1}^{n}\log(1+\delta_i),$$

with $\delta_i = \{f(X_i;\Psi^{(1)}(\boldsymbol{\beta}_0)) - f(X_i;\Psi_0)\}/f(X_i;\Psi_0)$. Using the inequality $\log(1+x) \leq x - x^2/2 + x^3/3$, we get

$$2\sum_{i=1}^{n}\log(1+\delta_i) \leq 2\sum_{i=1}^{n}\delta_i - \sum_{i=1}^{n}\delta_i^2 + 2/3\sum_{i=1}^{n}\delta_i^3. \tag{3.8}$$

We now expand every term in (3.8) in terms of $\mathbf{t}^{(1)}$. Note that

$$\begin{aligned} f(X_i;\Psi^{(1)}(\boldsymbol{\beta}_0)) - f(X_i;\Psi_0) &= \sum_{h=1}^{m_0}(\alpha_h^{(1)} - \alpha_{0h})f(X_i;\theta_{0h}) \\ &\quad + \sum_{h=1}^{m_0}\alpha_h^{(1)}\left\{\beta_h^{(1)}\left[f(X_i;\theta_{1h}^{(1)}) - f(X_i;\theta_{0h})\right]\right. \\ &\quad \left. + (1-\beta_h^{(1)})\left[f(X_i;\theta_{2h}^{(1)}) - f(X_i;\theta_{0h})\right]\right\}. \end{aligned} \tag{3.9}$$

Because $\sum_{h=1}^{m_0}\alpha_h^{(1)} = \sum_{h=1}^{m_0}\alpha_{0h} = 1$, the first term in (3.9) can be expressed as

$$\begin{aligned} \sum_{h=1}^{m_0}(\alpha_h^{(1)} - \alpha_{0h})f(X_i;\theta_{0h}) &= \sum_{h=1}^{m_0}(\alpha_h^{(1)} - \alpha_{0h})[f(X_i;\theta_{0h}) - f(X_i;\Psi_0)] \\ &= \sum_{h=1}^{m_0-1}(\alpha_h^{(1)} - \alpha_{0h})\Delta_{ih}f(X_i;\Psi_0). \end{aligned}$$

By Lemma 1, $\Psi^{(1)}$ and Ψ_0 have nearly the same support, which supports the following expansion with a well-behaved remainder term ϵ_{ih}:

$$f(X_i;\theta_{1h}^{(1)}) - f(X_i;\theta_{0h}) = (\theta_{1h}^{(1)} - \theta_{0h})f'(X_i;\theta_{0h}) + \frac{1}{2}(\theta_{1h}^{(1)} - \theta_{0h})^2 f''(X_i;\theta_{0h}) + \epsilon_{ih}$$
$$= \left[(\theta_{1h}^{(1)} - \theta_{0h})Y_i(\theta_{0h}) + (\theta_{1h}^{(1)} - \theta_{0h})^2 Z_i(\theta_{0h})\right] f(X_i;\Psi_0) + \epsilon_{ih}.$$

Combining the two expansions for (3.9) leads to

$$\delta_i = \sum_{h=1}^{m_0-1}(\alpha_h^{(1)} - \alpha_{0h})\Delta_{ih} + \sum_{h=1}^{m_0}\left[\alpha_h^{(1)} m_{1h}^{(1)} Y_i(\theta_{0h}) + \alpha_h^{(1)} m_{2h}^{(1)} Z_i(\theta_{0h})\right] + \sum_{h=1}^{m_0}\epsilon_{ih}$$
$$= (\mathbf{t}^{(1)})^\tau \mathbf{b}_i + \sum_{h=1}^{m_0}\epsilon_{ih} \tag{3.10}$$

and letting $\epsilon_n = \sum_{i=1}^n \sum_{h=1}^{m_0} \epsilon_{ih}$, we get

$$\sum_{i=1}^n \delta_i = (\mathbf{t}^{(1)})^\tau \sum_{i=1}^n \mathbf{b}_i + \epsilon_n. \tag{3.11}$$

The remainder term ϵ_n has the following order assessment:

$$\left|\sum_{i=1}^n \epsilon_{ih}\right| = n^{1/2}\alpha_h^{(1)}\left|\beta_h^{(1)}(\theta_{1h}^{(1)} - \theta_{0h})^3\left[n^{-1/2}\sum_{i=1}^n U_{ih}(\theta_{1h}^{(1)})\right]\right.$$
$$\left. + (1-\beta_h^{(1)})(\theta_{2h}^{(1)} - \theta_{0h})^3\left[n^{-1/2}\sum_{i=1}^n U_{ih}(\theta_{2h}^{(1)})\right]\right|$$
$$= O_p(n^{1/2})\left[\beta_h^{(1)}|\theta_{1h}^{(1)} - \theta_{0h}|^3 + (1-\beta_h^{(1)})|\theta_{2h}^{(1)} - \theta_{0h}|^3\right]$$
$$= o_p\left(n^{1/2} m_{2h}^{(1)}\right) = o_p\left(n^{1/2}\|\mathbf{t}^{(1)}\|\right), \tag{3.12}$$

where $\|\mathbf{t}\|^2 = \mathbf{t}^\tau\mathbf{t}$. Note that the second-last equality holds because $\theta_{jh}^{(1)} - \theta_{0h} = o_p(1)$ according to Lemma 1. Thus, (3.11) gives us a key intermediate order assessment or expansion,

$$\sum_{i=1}^n \delta_i = (\mathbf{t}^{(1)})^\tau \sum_{i=1}^n \mathbf{b}_i + o_p(n^{1/2}\|\mathbf{t}^{(1)}\|). \tag{3.13}$$

This assessment allows us to directly expand the first term in (3.8).

We now focus on expanding the second term in (3.8). It can be seen that

$$\sum_{i=1}^n \delta_i^2 = (\mathbf{t}^{(1)})^\tau \sum_{i=1}^n (\mathbf{b}_i \mathbf{b}_i^\tau)(\mathbf{t}^{(1)}) + 2(\mathbf{t}^{(1)})^\tau \sum_{h=1}^{m_0}\sum_{i=1}^n \epsilon_{ih}\mathbf{b}_i + \sum_{i=1}^n\left(\sum_{h=1}^{m_0}\epsilon_{ih}\right)^2. \tag{3.14}$$

By the law of large numbers,

$$(\mathbf{t}^{(1)})^\tau \sum_{i=1}^{n} (\mathbf{b}_i \mathbf{b}_i^\tau)(\mathbf{t}^{(1)}) = n(\mathbf{t}^{(1)})^\tau \mathbf{B}(\mathbf{t}^{(1)})[1 + o_p(1)]. \tag{3.15}$$

Similar to the derivation for (3.12), we have

$$\begin{aligned}\sum_{i=1}^{n} \epsilon_{ih}^2 &\le \left[\beta_h^{(1)} |\theta_{1h}^{(1)} - \theta_{0h}|^3 + \left(1 - \beta_h^{(1)}\right) |\theta_{2h}^{(1)} - \theta_{0h}|^3\right] \sum_{i=1}^{n} \left[U_{ih}^2(\theta_{1h}^{(1)}) + U_{ih}^2(\theta_{2h}^{(1)})\right] \\ &= o_p(n\|\mathbf{t}^{(1)}\|^2) \end{aligned} \tag{3.16}$$

because condition C4 implies $\sum_{i=1}^{n} U_{ih}^2(\theta_{1h}^{(1)}) = O_p(n)$ and so on. By the Cauchy inequality, the second term in (3.14) is bounded by the geometric mean of (3.15) and (3.16) which also has order $o_p(n\|\mathbf{t}^{(1)}\|^2)$. Thus, we get the second intermediate result

$$\sum_{i=1}^{n} \delta_i^2 = (\mathbf{t}^{(1)})^\tau \mathbf{B}(\mathbf{t}^{(1)})\{1 + o_p(1)\}. \tag{3.17}$$

After going over the above derivation, it becomes easy to see that

$$\sum_{i=1}^{n} \delta_i^3 = o_p(n\|(\mathbf{t}^{(1)})\|^2). \tag{3.18}$$

Substituting (3.13), (3.17) and (3.18) into (3.8) and then connecting it to $R_{1n}[\Psi^{(1)}(\boldsymbol{\beta}_0)]$, we get the upper bound (3.7). This completes the proof.

Lemma 3: Suppose that $f(x;\theta)$ and $p(\beta)$ satisfy conditions C0 to C5. If for some $k \ge 1$, under the null distribution $f(x;\Psi_0)$ and for each given $\boldsymbol{\beta}_0 \in \mathsf{B}^{m_0}$,

$$\begin{aligned}\boldsymbol{\alpha}^{(k)} - \boldsymbol{\alpha}_0 = O_p(n^{-1/2}),\ \ \boldsymbol{\beta}^{(k)} - \boldsymbol{\beta}_0 = O_p(n^{-1/6}),\\ \boldsymbol{\theta}_1^{(k)} - \boldsymbol{\theta}_0 = O_p(n^{-1/4}),\ \ \boldsymbol{\theta}_2^{(k)} - \boldsymbol{\theta}_0 = O_p(n^{-1/4}),\end{aligned}$$

and $\boldsymbol{m}_1^{(k)} = O_p(n^{-1/2})$, then we have $\boldsymbol{m}_1^{(k+1)} = O_p(n^{-1/2})$ and

$$\begin{aligned}\boldsymbol{\alpha}^{(k+1)} - \boldsymbol{\alpha}_0 = O_p(n^{-1/2}),\ \ \boldsymbol{\beta}^{(k+1)} - \boldsymbol{\beta}_0 = O_p(n^{-1/6}),\\ \boldsymbol{\theta}_1^{(k+1)} - \boldsymbol{\theta}_0 = O_p(n^{-1/4}),\ \ \boldsymbol{\theta}_2^{(k+1)} - \boldsymbol{\theta}_0 = O_p(n^{-1/4}).\end{aligned}$$

PROOF: We divide the proof into two major steps. The first step shows that $\beta_h^{(k+1)} - \beta_{0h} = O_p(n^{-1/6})$. This step is a preparatory step. It enables us to employ the same techniques used for proving Lemmas 1 and 2. In the second step, we apply the same ideas used in the proofs of Lemmas 1 and 2 except for omitting some repetitive details.

Step 1: We show that $\beta_h^{(k+1)} - \beta_{0h} = O_p(n^{-1/6})$ by pre-assuming

$$\sum_{i=1}^{n} \frac{f(X_i; \theta_{1h}^{(k)})}{f(X_i; \Psi^{(k)}(\boldsymbol{\beta}_0))} = n[1 + O_p(n^{-1/6})] \tag{3.19}$$

and defer its proof slightly. Let

$$H_{nh}(\beta) = \sum_{i=1}^{n} w_{i1h}^{(k)} \log(\beta) + \sum_{i=1}^{n} w_{i2h}^{(k)} \log(1 - \beta),$$

where the $w_{ijh}^{(k)}$ have been defined in EM iterations and they satisfy, if (3.19) is true,

$$\sum_{i=1}^{n} w_{i1h}^{(k)} = \alpha_h^{(k)} \beta_h^{(k)} \sum_{i=1}^{n} \frac{f(X_i; \theta_{1h}^{(k)})}{f(X_i; \Psi^{(k)}(\boldsymbol{\beta}_0))} = n\alpha_h^{(k)} \beta_h^{(k)} [1 + O_p(n^{-1/6})],$$

$$\sum_{i=1}^{n} w_{i2h}^{(k)} = n\alpha_n^{(k)} - \sum_{i=1}^{n} w_{i1h}^{(k)} = n\alpha_h^{(k)}(1 - \beta_h^{(k)})[1 + O_p(n^{-1/6})].$$

Note that $H_{nh}(\beta)$ is maximised at $\hat{\beta}_h = (n\alpha_n^{(k)})^{-1} \sum_{i=1}^{n} w_{i1h}^{(k)} = \beta_h^{(k)}[1 + O_p(n^{-1/6})]$ which is in the $O_p(n^{-1/6})$ neighbourhood of β_{0h} by assumption. The quantity of concern, $\beta_h^{(k+1)}$, is made by EM iteration to maximise a slightly different function

$$Q_{nh}(\beta) = H_{nh}(\beta) + p(\beta).$$

So $\beta_h^{(k+1)}$ should be nearly the same as $\hat{\beta}_h$.

Because $\hat{\beta}_h$ maximises the smooth function H_{nh}, we must have, when β^* is a value such that $|\beta^* - \hat{\beta}_h|$ is small enough,

$$H_{nh}(\hat{\beta}_h) - H_{nh}(\beta^*) \geq \epsilon \alpha_h^{(k)} n (\beta^* - \hat{\beta}_h)^2$$

for some $\epsilon > 0$. In particular, when $|\beta^* - \hat{\beta}_h| \geq n^{-1/6}$ but is still small enough,

$$H_{nh}(\hat{\beta}_h) - H_{nh}(\beta^*) \geq \epsilon \alpha_h^{(k)} n^{2/3}.$$

This implies that for such β^*,

$$Q_{nh}(\beta^*) - Q_{nh}(\hat{\beta}_h) \leq p(\beta^*) - p(\hat{\beta}_h) - \alpha_h^{(k)} \epsilon n^{2/3} < 0$$

when n is large enough. Consequently, the maximum point of Q_{nh}, $\hat{\beta}_h^{(k+1)}$, must be within an $O_p(n^{-1/6})$ neighbourhood of $\hat{\beta}_h$ or of β_{0h}. This is the first claim we prove in this lemma, and we need to return to (3.19).

Using an expansion of δ_i that we obtained in the proof of Lemma 1,

$$\frac{f(X_i;\Psi^{(k)}(\boldsymbol{\beta}_0)) - f(X_i;\Psi_0)}{f(X_i;\Psi_0)} = \sum_{h=1}^{m_0-1}(\alpha_h^{(k)} - \alpha_0)\Delta_{ih} + \sum_{h=1}^{m_0}\alpha_h^{(k)} m_{1h}^{(k)} Y_i(\theta_{0h})$$
$$+\sum_{h=1}^{m_0}\alpha_h^{(k)}\beta_h^{(k)}(\theta_{1h}^{(k)} - \theta_{0h})^2 Z_i(\tilde{\theta}_{1h}) + \sum_{h=1}^{m_0}\alpha_h^{(k)}(1-\beta_h^{(k)})(\theta_{2h}^{(k)} - \theta_{0h})^2 Z_i(\tilde{\theta}_{2h}). \tag{3.20}$$

The first term in (3.20) is $O_p(n^{-1/2})$ because $|\Delta_{ih}| < 1$ and $\alpha_h^{(k)} - \alpha_0 = O_p(n^{-1/2})$ by assumption.

The uniform third moment condition C4 implies that $\max_{1\le i\le n}|Y_i(\theta_{0h})| = O_p(n^{1/3})$. Hence, under the lemma assumption $m_{1h}^{(k)} = O_p(n^{-1/2})$, the second term in (3.20) is of order $O_p(n^{-1/2+1/3}) = O_p(n^{-1/6})$.

The uniform third moment condition C4 also implies that, for $j = 1, 2$, $\max_{1\le i\le n}|Z_i(\tilde{\theta}_{jh}^{(k)})| = O_p(n^{1/3})$. Combined with the assumption $\theta_{jh}^{(k)} - \theta_{0h} = O_p(n^{-1/4})$, we get $(\theta_{jh}^{(k)} - \theta_{0h})^2 = O_p(n^{-1/2})$. Therefore, the third and fourth terms are also $O_p(n^{-1/6})$.

The above assessments imply that

$$\max_i[|f(X_i;\Psi^{(k)}(\boldsymbol{\beta}_0))/f(X_i;\Psi_0) - 1|] = O_p(n^{-1/6}),$$

which implies

$$\max_i[|f(X_i;\Psi_0)/f(X_i;\Psi^{(k)}(\boldsymbol{\beta}_0)) - 1|] = O_p(n^{-1/6}).$$

Similarly, since $\theta_{1h}^{(k)} - \theta_{0h} = O_p(n^{-1/4})$ by assumption, we have

$$\max_i\{|f(X_i;\theta_{1h}^{(k)})/f(X_i;\theta_{0h}) - 1|\} = O_p(n^{-1/6}).$$

Consequently, the fact that $E\{f(X;\theta_{0h})/f(X;\Psi_0)\} = 1$ together with the above order assessments implies that

$$\begin{aligned}\frac{1}{n}\sum_{i=1}^{n}\frac{f(X_i;\theta_{1h}^{(k)})}{f(X_i;\Psi^{(k)}(\boldsymbol{\beta}_0))} &= \frac{1}{n}\sum_{i=1}^{n}\frac{f(X_i;\theta_{0h})}{f(X_i;\Psi_0)}\times\frac{f(X_i;\theta_{1h}^{(k)})}{f(X_i;\theta_{0h})}\times\frac{f(X_i;\Psi_0)}{f(X_i;\Psi^{(k)}(\boldsymbol{\beta}_0))}\\ &= \frac{1}{n}\sum_{i=1}^{n}\frac{f(X_i;\theta_{0h})}{f(X_i;\Psi_0)}[1 + O_p(n^{-1/6})]\\ &= 1 + O_p(n^{-1/6}),\end{aligned}$$

which is (3.19). This completes the first step of the proof.

Step 2: Although the proof in this step is tedious, the idea is rather simple. The EM iteration always increases the value of the likelihood function or that of the modified likelihood (Dempster *et al.*, 1977). Asymptotically, however, the likelihood can attain its maximum only in a small neighbourhood of the true parameter. For regular models, the size of this neighbourhood is $n^{-1/2}$. For mixture models, the size is $n^{-1/2}$ in terms of $\mathbf{t}^{(k+1)}$. Because $\beta_h^{(k+1)}$ is in $n^{-1/6}$ distance from β_{0h} which is not zero, $\mathbf{t}^{(k+1)}$ is of order $n^{-1/2}$ only if $\boldsymbol{\alpha}^{(k+1)}$ and $\boldsymbol{\theta}^{(k+1)}$ are in $n^{-1/2}$ and $n^{-1/4}$ neighbourhoods of $\boldsymbol{\alpha}_0$ and $\boldsymbol{\theta}_0$, respectively. This leads to the result of the lemma.

To provide a rigorous proof, we invent the notion of partial EM iteration. In regular EM iteration, the Q function in the M-step is maximised with respect to all the parameters in the model. In this proof, we examine the effect of updating parameter values one subpopulation at a time. More specifically, we define

$$\Psi_l^{(k+1)}(\theta) = \sum_{1 \le h \le m_0; h \neq l} \alpha_h^{(k)} \left[\beta_h^{(k)} I(\theta_{1h}^{(k)} \le \theta) + (1 - \beta_h^{(k)}) I(\theta_{2h}^{(k)} \le \theta) \right]$$
$$+ \alpha_l^{(k)} \left[\beta_l^{(k)} I(\theta_{1l}^{(k+1)} \le \theta) + (1 - \beta_l^{(k)}) I(\theta_{2l}^{(k+1)} \le \theta) \right].$$

This mixing distribution maximises the Q function with respect to the parameters in the lth subpopulation only. Note that $\Psi_l^{(k+1)}(\theta)$ is identical to $\Psi^{(k)}(\theta)$ except for the lth pair of support points. Since the Q function at $\Psi_l^{(k+1)}(\theta)$ is larger than when it is evaluated at $\Psi^{(k)}(\theta)$, the value of log-likelihood or modified log-likelihood also increases according to Dempster *et al.* (1977). Therefore we have

$$p\ell_n(\Psi_l^{(k+1)}) \ge p\ell_n(\Psi^{(k)}) \ge p\ell_n(\hat{\Psi}_0) = \ell_n(\hat{\Psi}_0) + p(\boldsymbol{\beta}_0) \ge \ell_n(\Psi_0) + p(\boldsymbol{\beta}_0).$$

Because the penalty is always nonpositive, it further implies that

$$\ell_n(\Psi_l^{(k+1)}) \ge \ell_n(\Psi_0) + p(\boldsymbol{\beta}_0).$$

The above result implies the consistency of $\Psi_l^{(k+1)}$ for Ψ_0. Since $\boldsymbol{\alpha}^{(k)}$, $\boldsymbol{\theta}_1^{(k)}$ and $\boldsymbol{\theta}_2^{(k)}$ are consistent as assumed, and $\beta_h^{(k+1)} = \beta_{0h} + O_p(n^{-1/6})$, the consistency of $\Psi_l^{(k+1)}$ is possible only if both $\theta_{1l}^{(k+1)} = \theta_{0l} + o_p(1)$ and $\theta_{2l}^{(k+1)} = \theta_{0l} + o_p(1)$. Because this result is applicable to each $l = 1, 2, \ldots, m_0$, it extends to vector form:

$$\boldsymbol{\theta}_1^{(k+1)} = \boldsymbol{\theta}_0 + o_p(1); \;\; \boldsymbol{\theta}_2^{(k+1)} = \boldsymbol{\theta}_0 + o_p(1). \tag{3.21}$$

Now we apply the same logic to $\Psi^{(k+1)}$, which is obtained by updating the parameters in all the subpopulations from $\Psi^{(k)}$. This logic implies that $\Psi^{(k+1)}$ is a consistent estimator of Ψ_0. Given the consistency of the support point (3.21) that we have just shown, we further conclude that $\boldsymbol{\alpha}^{(k+1)} = \boldsymbol{\alpha}_0 + o_p(1)$. That is, $\Psi^{(k+1)}$ is consistent for Ψ_0 not only as a mixing distribution but also in terms of supporting points and corresponding mixing proportions.

The last task of the proof is to strengthen the consistency result with the particular rates of convergence stated in this lemma. This part of the proof is logically the same as the proof of Lemma 2: an all-round consistent estimator $\Psi^{(k+1)}$ of Ψ_0 maintaining a high likelihood value must have its $\mathbf{t}^{(k+1)} = O_p(n^{-1/2})$. Interpreting this rate result gives the rate assessments of this lemma. This completes the proof.

In the following proof, we need a few standard results from the literature. We will introduce them as lemmas within the proof rather than present them now. This is because these results do not appear relevant unless they are in the right context.

Proof of Theorem 1: Since the null model Ψ_0 is regular when its order m_0 is known, the following classical expansion is applicable (Serfling, 1980):

$$R_{0n} = 2[\ell_n(\hat{\Psi}_0) - \ell_n(\Psi_0)] = \left(\sum_{i=1}^{n} \mathbf{b}_{1i}\right)^{\tau} (n\mathbf{B}_{11})^{-1} \left(\sum_{i=1}^{n} \mathbf{b}_{1i}\right) + o_p(1).$$

At the same time, let

$$R_{1n}[\Psi^{(k)}(\boldsymbol{\beta}_0)] = 2\{p\ell_n[\Psi^{(k)}(\boldsymbol{\beta}_0)] - \ell_n(\Psi_0)\}.$$

The upper bound established in the proof of Lemma 2 on $R_{1n}(\Psi^{(1)})$ is generally applicable. That is,

$$R_{1n}[\Psi^{(k)}(\boldsymbol{\beta}_0)] \leq 2(\mathbf{t}^{(k)})^{\tau} \sum_{i=1}^{n} \mathbf{b}_i - n(\mathbf{t}^{(k)})^{\tau}\mathbf{B}(\mathbf{t}^{(k)})[1 + o_p(1)] + o_p(1).$$

Now we define a generic $\mathbf{t} = (\mathbf{t}_1^{\tau}, \mathbf{t}_2^{\tau})^{\tau}$ so that $\mathbf{t}_1$ and $\mathbf{t}_2$ have lengths $2m_0 - 1$ and m_0. For any finite k,

$$\begin{aligned} R_{1n}[\Psi^{(k)}(\boldsymbol{\beta}_0)] - R_{0n} \leq \sup_{\mathbf{t}} \left(2\mathbf{t}^{\tau} \sum_{i=1}^{n} \mathbf{b}_i - n\mathbf{t}^{\tau}\mathbf{B}\mathbf{t}\right) \\ - \left(\sum_{i=1}^{n} \mathbf{b}_{1i}\right)^{\tau} (n\mathbf{B}_{11})^{-1} \left(\sum_{i=1}^{n} \mathbf{b}_{1i}\right) + o_p(1). \end{aligned}$$

We note that the above upper bound also serves as the upper bound for $EM_n^{(k)}$. Letting $\tilde{\mathbf{t}}_1 = \mathbf{t}_1 + \mathbf{B}_{11}^{-1}\mathbf{B}_{12}\mathbf{t}_2$, we find

$$2\mathbf{t}^{\tau}\left(\sum_{i=1}^{n} \mathbf{b}_i\right) - n\mathbf{t}^{\tau}\mathbf{B}\mathbf{t} = 2\tilde{\mathbf{t}}_1^{\tau}\left(\sum_{i=1}^{n} \mathbf{b}_{1i}\right) - n\tilde{\mathbf{t}}_1^{\tau}\mathbf{B}_{11}\tilde{\mathbf{t}}_1 + 2\mathbf{t}_2^{\tau}\left(\sum_{i=1}^{n} \tilde{\mathbf{b}}_{2i}\right) - n\mathbf{t}_2^{\tau}\tilde{\mathbf{B}}_{22}\mathbf{t}_2.$$

Therefore,

$$
\begin{aligned}
EM_n^{(k)} &\le \sup_{\tilde{\mathbf{t}}_1}\left[2\tilde{\mathbf{t}}_1^\tau\left(\sum_{i=1}^n \mathbf{b}_{1i}\right) - n\tilde{\mathbf{t}}_1^\tau \mathbf{B}_{11}\tilde{\mathbf{t}}_1\right] + \sup_{\mathbf{t}_2}\left[2\mathbf{t}_2^\tau\left(\sum_{i=1}^n \tilde{\mathbf{b}}_{2i}\right) - n\mathbf{t}_2^\tau \tilde{\mathbf{B}}_{22}\mathbf{t}_2\right] \\
&\quad - \left(\sum_{i=1}^n \mathbf{b}_{1i}\right)^\tau (n\mathbf{B}_{11})^{-1}\left(\sum_{i=1}^n \mathbf{b}_{1i}\right) + o_p(1) \\
&= \left(\sum_{i=1}^n \mathbf{b}_{1i}\right)^\tau (n\mathbf{B}_{11})^{-1}\left(\sum_{i=1}^n \mathbf{b}_{1i}\right) + \sup_{\mathbf{t}_2}\left[2\mathbf{t}_2^\tau\left(\sum_{i=1}^n \tilde{\mathbf{b}}_{2i}\right) - n\mathbf{t}_2^\tau \tilde{\mathbf{B}}_{22}\mathbf{t}_2\right] \\
&\quad - \left(\sum_{i=1}^n \mathbf{b}_{1i}\right)^\tau (n\mathbf{B}_{11})^{-1}\left(\sum_{i=1}^n \mathbf{b}_{1i}\right) + o_p(1) \\
&\le \sup_{\mathbf{t}_2}\left[2\mathbf{t}_2^\tau\left(\sum_{i=1}^n \tilde{\mathbf{b}}_{2i}\right) - n\mathbf{t}_2^\tau \tilde{\mathbf{B}}_{22}\mathbf{t}_2\right] + o_p(1). \qquad (3.22)
\end{aligned}
$$

The transformation from $\mathbf{t}_1$ to $\tilde{\mathbf{t}}_1$ does not change the domain of $\mathbf{t}_2$: all of its elements remain nonnegative.

Now we show that $EM_n^{(k)}$ asymptotically attains this upper bound for all $k \le K$. Since the EM iteration increases the modified likelihood, we need only show that this is the case when $k = 1$. To prove this for $k = 1$, we need only find a set of parameter values satisfying the conditions of $\Psi^{(1)}(\boldsymbol{\beta}_0)$ at which the upper bound (3.22) is attained. Recall that $\mathbf{t}^{(1)}$ is defined through parameters in $\Psi^{(1)}(\boldsymbol{\beta}_0)$. With the $2m_0$ support points built in, there are $3m_0 - 1$ free parameters in $\Psi^{(1)}(\boldsymbol{\beta}_0)$ that match the dimension of $\mathbf{t}$. Hence, we can easily find $\alpha_h, \theta_{1h}, \theta_{2h}, h = 1, 2, \ldots, m_0$, such that $\mathbf{t}_1^{(1)} = (n\mathbf{B}_{11})^{-1}\sum_{i=1}^n \mathbf{b}_{1i} - \mathbf{B}_{11}^{-1}\mathbf{B}_{12}\mathbf{t}_2^{(1)}$ and

$$
\mathbf{t}_2^{(1)} = \arg\sup_{\mathbf{t}_2 \ge 0}\left[2\mathbf{t}_2^\tau\left(\sum_{i=1}^n \tilde{\mathbf{b}}_{2i}\right) - n\mathbf{t}_2^\tau \tilde{\mathbf{B}}_{22}\mathbf{t}_2\right].
$$

We do not have an explicit expression for $\mathbf{t}_2^{(1)}$ because of the complication arising from the range restriction $\mathbf{t}_2 \ge 0$. Yet the existence of such a solution is obvious. It is easy to verify that such a choice satisfies

$$
\boldsymbol{\alpha}^{(1)} - \boldsymbol{\alpha}_0 = O_p(n^{-1/2}); \quad \boldsymbol{\theta}_j^{(1)} - \boldsymbol{\theta}_0 = O_p(n^{-1/4})
$$

for $j = 1, 2$. This order information, in turn, allows us to get the expansion

$$
EM_n^{(1)} \ge R_{1n}[\Psi^{(1)}(\boldsymbol{\beta}_0)] - R_{0n} \ge \sup_{\mathbf{t}_2 \ge 0}\left[2\mathbf{t}_2^\tau\left(\sum_{i=1}^n \tilde{\mathbf{b}}_{2i}\right) - n\mathbf{t}_2^\tau \tilde{\mathbf{B}}_{22}\mathbf{t}_2\right] + o_p(1).
$$

Combining the lower and upper bounds, we arrive at

$$
EM_n^{(1)} = \sup_{\mathbf{t}_2 \ge 0}\left[2\Big(\sum_{i=1}^n \tilde{\mathbf{b}}_{2i}\Big)^\tau \mathbf{t}_2 - n\mathbf{t}_2^\tau \tilde{\mathbf{B}}_{22}\mathbf{t}_2\right] + o_p(1).
$$

Because of the multivariate asymptotic normality of $n^{-1}\sum_{i=1}^{n}\tilde{\mathbf{b}}_{2i}$, the above expansion implies that the limiting distribution of $EM_n^{(K)}$ is the same as the distribution of

$$\sup_{\mathbf{v}\geq 0}(2\mathbf{v}^{\tau}\mathbf{w}-\mathbf{v}^{\tau}\tilde{\mathbf{B}}_{22}\mathbf{v}),$$

where $\mathbf{w}$ is a multivariate normally distributed random vector with variance–covariance matrix $\tilde{\mathbf{B}}_{22}$. This completes the proof of the first part of the Theorem.

The remaining task is to obtain the chi-square representation. For this purpose, we first introduce a lemma from Kudo (1963) or Nüesch (1966).

Lemma 4: Let $\mathbf{w}$ be a vector of dimension m and $\mathbf{W}$ be a positive definite matrix of dimension $m \times m$. The vector $\hat{\mathbf{v}}$ maximises $2\mathbf{v}^{\tau}\mathbf{w}-\mathbf{v}^{\tau}\mathbf{W}\mathbf{v}$ under the constraint $\mathbf{v}\geq 0$ only if both (a) $\hat{\mathbf{v}}^{\tau}\mathbf{W}-\mathbf{w}^{\tau}\geq 0$ and (b) $\hat{\mathbf{v}}^{\tau}\mathbf{W}_h-\mathbf{w}_h>0$ implies $\hat{\mathbf{v}}_h=0$ where subindex h means the hth element.

PROOF: Now we use the lemma to continue the proof. Let $\mathbf{W}=\tilde{\mathbf{B}}_{22}$ and let $\hat{\mathbf{v}}$ be the maximum point. By Lemma 4, we have $(\mathbf{w}^{\tau}-\hat{\mathbf{v}}^{\tau}\mathbf{W})\hat{\mathbf{v}}=0$ and therefore

$$\sup_{\mathbf{v}\geq 0}(2\mathbf{v}^{\tau}\mathbf{w}-\mathbf{v}^{\tau}\mathbf{W}\mathbf{v})=\hat{\mathbf{v}}^{\tau}W\hat{\mathbf{v}}.$$

Given $\mathbf{W}$, the solution $\hat{\mathbf{v}}$ is a function of $\mathbf{w}$. Depending on the positiveness of the elements of $\hat{\mathbf{v}}$, we can partition the range of $\mathbf{w}$ into 2^{m_0} disjoint regions. There are $\binom{m_0}{p}$ regions of $\mathbf{w}$ whose corresponding solution $\hat{\mathbf{v}}$ has p positive elements. One of them is

$$\mathcal{A}=\{\mathbf{w}:\hat{\mathbf{v}}_1=0,\hat{\mathbf{v}}_2>0\},\tag{3.23}$$

where $(\hat{\mathbf{v}}_1^{\tau},\hat{\mathbf{v}}_2^{\tau})$ is a partition of $\mathbf{v}^{\tau}$ of lengths m_0-p and p. Let $\mathbf{w}$ and $\mathbf{W}$ be partitioned accordingly:

$$\mathbf{w}^{\tau}=(\mathbf{w}_1^{\tau},\mathbf{w}_2^{\tau}),\quad \mathbf{W}=\begin{pmatrix}\mathbf{W}_{11} & \mathbf{W}_{12}\\ \mathbf{W}_{21} & \mathbf{W}_{22}\end{pmatrix}.$$

By Lemma 4 and because $\hat{\mathbf{v}}_1=0$ and $\hat{\mathbf{v}}_2>0$, we have

$$\hat{\mathbf{v}}_2^{\tau}\mathbf{W}_{21}-\mathbf{w}_1^{\tau}>0 \quad\text{and}\quad \hat{\mathbf{v}}_2^{\tau}\mathbf{W}_{22}-\mathbf{w}_2^{\tau}=0$$

or equivalently

$$\hat{\mathbf{v}}_2=\mathbf{W}_{22}^{-1}\mathbf{w}_2>0\quad\text{and}\quad \mathbf{W}_{12}\mathbf{W}_{22}^{-1}\mathbf{w}_2-\mathbf{w}_1>0.$$

Therefore $\hat{\mathbf{v}}^{\tau}\mathbf{W}\hat{\mathbf{v}}=\mathbf{w}_2^{\tau}\mathbf{W}_{22}^{-1}\mathbf{w}_2$ and the region can equivalently be defined as

$$\mathcal{A}=\{\mathbf{w}:\mathbf{W}_{22}^{-1}\mathbf{w}_2>0\quad\text{and}\quad \mathbf{W}_{12}\mathbf{W}_{22}^{-1}\mathbf{w}_2-\mathbf{w}_1>0\}.\tag{3.24}$$

Using the law of total probability, we have

$$\Pr(\hat{\mathbf{v}}^\tau \mathbf{W}\hat{\mathbf{v}} \le x) = \sum_{p=0}^{m_0} \sum_{\mathcal{A}\in\mathcal{F}_p} \Pr(\hat{\mathbf{v}}^\tau \mathbf{W}\hat{\mathbf{v}} \le x|\mathbf{w} \in \mathcal{A})\Pr(\mathbf{w} \in \mathcal{A}),$$

where $\mathcal{F}_p$ is the collection of $\binom{m_0}{p}$ regions of $\mathbf{w}$ in the form of (3.23). By (3.24),

$$\Pr(\hat{\mathbf{v}}^\tau \mathbf{W}\hat{\mathbf{v}} \le x|\mathbf{w} \in \mathcal{A}) = \Pr(\mathbf{w}_2^\tau \mathbf{W}_{22}^{-1}\mathbf{w}_2 \le x|\mathbf{W}_{22}^{-1}\mathbf{w}_2 > 0, \mathbf{W}_{12}\mathbf{W}_{22}^{-1}\mathbf{w}_2 - \mathbf{w}_1 > 0).$$

Because of normality, we find that $\mathbf{w}_2$ and $\mathbf{W}_{12}\mathbf{W}_{22}^{-1}\mathbf{w}_2 - \mathbf{w}_1$ are independent. Therefore

$$\Pr(\hat{\mathbf{v}}^\tau \mathbf{W}\hat{\mathbf{v}} \le x|\mathbf{w} \in \mathcal{A}) = \Pr(\mathbf{w}_2^\tau \mathbf{W}_{22}^{-1}\mathbf{w}_2 \le x|\mathbf{W}_{22}^{-1}\mathbf{w}_2 > 0).$$

Clearly, $\mathbf{w}_2^\tau \mathbf{W}_{22}^{-1}\mathbf{w}_2$ has a χ_p^2 distribution when $\mathcal{A} \in \mathcal{F}_p$. In addition, $\mathbf{w}_2^\tau \mathbf{W}_{22}^{-1}\mathbf{w}_2$ and $\mathbf{W}_{22}^{-1/2}\mathbf{w}_2/\sqrt{\mathbf{w}_2^\tau \mathbf{W}_{22}^{-1}\mathbf{w}_2}$ are independent. Hence,

$$\begin{aligned}\Pr(\hat{\mathbf{v}}^\tau \mathbf{W}\hat{\mathbf{v}} \le x|\mathbf{w} \in \mathcal{A}) &= \Pr(\mathbf{w}_2^\tau \mathbf{W}_{22}^{-1}\mathbf{w}_2 \le x|\mathbf{W}_{22}^{-1}\mathbf{w}_2 > 0)\\ &= \Pr(\mathbf{w}_2^\tau \mathbf{W}_{22}^{-1}\mathbf{w}_2 \le x|\mathbf{W}_{22}^{-1/2}\mathbf{w}_2/\sqrt{\mathbf{w}_2^\tau \mathbf{W}_{22}^{-1}\mathbf{w}_2} > 0)\\ &= \Pr(\mathbf{w}_2^\tau \mathbf{W}_{22}^{-1}\mathbf{w}_2 \le x) = \Pr(\chi_p^2 \le x).\end{aligned}$$

The above derivation applies to each region in $\mathcal{F}_p$. Let $a_p = \sum_{\mathcal{A}\in\mathcal{F}_p} \Pr(\mathbf{w} \in \mathcal{A})$; then

$$\Pr(\hat{\mathbf{v}}^\tau W\hat{\mathbf{v}} \le x) = \sum_{p=0}^{m_0} a_p \Pr(\chi_p^2 \le x).$$

This completes the proof.

The final task is to find the analytical expressions of a_p. This is achieved by employing the lemma of David (1953).

Lemma 5: Let $\mathbf{w} = (w_1, w_2, w_3)$ be multivariate normal distributed with mean 0 and correlations $\mathrm{cor}(w_i, w_j) = \omega_{ij}$ for $i, j = 1, 2, 3$. Then

$$\Pr(w_1 > 0, w_2 > 0) = (\pi - \arccos\omega_{12})/(2\pi)$$

and

$$\begin{aligned}&\Pr(w_1 > 0, w_2 > 0, w_3 > 0)\\ &\quad = [2\pi - (\arccos\omega_{12} + \arccos\omega_{13} + \arccos\omega_{23})]/(4\pi).\end{aligned}$$

We now derive the formulae presented in the last section.

When $m_0 = 1$, we have

$$a_0 = \Pr(-w_1 > 0) = \Pr(w_1 < 0) = 0.5$$

and $a_1 = 1 - a_0 = 0.5$.

When $m_0 = 2$, because of distributional symmetry,

$$a_0 = \Pr(-\mathbf{w} > 0) = \Pr(w_1 > 0, w_2 > 0) = (\pi - \arccos \omega_{12})/(2\pi)$$

and because the correlation between two elements of $\mathbf{W}^{-1}\mathbf{w}$ is $-\omega_{12}$,

$$a_2 = \Pr(\mathbf{W}^{-1}\mathbf{w} > 0) = 0.5 - \arccos(-\omega_{12})/(2\pi) = \arccos(\omega_{12})/(2\pi).$$

Since $a_0 + a_2 = 0.5$, we get $a_1 = 0.5$.

When $m_0 = 3$, due to distributional symmetry,

$$\begin{aligned} a_0 &= \Pr(-\mathbf{w} > 0) = \Pr(w_1 > 0, w_2 > 0, w_3 > 0) \\ &= [2\pi - (\arccos \omega_{12} + \arccos \omega_{13} + \arccos \omega_{23})]/(4\pi) \end{aligned}$$

and $a_3 = \Pr(\mathbf{W}^{-1}\mathbf{w} > 0)$. Note that the covariance matrix of $\mathbf{W}^{-1}\mathbf{w}$ is $\mathbf{W}^{-1}$. It is easy to check that the correlation between the ith and jth elements of $\mathbf{W}^{-1}\mathbf{w}$ is $-\omega_{ij:k}$ for $\{i, j, k\} = \{1, 2, 3\}$. Hence,

$$\begin{aligned} a_3 &= 0.5 - [\arccos(-\omega_{12:3}) + \arccos(-\omega_{13:2}) + \arccos(-\omega_{23:1})]/(4\pi) \\ &= [\arccos \omega_{12:3} + \arccos \omega_{13:2} + \arccos \omega_{23:1} - \pi]/(4\pi) \end{aligned}$$

and

$$\begin{aligned} a_2 &= \sum_{\mathcal{F}_2} \Pr(\mathbf{w} \in \mathcal{F}_2) \\ &= \sum_{\mathcal{F}_2} \Pr(\mathbf{W}_{22}^{-1}\mathbf{w}_2 > 0, \mathbf{W}_{12}\mathbf{W}_{22}^{-1}\mathbf{w}_2 - \mathbf{w}_1 > 0) \\ &= \sum_{\mathcal{F}_2} \Pr(\mathbf{W}_{22}^{-1}\mathbf{w}_2 > 0)\Pr(\mathbf{W}_{12}\mathbf{W}_{22}^{-1}\mathbf{w}_2 - \mathbf{w}_1 > 0) \\ &= \frac{1}{2}\sum_{\mathcal{F}_2} \Pr(\mathbf{W}_{22}^{-1}\mathbf{w}_2 > 0). \end{aligned}$$

The second-last equality is a result of the independence between $\mathbf{W}_{22}^{-1}\mathbf{w}_2$ and $\mathbf{W}_{12}\mathbf{W}_{22}^{-1}\mathbf{w}_2 - \mathbf{w}_1$, and the last equality is true because $\mathbf{W}_{12}\mathbf{W}_{22}^{-1}\mathbf{w}_2 - \mathbf{w}_1$ is a 0-mean normal random variable.

Note that $\mathcal{F}_2$ contains three partitions and for each partition the form of the covariance matrix of $\mathbf{W}_{22}^{-1}\mathbf{w}_2$ remains the same: $\mathbf{W}_{22}^{-1}$.

When $\mathbf{w}_2 = (w_1, w_2)$ the correlation $\mathrm{cor}(w_1, w_2) = -\omega_{12}$; when $\mathbf{w}_2 = (w_1, w_3)$, $\mathrm{cor}(w_1, w_3) = -\omega_{13}$; and when $\mathbf{w}_2 = (w_2, w_3)$, $\mathrm{cor}(w_2, w_3) = -\omega_{23}$. Consequently, Lemma 5 implies that

$$\begin{aligned} a_2 &= \frac{1}{4\pi}\{[\pi - \arccos(-\omega_{12})] + [\pi - \arccos(-\omega_{13})] + [\pi - \arccos(-\omega_{23})]\} \\ &= (\arccos \omega_{12} + \arccos \omega_{13} + \arccos \omega_{23})/(4\pi). \end{aligned}$$

Since $a_0 + a_2 = 0.5$, we also get $a_1 + a_3 = 0.5$.

3.4 Discussion

The notion of the partial EM iteration is new, and the use of the results from Kudo (1963) is innovative in this paper. The technique used for expanding the EM test statistic has its origin in Chen *et al.* (2001). Implementing the idea in the current context was nevertheless a challenging task.

References

Billingsley, P. (1968) *Convergence of Probability Measures*. John Wiley & Sons, Ltd.

Chen, H. and Chen, J. (2001) The likelihood ratio test for homogeneity in finite mixture models. *The Canadian Journal of Statistics*, **29**, 201–215.

Chen, H., Chen, J. and Kalbfleisch, J. D. (2001) A modified likelihood ratio test for homogeneity in finite mixture models. *Journal of the Royal Statistical Society, Series B*, **63**, 19–29.

Chen, H., Chen, J. and Kalbfleisch, J. D. (2004) Testing for a finite mixture model with two components. *Journal of the Royal Statistical Society, Series B*, **66**, 95–115.

Chen, J. (1998) Penalized likelihood ratio test for finite mixture models with multinomial observations. *The Canadian Journal of Statistics*, **26**, 583–599.

Chen, J. and Cheng, P. (1995) The limit distribution of the restricted likelihood ratio statistic for finite mixture models. *Northeast Mathematics Journal*, **11**, 365–374.

Chen, J. and Li, P. (2009) Hypothesis test for normal mixture models: the EM approach. *The Annuals of Statistics*, **37**, 2523–2542.

Chernoff, H. and Lander, E. (1995) Asymptotic distribution of the likelihood ratio test that a mixture of two binomials is a single binomial. *Journal of Statistical Planning and Inference*, **43**, 19–40.

Dacunha-Castelle, D. and Gassiat, E. (1999) Testing the order of a model using locally conic parametrization: population mixtures and stationary ARMA processes. *The Annals of Statistics*, **27**, 1178–1209.

David, F. N. 1953 A note on the evaluation of the multivariate normal integral. *Biometrika*, **40**, 458–459.

Dempster, A. P., Laird, N. M. and Rubin, D. B. (1977) Maximum likelihood from incomplete data via EM algorithm (with discussion). *Journal of the Royal Statistical Society, Series B*, **39**, 1–38.

Hartigan, J. A. (1985) A failure of likelihood asymptotics for normal mixtures. In *Proceedings of the Berkeley Conference in Honor of J. Neyman and Kiefer*, Vol. 2, (eds L. LeCam and R. A. Olshen), pp. 807–810.

Kudo, A. (1963) A multivariate analogue of the one-sided test. *Biometrika*, **50**, 403–418.

Lemdani, M. and Pons, O. (1995) Tests for genetic linkage and homogeneity. *Biometrics*, **51**, 1033–1041.

Li, P. and Chen, J. (2010) Testing the order of a finite mixture. *Journal of the American Statistical Association*, **105**, 1084–1092.

Li, P., Chen, J. and Marriott, P. (2009) Non-finite Fisher information and homogeneity: the EM approach. *Biometrika*, **96**, 411–426.

Liu, X. and Shao, Y. (2003) Asymptotics for likelihood ratio tests under loss of identifiability. *The Annals of Statistics*, **31**, 807–832.

Nüesch, P. (1966) On the problem of testing location in multivariate populations for restricted alternatives. *Annals of Mathematical Statistics*, **37**, 113–119.

Serfling, R. J. (1980) *Approximation Theorems of Mathematical Statistics*, John Wiley & Sons, Ltd.

Titterington, D. M., Smith, A. F. M., and Makov, U. E. (1985) *Statistical Analysis of Finite Mixture Distributions*, John Wiley & Sons, Ltd.

Wald, A. (1949) Note on the consistency of the maximum likelihood estimate. *The Annals of Mathematical Statistics*, **20**, 595–601.

4

Comparing Wald and likelihood regions applied to locally identifiable mixture models

Daeyoung Kim and Bruce G. Lindsay

4.1 Introduction

Although finite mixture models have been used in many areas for over 100 years, interest in them as an effective tool for density estimation and model-based clustering greatly increased due to modern computing power. Nowadays statistical inference for finite mixture models can be done by the method of maximum likelihood (ML), which provides procedures for point estimation, hypothesis testing and construction of confidence sets. This chapter is concerned with comparison between the confidence sets based on the Wald method and the likelihood method in a finite mixture model.

There are several entries in the literature that suggest superiority of likelihood confidence sets over Wald confidence sets (for example, see Cox and Hinkley, 1974; Meeker and Escobar, 1995; Agresti, 2002; Kalbfleisch and Prentice, 2002; Lang, 2008).

The likelihood confidence sets generally have better coverage properties than the Wald sets, especially when the sample size is not large. The likelihood confidence sets are also invariant to parameterisation, unlike the Wald sets. A major challenge

Mixtures: Estimation and Applications, First Edition. Edited by Kerrie L. Mengersen, Christian P. Robert and D. Michael Titterington.

with the application of the likelihood sets to real data arises from the fact that finding the boundaries of the targeted likelihood confidence set requires more complicated computations, compared to the Wald set approach.

To address this computational issue, Kim and Lindsay (2011) created a simulation-based visualisation method, *modal simulation*, as a way of describing the (profile) likelihood confidence region. The modal simulation method generates sample points on the boundaries of the relevant sets and the resulting dataset can be used to display every (profile) likelihood confidence set of interest without further numerical optimisation. Their method assumed that one is simulating from a likelihood with a single mode. Kim and Lindsay (2010) adapted the simulation method to multimodal likelihood such as likelihood in a finite mixture model.

The visual analyses in Kim and Lindsay (2011) showed that there exist model/data situations where the shape of likelihood regions is not close to the ellipsoidal shape of the Wald regions. In particular, Kim and Lindsay (2010) illustrated that the carefully constructed likelihood set in a finite mixture model appears to bear less resemblance to the elliptical Wald confidence set when the components are not well separated relative to the sample size. We consider it important from a practical point of view to develop guidance to assess the adequacy of using the Wald confidence sets for the parameters and Fisher information matrix for the ML estimator. This is because the inverse of the Fisher information matrix provides a much more succinct description of the confidence region through the variances and correlations of the parameter estimates. Moreover, if the Wald regions and likelihood regions are quite similar, the use of normal theory-based confidence region is arguably more justified.

In this chapter we propose two diagnostics to assess the similarity between the likelihood and the Wald regions in any parametric model. One diagnostic is based on estimating the volumes of the two regions and comparing them. The second diagnostic is based on evaluation of the worst case differences between the Wald and likelihood univariate intervals over a class of parameters. Both diagnostics will use the outputs from the modal simulation method that display the boundaries of the targeted likelihood set. Note that the proposed diagnostics in this chapter are general in the sense that they can be applied to any statistical model and data where likelihood is bounded and the confidence set around the MLE has only one critical point, the MLE itself. Our application to the mixture problem shows how it can be adapted to a multimodal problem when the local topology of the mixture likelihood is simple.

Before applying the proposed diagnostics to a finite mixture model, one needs to consider that the likelihood sets for the mixture parameters have a potentially complex topological structure as a consequence of two types of nonidentifiability inherent in the mixture parameters, *labelling nonidentifiability* (Redner and Walker, 1984) and *degenerate nonidentifiability* (Crawford, 1994, and Lindsay, 1995). Based on a detailed study of topology of the likelihood region, Kim and Lindsay (2010) demonstrated that there exists a limited range of confidence levels at which one can construct a reasonable likelihood confidence region, where by reasonable we mean that the confidence set is a locally identifiable subset of the full parameter space. Beyond that range there is no natural way of constructing identifiable confidence sets, at least in a way that corresponds to the natural structure of the likelihood region. In this chapter we will employ the two proposed diagnostic tools only at reasonable confidence levels.

The rest of the chapter is organized as follows. Section 4.2 overviews a topological structure of the (profile) likelihood region described in Kim and Lindsay (2010). Section 4.3 reviews the modal simulation method developed by Kim and Lindsay (2011) for a case where the targeted likelihood confidence region contains a single critical point, the MLE. In this section we illustrate, with a regression example, how one can employ the modal simulation to reconstruct (profile) likelihood confidence regions for the parameters of interest, and visually detect discrepancies between the two types of confidence region. In Section 4.4 we propose two diagnostics designed to assess the similarity between the Wald sets and likelihood sets for the parameters. Section 4.5 shows how one can apply the modal simulation and two proposed diagnostics to a locally identifiable likelihood confidence set in a finite mixture model. Section 4.6 evaluates the performance of our proposed diagnostics in two real datasets.

4.2 Background on likelihood confidence regions

In this section we review the topological structure of the likelihood regions for a parameter vector and its relation to profile likelihood regions described in Kim and Lindsay (2010).

4.2.1 Likelihood regions

Given independent data vector $\mathbf{Y}_i$ $(i = 1, \ldots, n)$ with a density in the parametric family $\{p(\mathbf{y}_i \mid \boldsymbol{\theta}), \boldsymbol{\theta} \in \boldsymbol{\Omega} \subset R^p\}$, one can construct the likelihood of a p-dimensional parameter vector $\boldsymbol{\theta}$:

$$L(\boldsymbol{\theta}) = \prod_{i=1}^{n} p(\mathbf{y}_i \mid \boldsymbol{\theta}). \tag{4.1}$$

We will denote the maximum-likelihood estimator (MLE) for the parameter maximising the likelihood in Equation (4.1) by $\hat{\boldsymbol{\theta}}$. We assume the model is regular and the likelihood is bounded, but we do not assume it is unimodal.

One constructs a test of the null hypothesis $H_0 : \boldsymbol{\theta} = \boldsymbol{\theta}_0$ based on the likelihood-ratio statistic, $T_1 = -2 \log L(\boldsymbol{\theta}) + 2 \log L(\hat{\boldsymbol{\theta}})$. Note that we assume that the asymptotic distribution of T_1 does not depend on $\boldsymbol{\theta}_0$ and that asymptotic critical values are available for T_1 although one can do adjustments to these values for finite sample sizes. Then inversion of this test provides us with the general form of the likelihood confidence region for $\boldsymbol{\theta}$,

$$C_c^{LR} = \{\boldsymbol{\theta} : T_1 \leq q_{1-\alpha}\} = \{\boldsymbol{\theta} : L(\boldsymbol{\theta}) \geq c\}, \tag{4.2}$$

where $q_{1-\alpha}$ is the $1 - \alpha$ quantile of the (asymptotic) distribution of T_1 and $c = L(\hat{\boldsymbol{\theta}})e^{-q_{1-\alpha}/2}$. We refer to C_c^{LR} in Equation (4.2) as the *elevation c likelihood confidence region*. The value of c is usually interpreted via the confidence level of the corresponding likelihood-ratio statistic in the p-dimensional parameter space, which we denote by $Conf_p(c)$. The shape of C_c^{LR} is determined by the *elevation c likelihood contour set*, $\{\boldsymbol{\theta} : L(\boldsymbol{\theta}) = c\}$.

Kim and Lindsay (2010) used a sea-level analogy to describe a topological structure of the likelihood regions C_c^{LR}. Picture the parameter $\boldsymbol{\theta}$ as the coordinate system for a flat plain and the likelihood as the elevation of the land surface above the plain. We assume that the likelihood is smooth in the sense that it is twice continuously differentiable, and that the Hessian matrix, the matrix of second-order partial derivatives of the likelihood, is nonsingular at all critical points. Suppose we flood the land surface up to elevation level c. If one floods the surface to elevation c, the contour sets $\{\boldsymbol{\theta} : L(\boldsymbol{\theta}) = c\}$ can be thought of as shorelines on the likelihood surface. The confidence regions C_c^{LR} describe the regions of the coordinate space that define the land above the water level c. Based on results from Morse theory (Klemelä, 2009; Matsumoto, 2002), the topology of the likelihood regions can only change as the water level passes through the elevation of a critical point. If the likelihood has only one critical point corresponding to the MLE (i.e. the mode), the confidence region corresponds to a single connected island, with a peak at the single mode, for any water level c. Note that what we mean by an *island* is a *path-connected set* in topology where any two points in the set can be connected by a continuous path between them that lies entirely in the set.

When there are more than two modes whose elevations are different, the description of the likelihood topology becomes more complex. As the water level goes down, the islands associated with the modes appear and coalesce, and their coalescence is determined by the elevations and the connecting properties of the saddlepoints on the likelihood.

4.2.2 Profile likelihood regions

When one is interested in a function of the parameters, defined by an $r(\leq p)$-dimensional vector of parameters $\beta(\boldsymbol{\theta})$, one defines a profile likelihood $L_{prof}(\beta)$ and constructs profile confidence regions C_c^{PLR} for inference:

$$L_{prof}(\beta) = \sup\left[L(\boldsymbol{\theta}) : \beta(\boldsymbol{\theta}) = \beta\right], \tag{4.3}$$

$$C_c^{PLR} = [\beta : L_{prof}(\beta) \geq c]. \tag{4.4}$$

Note that C_c^{PLR} is obtained by the inversion of tests of $H_0 : \beta(\boldsymbol{\theta}) = \beta_0$ based on $T_2 = -2\log L_{prof}(\beta_0) + 2\log L_{prof}(\hat{\beta})$ where $\hat{\beta} = \beta(\hat{\boldsymbol{\theta}})$. Note that c is usually interpreted via a confidence level in the r-dimensional parameter space, which we denote by $Conf_r(c)$.

Kim and Lindsay (2010) pointed out an important property describing the relationship between the profile confidence region and the full-likelihood confidence region, the *nesting property*: if $\boldsymbol{\theta}$ is a point in the likelihood confidence region C_c^{LR} of Equation (4.2), $\beta(\boldsymbol{\theta})$ must also be in the profile likelihood confidence region C_c^{PLR} of Equation (4.4) at the same elevation c. Note, however, that the value of c has a different confidence interpretation depending on whether we are looking at the full-likelihood set or the profile one. A particular value of c generates a confidence set with level $Conf_p(c)$ in C_c^{LR}, but, for C_c^{PLR} in an r-dimensional space, it corresponds to a larger level $Conf_r(c)$.

If one wants to use profile confidence regions to view important features of the full-likelihood region, it is very critical to select the right profiles because the profile regions need not reveal the full structure of the full likelihood for $\boldsymbol{\theta}$. Suppose that the profile set is based on a suitably regular function $\beta(\boldsymbol{\theta})$ and there exist M separated islands in the full dimensional likelihood. Then it is not necessarily true that a profile plot shows the M separated islands. However, the good news is that the converse is true: if one can identify M separated islands in a profile plot, then there exist at least that many in the full dimensional likelihood. We will later use this 'island counting' property of the profile. This is important in the mixture model because, as we shall see, the labelling problem creates multiple modes that may or may not be visible in the profile plots.

4.2.3 Alternative methods

There exist many alternative methods for constructing confidence sets. Here we mention two methods due to their relationship to our topic.

The first alternative method is the Wald confidence set based on the Wald statistic (the asymptotic normality of the MLE $\hat{\boldsymbol{\theta}}$ and associated Fisher information matrix $\mathbf{I}$), that is the set of all the parameters in the ellipsoid

$$C^W = \{\boldsymbol{\theta} : (\hat{\boldsymbol{\theta}} - \boldsymbol{\theta})^T \boldsymbol{I}(\hat{\boldsymbol{\theta}} - \boldsymbol{\theta}) \leq w\}, \tag{4.5}$$

where w is chosen to achieve a desired asymptotic confidence region. One nice feature of the Wald set is that the inverse of Fisher information matrix is the asymptotic covariance matrix that summarises the structure of all the various confidence sets for the parameters of interest. For example, for the scalar parameter, they are plus/minus standard errors and, for the multidimensional parameter, they are elliptical regions. Thus, for one- or two-dimensional parameters of interest, it is easy to construct the Wald sets for them.

However, it is known that the Wald sets are inferior to the likelihood sets (Cox and Hinkley, 1974; Meeker and Escobar, 1995; Agresti, 2002; Kalbfleisch and Prentice, 2002; Lang, 2008). When the sample size is not large, the likelihood sets generally have better coverage than the Wald sets. The Wald sets also require estimation of the covariance matrix for the ML estimator and so different estimation methods may produce quite different results. In addition, the Wald set is not invariant to parameterisation, unlike the likelihood set, so that a search for good parameterisation is sometimes required.

Another approach for confidence set construction is the parametric bootstrap, accounting for finite sample effects (Hall, 1992; Efron and Tibshirani, 1993; Davison and Hinkley, 2006). For using this method one needs to simulate data repeatedly from the fitted model, compute the estimates for the parameters in each replicate and then use these samples to construct the confidence sets. This approach has been proposed for use in mixture models (McLachlan, 1987; Basford *et al.*, 1997; McLachlan and Peel, 2000). However, there are the theoretical drawbacks of the bootstrap method compared to the likelihood confidence sets (Owen, 2001; Lang, 2008). In terms of the computational viewpoint, there are also some issues. First, the repeated computation of the parameter estimates from new datasets could be computationally expensive. Second, the parametric bootstrap approach has an inherent sparsity problem in describing the boundaries of the targeted confidence sets (Kim and Lindsay, 2011). Finally, in the mixture problem, the sample output does not have inherent labels, so the labelling problem is only solved by using an ad hoc label assignment method.

4.3 Background on simulation and visualisation of the likelihood regions

As confidence sets based on ML estimation, the likelihood confidence set and the Wald confidence set are both commonly used over a wide class of parametric models. In Section 4.2.3 we discussed advantages of the (profile) likelihood set over the Wald set. However, a major challenge in using the (profile) likelihood sets is a description of the boundaries of these sets. This requires a numerical optimisation that involves repeated computation of the (profile) likelihood-ratio statistic over different constraint parameter sets. This could be a daunting and time-consuming task for the likelihood sets with complex boundaries and high-dimensional parameter space.

In order to overcome this computational issue, Kim and Lindsay (2010, 2011) created the *modal simulation* method designed to generate samples on the boundaries of the confidence region generated by an inference function such as the likelihood. They showed by examples that modal simulation along with data analysis gives a more useful set of tools than standard numerical analysis in describing an inference function region. In this section we review modal simulation for displaying the targeted likelihood region that contains just a single critical point, the MLE. We also illustrate how one can employ this methodology to reconstruct the targeted (profile) likelihood region in a multiple linear regression example.

4.3.1 Modal simulation method

Given the observed data and the MLE $\hat{\boldsymbol{\theta}}$ for the parameters in an assumed model, the modal simulation methodology assumes that one is simulating a sample of points from the elevation c likelihood contour set, $\{\boldsymbol{\theta} : L(\boldsymbol{\theta}) = c\}$, that contains $\hat{\boldsymbol{\theta}}$. A strategy of generating the sample on $\{\boldsymbol{\theta} : L(\boldsymbol{\theta}) = c\}$ can be described as follows.

Suppose one has found the MLE mode, $\hat{\boldsymbol{\theta}}$, and the covariance estimator, $V_{\hat{\boldsymbol{\theta}}}$. Suppose one also defines a ray generated by a vector $\mathbf{z} \in R^p$ to be $\boldsymbol{\theta}(\epsilon) = \hat{\boldsymbol{\theta}} + \epsilon V_{\hat{\boldsymbol{\theta}}}{}^{1/2}\mathbf{z}$ where $\epsilon \in R$.

Step 1: Generate $\mathbf{z}$ from the p-dimensional standard normal distribution and find a targeted elevation c.

Step 2: Determine ϵ satisfying $L[\boldsymbol{\theta}(\epsilon)] = c$. Denote the computed ϵ by $\tilde{\epsilon} = \tilde{\epsilon}(c, \mathbf{z})$. We let $\tilde{\boldsymbol{\theta}} = \boldsymbol{\theta}(\tilde{\epsilon})$ be the simulated value generated by $(c, \mathbf{z})$. We compute $\tilde{\boldsymbol{\theta}}$ for both the positive and negative ϵ solutions.

For a targeted elevation c and given $\mathbf{z}$, the ray starts from a chosen mode $\hat{\boldsymbol{\theta}}$ and heads in direction $\mathbf{z}/||\mathbf{z}||$. The goal is to determine when this ray reaches the boundary of the targeted $\{\boldsymbol{\theta} : L(\boldsymbol{\theta}) = c\}$, as this point will become the sampled value of $\boldsymbol{\theta}$, a random point depending on the random direction $\mathbf{z}/||\mathbf{z}||$. The sampling proceeds along the ray to find solutions to $L(\boldsymbol{\theta}) = c$. One can obtain such an ϵ using a one-dimensional root finding algorithm. If one repeats the calculation of $\tilde{\boldsymbol{\theta}}$ for many generated $\mathbf{z}$, one can obtain a large set of simulated parameter values on the boundary of $\{\boldsymbol{\theta} : L(\boldsymbol{\theta}) = c\}$. Note that, if c_w is an elevation value for the Wald confidence set, setting ϵ equal to $\sqrt{c_w}/||\mathbf{z}||$ gives a point on the boundary of the Wald set with critical value c_w.

Note that, regarding the form of $V_{\boldsymbol{\theta}}$ in a ray, one can use the Fisher information or one of its asymptotic equivalents. As for the elevation c corresponding to the confidence level of interest, one can use the limiting distribution of the likelihood-ratio statistic for an easy transition between the full confidence set and the profile confidence set. If one wishes to use an elevation that accounts for having a small sample size, one can do an adjustment by doing a single parametric bootstrap simulation to estimate the distribution of the likelihood-ratio statistic under sampling from $\hat{\boldsymbol{\theta}}$ (Hall, 1992; Efron and Tibshirani, 1993; Davison and Hinkley, 2006; Kim and Lindsay, 2011).

There are three important features of the modal simulation method. First, this method sharply and efficiently defines the boundaries of likelihood confidence regions. Second, from the results of a single simulation run, one can display the profile confidence sets for any function of the parameters in an efficient manner. This feature arises from the nesting property mentioned in Section 4.2.2. That is, if one has a sample of B points $(\tilde{\boldsymbol{\theta}}_1, \ldots, \tilde{\boldsymbol{\theta}}_B)$ from $\{\boldsymbol{\theta} : L(\boldsymbol{\theta}) = c\}$ and $\beta(\boldsymbol{\theta})$ is a function of the parameters of interest, then $(\beta(\tilde{\boldsymbol{\theta}}_1), \ldots, \beta(\tilde{\boldsymbol{\theta}}_B))$ are a set of points from the profile set $\{\beta : L_{prof}(\beta) \geq c\}$. After generating a large sample, therefore, one can picture the profile likelihood confidence sets for various functions of the parameters, all without further numerical optimisation. Third, the simulated points displaying $\{\boldsymbol{\theta} : L(\boldsymbol{\theta}) = c\}$ are automatically associated with the MLE mode $\hat{\boldsymbol{\theta}}$. This feature is very crucial in addressing the labelling problem in a finite mixture model, as the association found in the full-dimensional likelihood provides simulated points with inherent labels that are the same as that of $\hat{\boldsymbol{\theta}}$. This is not available in a conventional numerical profile approach.

As noted by Kim and Lindsay (2010, 2011), if the targeted set around $\hat{\boldsymbol{\theta}}$ is *star-shaped*, the modal simulation will have only a single solution on $L(\boldsymbol{\theta}) = c$ along the positive and negative rays because the likelihood is monotonically decreasing along every ray from $\hat{\boldsymbol{\theta}}$, regardless of elevation. A challenge emerges from a possibility that the set is not star-shaped and a ray from $\hat{\boldsymbol{\theta}}$ can have multiple solutions on $L(\boldsymbol{\theta}) = c$. The first solution on the ray starting from $\hat{\boldsymbol{\theta}}$ is in the star-shaped region for $\hat{\boldsymbol{\theta}}$. Solutions further out along the same ray lay outside the star-shaped region for $\hat{\boldsymbol{\theta}}$. Those authors called these extra solutions *simulation outliers*.

If there is only one MLE mode in the likelihood, the simulation outliers are still in the confidence region for $\hat{\boldsymbol{\theta}}$ but their presence provides useful information about the structure of the targeted confidence region. The existence of just one such outlier proves that the shape of the likelihood region is quite different from that of the Wald region.

If the likelihood has multiple modes, we recommend simulating separately from the different modes so that one can see better the various island regions associated with each mode. However, in a multimodal surface, such as occurs with a mixture model, the interpretation of the outliers from a single mode is not simple. Kim and Lindsay (2010) noted that an outlier from a modal simulation based on a particular mode $\hat{\boldsymbol{\theta}}$ cannot be in the star-shaped region for that island, but might still be in the island associated with that mode (a 'star-shaped region' outlier), but it could also be an element of the island region associated with another mode (a 'wrong-modal region' outlier). If one wishes to describe the modal region for the chosen mode $\hat{\boldsymbol{\theta}}$ then the star-shaped region outliers from $\hat{\boldsymbol{\theta}}$ are legitimate elements and should be used. For more details on the methods to determine the type of simulation outliers and remedy wrong-modal region outliers, see Kim and Lindsay (2010). Since Wald and likelihood regions can only be similar when the likelihood region is a single star-shaped island, the rest of this chapter focuses on the unimodal case.

4.3.2 Illustrative example

In this section we illustrate application of the modal simulation method to a case of a unimodal likelihood. We here consider real data with a multiple linear regression model. The data contain the taste of matured cheese and concentrations of several chemicals in 30 samples of mature cheddar cheese (Moore and McCabe, 1989). The goal was to study the relationships between the taste of matured cheese, denoted by Taste, and three chemicals deciding the taste of cheese, acetic acid, hydrogen sulfide and lactic acid, denoted by Acetic, H2S and Lactic, respectively. Note that the first two predictors are log-transformed.

Based on scatterplots and residual analysis, we decided that the regression of Taste on H2S and Lactic as the best regression model, Taste = β_0 + β_1 H2S + β_2 Lactic + τ, where $\tau \sim N(0, \sigma^2)$. Note that the maximum-likelihood estimate for $\boldsymbol{\theta} = (\beta_0, \beta_1, \beta_2, \sigma^2)'$ was $\hat{\boldsymbol{\theta}} = (-27.5918, 3.94627, 19.8872, 88.9655)'$.

For further inferences suppose one is interested in constructing confidence sets for (β_2, σ^2). In this case β_0 and β_1 are nuisance parameters. We constructed three 95 % numerical profile confidence regions for (β_2, σ^2), the profile likelihood confidence region, the Wald confidence region based on the original parameterisation $\boldsymbol{\theta}$ and the Wald confidence region based on a new parameterisation $\boldsymbol{\theta}^\star = (\beta_0, \beta_1, \beta_2, \log \sigma^2)$. Note that the log transformation for σ^2 was used to eliminate the nonnegativity constraint. We will calibrate the confidence level for the elevation c using the limiting distribution of the likelihood-ratio statistic. In other

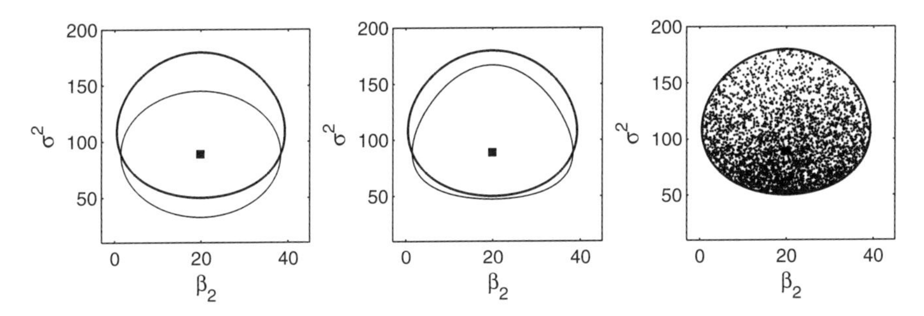

Figure 4.1 A 95 % numerical profile likelihood contour set (thick black) for (β_2, σ^2): superimposed on a numerical Wald set using $\boldsymbol{\theta}$ (thin black, left), superimposed on a numerical Wald set using $\boldsymbol{\theta}^{\star}$ (thin black, middle) and modal simulation-based profile plot (black dots, right).

words, $Conf_4(c) = 0.8002$ (corresponding to $Conf_2(c) = 0.95$) will be based on the chi-square distribution with 4 degrees of freedom.

The left and middle plots in Figure 4.1 are a 95 % numerical Wald set for (β_2, σ^2) based on $\boldsymbol{\theta}$ (thin black, left) and based on $\boldsymbol{\theta}^{\star}$ (thin black, middle), respectively, that are superimposed on the profile likelihood contour set (thick black). We observe a discrepancy between the likelihood set and both Wald sets, and the shape and spread of the Wald sets depend on parameterisations. The Wald set based on $\boldsymbol{\theta}$ is elliptical, unlike the profile likelihood set, and its upper limit of range of σ^2 is much lower than that of the profile likelihood set. It appears that the intersection of the profile likelihood set with the Wald set based on $\boldsymbol{\theta}^{\star}$ is much larger than the intersection of the profile likelihood set with the Wald set based on $\boldsymbol{\theta}$ because the log parameterisation in $\boldsymbol{\theta}^{\star}$ gets the spread of σ^2 better, even though it still loses in shape.

In order to reconstruct the 95 % numerically calculated profile likelihood set for (β_2, σ^2) using the modal simulation, we first generated 2000 rays from the $4(= p)$-dimensional standard normal distribution and used the same elevation c corresponding to $Conf_4(c) = 0.8002$ in the full parameter space. We found that there was no simulation outlier. The right-hand plot in Figure 4.1 shows 95 % modal simulation-based profile plots (black dots) for (β_2, σ^2) overlaid on a numerical profile likelihood contour set (thick black). We see that the points for (β_2, σ^2) in $\tilde{\boldsymbol{\theta}}$ simulated from the modal simulation succeeded in recovering the numerical profile likelihood set in terms of shape and spread.

4.4 Comparison between the likelihood regions and the Wald regions

As shown in visual analysis of Kim and Lindsay (2011) and Section 4.3.2, there appears to exist model/data situations where the likelihood regions are not similar to the ellipsoidal Wald regions. Unfortunately, there is no published guidance to assess the adequacy of using the Wald sets and the Fisher information for finite samples. In this section we propose two diagnostics designed to assess the difference between the likelihood set and the Wald set for the parameters in the full-dimensional parameter space. Note that these diagnostics are based on a set of samples simulated from the modal simulation introduced in Section 4.3.

4.4.1 Volume/volume error of the confidence regions

An idea of the first diagnostic is to estimate the volumes of the Wald set and the likelihood set, and to measure similarity/discrepancy between the two sets. To do so, we use an important feature of the modal simulation method: we are sampling uniformly on the sphere in the information-transformed coordinate system. Note that by information-transformed coordinate system we mean a new parameter space η from the original parameter space $\boldsymbol{\theta}$ given by $\eta \equiv \epsilon \mathbf{z} = V_{\hat{\theta}}^{-1/2}[\boldsymbol{\theta}(\epsilon) - \hat{\boldsymbol{\theta}}]$. That is, for given $\mathbf{z}$ and elevation c, the modal simulation provides sampled rays $\tilde{\epsilon}\mathbf{z}$ that reach the boundary of the targeted likelihood set in this new coordinate system. Thus, we can estimate the volume of the likelihood set in this coordinate system simply by using the sampled lengths of our rays, $||\tilde{\epsilon}\mathbf{z}||$. It will be useful in the following to rewrite the sampled rays in terms of their directions expressed as unit vectors, $u = \tilde{\epsilon}\mathbf{z}/||\tilde{\epsilon}\mathbf{z}||$. In the star-shaped case, the boundary point of the region is found when the ray has length $L(u) = ||\tilde{\epsilon}\mathbf{z}||$, so that the point on the boundary can be written as $\tilde{\epsilon}\mathbf{z} = L(u)u$.

Assume for a moment that the parameter space is just two-dimensional. If the likelihood region is star-shaped, we can calculate its volume using polar coordinates (r, ϕ). We first convert the unit vector to polar coordinates as $u = (\cos(\phi), \sin(\phi))$, and then write the length as a function of the angle as $L(\phi)$ instead of $L(u)$. Then, the volume of the star-shaped set can be computed by integration in these coordinates as

$$V(L) \equiv \int_0^{2\pi} \int_0^{L(\phi)} r \mathrm{d}r \mathrm{d}\phi = \frac{1}{2} \int_0^{2\pi} L(\phi)^2 \mathrm{d}\phi. \tag{4.6}$$

We can write the last integral in Equation (4.6) as $2\pi E[L(\Phi)^2]$, where Φ is uniformly distributed on zero to 2π. Thus, we can estimate this volume in the new coordinate system as

$$\hat{V}(L) = \frac{2\pi}{2} \frac{1}{B} \sum_{b=1}^{B} ||(\tilde{\epsilon}\mathbf{z})_b||^2, \tag{4.7}$$

for B sampled rays $(\tilde{\epsilon}\mathbf{z})_b$. For the Wald set in the new coordinates, the lengths would be a constant based on the targeted confidence level and thus the Wald set is just a ball, with easily computed volume: $\hat{V}(W) = (2\pi/2)\sqrt{c_w^2}$, where c_w is the critical value corresponding to the confidence level of interest.

This idea extends easily to the $p(> 2)$-dimensional case by using hyperspherical coordinates. Suppose there are B sampled p-dimensional rays at an elevation c, $(\tilde{\epsilon}\mathbf{z})_b$, where $b = 1, \ldots, B$, and c_w is the critical value of the Wald set corresponding to $Conf_p(c)$. Then the estimated volumes of the likelihood set and the Wald set, denoted by $\hat{V}(L)$ and $\hat{V}(W)$, are

$$\hat{V}(L) = v_p \frac{1}{B} \sum_{b=1}^{B} ||(\tilde{\epsilon}\mathbf{z})_b||^p, \quad \hat{V}(W) = v_p \sqrt{c_w^p}, \tag{4.8}$$

where v_p is the volume of the unit hypersphere in p-dimensional space.

Once one obtains the estimated volume of both sets for a targeted confidence level, one can measure similarity in the volumes between the two sets. To do so, we first estimate the volume of the set I that equals the intersection of the Wald and likelihood sets, denoted by $\hat{V}(I)$, by averaging the minimum of $\{||\tilde{\epsilon}\mathbf{z}||^p, \sqrt{c_w^p}\}$ over B sampled rays for a targeted confidence level. Then we can use relative volumes $R_{IL} = \hat{V}(I)/\hat{V}(L)$ and $R_{IW} = \hat{V}(I)/\hat{V}(W)$ as measures of agreement between the two confidence sets.

Given a value of $\hat{V}(I)$, we can calculate the discrepancy in the estimated volumes between the two sets by computing both one-sided volumetric errors. One is the error in the region of points that are in the likelihood region but not the Wald region, denoted by VE_+, and the other the error in the region of points in the Wald region but not the likelihood region, denoted by VE_-. These two volumetric errors can give a clue to the basic shape differences between the sets. One can calculate them as follows: $VE_+ = \hat{V}(L) - \hat{V}(I)$ and $VE_- = \hat{V}(W) - \hat{V}(I)$.

If the likelihood set and the Wald set are similar, both volumetric errors would be negligibly small so that the two agreement measures, R_{IL} and R_{IW}, would be close to 1. If the likelihood set is strictly outside the Wald set in all the coordinates, R_{IL} would be less than 1 but R_{IW} would be exactly 1. If both R_{IL} and R_{IW} are less than 1, the one set would not contain the other set in every direction; for some coordinates the likelihood set would be larger than the Wald set, but for other coordinates the other way around.

4.4.2 Differences in univariate intervals via worst case analysis

The motivation of the second diagnostic is to evaluate the worst case differences between the one-dimensional intervals over a class of parameters. Suppose we consider the class of parameters of interest to be all linear combinations of the normalised parameters $\boldsymbol{\eta}$, which are themselves linear combinations of the original parameter $\boldsymbol{\theta}$. If one considers only half-intervals, by which we mean the length from the MLE $\hat{\boldsymbol{\theta}}$ to the endpoint, a profile confidence interval for a linear combination $\mathbf{a}^T\eta$, where $\mathbf{a}$ is a vector with length 1, can be viewed as the projection of the likelihood set onto a particular axis. Its endpoints are determined by an interval based on the ray of the modal simulation in the information-transformed coordinate system, $[\inf \mathbf{a}^T(\epsilon\mathbf{z}), \sup \mathbf{a}^T(\epsilon\mathbf{z})]$. Thus the most extreme cases for such an interval involve finding the vectors $\mathbf{a}$ that maximise and minimise these endpoints.

One considers an arbitrary ray $\mathbf{z}_r$ going to the boundary of the likelihood set from $\hat{\boldsymbol{\theta}}$, and find the ray with the longest length, $\mathbf{z}_r^*$. In practice we find the sampled ray $\epsilon\mathbf{z}$ with the largest $||\epsilon\mathbf{z}||$ in the modal simulation. If one considers the linear combination with $\mathbf{a} = \mathbf{z}_r^*/||\mathbf{z}_r^*||$, this gives the largest endpoint, namely $||\mathbf{z}_r^*||$. Similarly, the shortest length is found by minimising $||\epsilon\mathbf{z}||$ in the modal simulation. Finally, these numbers can be divided by $\sqrt{c_w}$ to see how much shorter or longer the likelihood half-interval would be in the worst case. We denote a ratio of the shortest (longest) likelihood half-interval to that of the Wald set in the worst case by *S_HL* (*L_HL*). Note that one can use the linear combinations **a** that produce the maximum and minimum to identify the most troublesome parameters, which then might provide guidance on parameter transformations.

4.4.3 Illustrative example (revisited)

In the multiple linear regression example of Section 4.3.2 we visually showed that the profile likelihood set and the Wald sets could be different in terms of shape and spread, and different parameterisations could produce different shapes of the Wald confidence sets. In this section we apply both proposed diagnostics to the multiple linear regression example of Section 4.3.2 in order to investigate their performances in detecting differences between the confidence sets observed from the visual analysis.

In order to obtain the simulation outputs for both diagnostics we used 2000 **z**s from the four (=p)-dimensional standard normal distribution and the following four elevations c corresponding to $Conf_4(c)$ = (0.1017, 0.3918, 0.5721, 0.8435) in the full parameter space. Note that the elevations c for the four confidence levels were obtained from the limiting distribution of the likelihood-ratio statistic (i.e. the chi-square distribution with 4 degrees of freedom). We found that there was no simulation outlier.

Since the second diagnostic is based on a one-dimensional confidence level, we use $Conf_1(c)$ as the nominal confidence of a one-dimensional profile using $Conf_4(c)$ in the full parameter space. Note that the four confidence levels above correspond to $Conf_1(c) = (0.7, 0.9, 0.95, 0.99)$ in a one-dimensional space.

The visual analysis in Section 4.3.2 showed that the Wald set based on $\boldsymbol{\theta}^\star$ was more similar to the profile likelihood set than the Wald set based on $\boldsymbol{\theta}$. However, it was based on a single profile, and our diagnostics target the full four-dimensional confidence set. Nonetheless, we might expect that the two diagnostics should be able to show that $\boldsymbol{\theta}^\star$ is superior to $\boldsymbol{\theta}$ in terms of similarity to the likelihood confidence set. Thus, given **z** and the elevations of interest, we obtained the sampled rays $\tilde{\epsilon}\mathbf{z}$ in the information-transformed coordinates for both parameterisations, $\boldsymbol{\theta} = (\beta_0, \beta_1, \beta_2, \sigma^2)$ and $\boldsymbol{\theta}^\star = (\beta_0, \beta_1, \beta_2, \log\sigma^2)$.

Table 4.1 shows the results from the volumetric analysis and worst case analysis. In the volumetric analysis based on both parameterisations we observe that the discrepancy between the two sets at each parameterisation resulted from mainly VE_+, an error in the regions where the likelihood set included points outside the Wald sets, especially for small/moderate confidence levels. Note that the parameterisation $\boldsymbol{\theta}^\star$ appeared to produce a larger agreement set I than $\boldsymbol{\theta}$ so that the two relative volumes $R_{IL} = \hat{V}(I)/\hat{V}(L)$ and $R_{IW} = \hat{V}(I)/\hat{V}(W)$ for $\boldsymbol{\theta}^\star$ were much closer to

Table 4.1 Regression data: volumetric analysis and worst case analysis.

$Conf_1(c)$		Volumetric analysis					Worst case analysis	
		$\hat{V}(I)$	$VE+$	$VE-$	$\frac{\hat{V}(I)}{\hat{V}(W)}$	$\frac{\hat{V}(I)}{\hat{V}(L)}$	S_HL	L_HL
0.70	$\boldsymbol{\theta}$	1.034	0.181	0.120	0.896	0.851	0.846	1.205
	$\boldsymbol{\theta}^{\star}$	1.071	0.110	0.083	0.928	0.907	0.941	1.068
0.90	$\boldsymbol{\theta}$	6.224	2.112	1.096	0.850	0.746	0.773	1.359
	$\boldsymbol{\theta}^{\star}$	6.540	1.222	0.781	0.893	0.843	0.911	1.114
0.95	$\boldsymbol{\theta}$	12.248	5.509	2.509	0.830	0.690	0.739	1.450
	$\boldsymbol{\theta}^{\star}$	12.947	3.092	1.810	0.877	0.807	0.895	1.139
0.99	$\boldsymbol{\theta}$	35.029	25.712	8.993	0.796	0.577	0.680	1.660
	$\boldsymbol{\theta}^{\star}$	37.384	13.510	6.638	0.849	0.735	0.868	1.193

1 than those for $\boldsymbol{\theta}$. This observation is in agreement with what the numerical profile likelihood and two Wald sets for (β_2, σ^2) showed in Figure 4.1.

The second diagnostic, the worst case analysis at $\boldsymbol{\theta}$ and $\boldsymbol{\theta}^{\star}$, provides similar results. As the confidence level increased, the likelihood half-interval became much longer or shorter than the Wald set in the worst case. In the case of the $\boldsymbol{\theta}^{\star}$ parameterisation the length of the likelihood half-interval was less than 20 % longer (or shorter) than that of the Wald set. However, when the parameterisation $\boldsymbol{\theta}$ was considered, the worst case likelihood half-interval was longer than the Wald set by over 30 % for moderate/large confidence levels.

From the analysis implemented above we find that the proposed diagnostics can give information in assessing differences between the likelihood confidence set and the Wald set, and identifying the reasons for these differences. These tools are also useful for finding better parameterisation in approximating the likelihood confidence set.

4.5 Application to a finite mixture model

Sections 4.3 and 4.4 illustrated application of the modal simulation and both proposed diagnostics to a case where the targeted likelihood confidence region contains a single critical point, the MLE itself. In this section we show how one can employ the modal simulation and both proposed diagnostics in a finite mixture model where one can construct a locally identifiable likelihood confidence region. Since the finite mixture model has a complex topology of the likelihood regions as a consequence of nonidentifiabilities on the parameters, we review nonidentifiabilities on the mixture parameters and theories relevant to the construction of the likelihood confidence regions presented in Kim and Lindsay (2010). For better understanding we use two simulated datasets generated from a two-component normal mixture model with equal variances.

4.5.1 Nonidentifiabilities and likelihood regions for the mixture parameters

We consider a two-component normal mixture model with equal variances:

$$p(y \mid \boldsymbol{\theta}) = \pi_1 N(y; \xi_1, \omega) + \pi_2 N(y; \xi_2, \omega), \tag{4.9}$$

where $\pi_1 + \pi_2 = 1$ and $\boldsymbol{\theta} = [(\pi_1, \xi_1, \omega)', (\pi_2, \xi_2, \omega)']$, each column corresponding to a component. Here ξ_1 and ξ_2 are component-specific parameters that represent the mean parameters and ω is a structural parameter that represents a common variance parameter.

We first simulated 500 observations from $\boldsymbol{\theta}_\tau = [(0.4, -1, 1)', (0.6, 1, 1)']$ and obtained the MLE for $\boldsymbol{\theta}$ using the expectation maximisation (EM) algorithm (Dempster *et al.*, 1977). The estimated MLE had parameters $\hat{\boldsymbol{\theta}} = [(0.588, 0.948, 0.89)', (0.411, -1.003, 0.89)']$. Note that we carried out systematic algorithmic searches to verify that $\hat{\boldsymbol{\theta}}$ is the MLE at this dataset, and we did not find any secondary modes.

In a standard mixture problem where the number of components is fixed, there is no identifiability of the parameters $\boldsymbol{\theta}$ due to *labelling nonidentifiability* (Redner and Walker, 1984). This nonidentifiability occurs because the mixture parameters $\boldsymbol{\theta}$ are only identifiable up to a column permutation of $\boldsymbol{\theta}$. For example, $\boldsymbol{\theta}_\tau$ and the MLE $\hat{\boldsymbol{\theta}}$ in Equation (4.9) have the same density and likelihood as $\boldsymbol{\theta}_\tau^\sigma = [(0.6, 1, 1)', (0.4, -1, 1)']$ and $\hat{\boldsymbol{\theta}}^\sigma = [(0.411, -1.003, 0.89)', (0.588, 0.948, 0.89)']$, respectively, where $\boldsymbol{\theta}^\sigma$ is the column permutation of $\boldsymbol{\theta}$. Thus $\hat{\xi}_1$ in $\hat{\boldsymbol{\theta}}$ could be interpreted as either 0.948 or -1.003.

More generally, in a K-component mixture there are $K!$ true values corresponding to all possible column permutations of $\boldsymbol{\theta}_\tau$. If $\hat{\boldsymbol{\theta}}$ is an MLE mode of the mixture likelihood, the modes of the mixture likelihood come in a *modal group* of $K!$ points because the likelihood has (at least) $K!$ MLE modes, corresponding to the column permutations of $\hat{\boldsymbol{\theta}}$.

Despite this problem, it is common to treat the parameters as identifiable when doing a likelihood-based analysis because (i) the point estimators for the parameters can be defined uniquely by the output of the likelihood maximisation algorithm and (ii) there is a forms of *asymptotic identifiability* (Redner and Walker, 1984), that one can appeal to in order to enable inference on identifiable parameters as the sample size increases. Note that asymptotic identifiability becomes a justification for using a Wald confidence set surrounding one of the $K!$ MLE modes (asymptotic normality for an MLE mode and the Fisher information to estimate standard errors).

However, if one wishes to do inference in the original space $\boldsymbol{\theta}$, there exists another problem, *degenerate nonidentifiability* (Crawford, 1994; Lindsay, 1995). This occurs at a *degenerate parameter* $\boldsymbol{\theta}_0$ for a K-component mixture model where a mixture density at $\boldsymbol{\theta}_0$ has the same density as the parameter with fewer than K components (columns). In our example of Equation (4.9), a degenerate parameter corresponds to a parameter $\boldsymbol{\theta}$ that generates the same density as a parameter with a single component. For instance, $\boldsymbol{\theta}_0 = [(\pi_1, \xi_0, \omega)', (\pi_2, \xi_0, \omega)']$,

$[(0, \xi, \omega)', (1, \xi_0, \omega)']$ and $[(1, \xi_0, \omega)', (0, \xi, \omega)']$ are in the boundary of the parameter space in Equation (4.9) and all generate a one-component density with parameter ξ_0 for arbitrary ξ, π_1 and ω.

In our example the number of components is known, so we are not in a boundary situation. However, as shown in Kim and Lindsay (2010), the degenerate class of parameters introduces some pathologies into the topological structure of the likelihood regions, and one still faces the problem that parameter values near this boundary will create challenges to asymptotic inference methods. Moreover, the parameters with $\xi_1 = \xi_2$ are in the interior of the cross-product space for ξ_1 and ξ_2, which is related to labelling nonidentifiability.

Based on a detailed study of the topology of the mixture likelihood, Kim and Lindsay (2010) showed that there exists a limited range of confidence levels at which one can construct a likelihood confidence region for the mixture parameters that is reasonable in the sense that it is an identifiable subset of the full parameter space (i.e. a locally identifiable confidence set located around one of the $K!$ MLE modes). Beyond that range there is no natural way of constructing the meaningful confidence sets because the likelihood set contains the parameters displaying both types of nonidentifiability. When the confidence level was suitable for constructing a locally identifiable confidence set, those authors said that the likelihood had induced *topological identifiability* for the parameters at that level. This, in turn, meant that one could, in a meaningful way, ascribe labels to the component parameters in the confidence set.

For our simulated example, the component-specific parameter was univariate and there was no secondary mode. In this case there is an identifiability rule: if the component-specific parameter is univariate and there is no secondary mode, then the mixture likelihood induces a locally identifiable confidence region for any confidence level below $Conf_p(\psi_{dgnt})$, where ψ_{dgnt} is the highest elevation in the degenerate class of parameters (i.e. the likelihood of the MLE for a $(K-1)$-component mixture model).

In order to apply the rule described above to our simulated example, we first calculated the maximum of the likelihood, ψ_{dgnt}, among the degenerate parameters, which here corresponded to having a single component, and found the confidence level for ψ_{dgnt} in the full dimensional parameter space, which is $Conf_4(\psi_{dgnt}) = 0.986$. This means that for an $n = 500$ simulated sample 0.986 represents the exact upper bound on the confidence levels one can use when constructing an identifiable confidence set.

In constructing the likelihood regions for the parameters of dimension 4 for our example, we can try to view the full likelihood structure through two-dimensional profiles. The left plot of Figure 4.2 is the numerical profile contour for (π_1, ξ_1) at the elevation c with $Conf_2(c) = 0.95$ (corresponding to $Conf_4(c) = 0.80$). Here we provide a numerical profile likelihood contour that was obtained by maximising $L(\boldsymbol{\theta})$ over (ξ_2, ω) for each fixed (π_1, ξ_1). We can see that the two separated regions, one corresponding to the MLE $\hat{\boldsymbol{\theta}}$ with $(\pi_1, \xi_1) = (0.588, 0.948)$ and one corresponding to the permuted mode $\hat{\boldsymbol{\theta}}^{\sigma}$ with $(\pi_1, \xi_1) = (0.411, -1.003)$, have appeared in the profile plot. These are the profile images of both modal regions in the full four-dimensional parameter space.

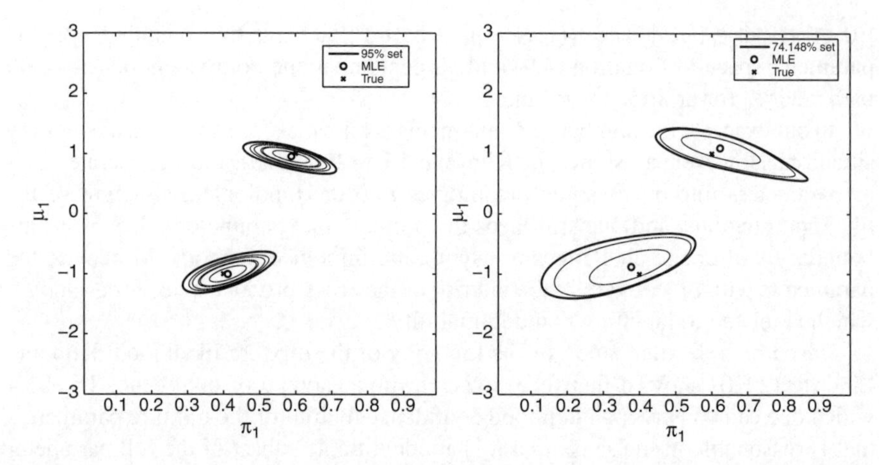

Figure 4.2 Numerical profile likelihood contours for (π_1, ξ_1) with $\hat{\boldsymbol{\theta}}(o)$ and $\boldsymbol{\theta}_\tau(x)$: $n = 500$ with $Conf_2(c) = 0.95$ (left) and $n = 100$ with $Conf_2(c) = 0.7415$ (right).

We observe that the likelihood confidence region is separable into two identifiable subsets, say C_1(upper region) and C_2 (lower region), one for each of the two modes of the likelihood. By the 'island counting property' described at the end of Section 4.2.2, it follows that the full four-dimensional parameter confidence region also separates into two identifiable sets at the same elevation, with $Conf_4(c) = 0.80$. This result agrees with the identifiability rule mentioned above, which applies here because $0.80 < Conf_4(\psi_{dgnt}) = 0.986$. Due to identifiability, one can use labels to describe the confidence sets. For example, one can call C_1 the confidence set for the first component (π_1, ξ_1) and C_2 the confidence set for the second component (π_2, ξ_2).

Moreover, it appears that two identifiable subsets C_1 and C_2 generated by the likelihood regions are approximately elliptical. In this case one might be confident in using the Wald set and the Fisher information for standard errors for the MLE mode.

In the same simulation model reducing sample size can lead to a dramatic change in the topological structure of the mixture likelihood (Kim and Lindsay, 2010). We simulated another data set with 100 observations from the same mixture model, $\boldsymbol{\theta}_\tau = [(0.4, -1, 1)', (0.6, 1, 1)']$, and obtained the MLE for $\boldsymbol{\theta}$: $\hat{\boldsymbol{\theta}} = [(0.623, 1.094, 0.83)', (0.377, -0.879, 0.83)']$. Here we again did not find any secondary modes. In this case the exact upper bound on the confidence levels generating a locally identifiable confidence set around $\hat{\boldsymbol{\theta}}$ was $Conf_4(\psi_{dgnt}) = 0.505$ (corresponding to $Conf_2(\psi_{dgnt}) = 0.816$). This means that the confidence regions for both modes at elevation c with $Conf_4(c) = 0.80$ (corresponding to $Conf_2(c) = 0.95$) must be connected through the degenerate set in the full parameter space. In a profile confidence set for (π_1, ξ_1) at any confidence level above 81.6 %, therefore, both modes at $(\pi_1, \xi_1) = (0.623, 1.094)$ and $(\pi_1, \xi_1) = (0.377, -0.879)$ will be in one connected island region.

In order to construct identifiable likelihood regions for a case of $n = 100$, one can increase the elevation, which corresponds to reducing the confidence level of interest. The right plot of Figure 4.2 is the numerical profile contour for (π_1, ξ_1) at the elevation with $Conf_2(c) = 0.7415$ (corresponding to $Conf_4(c) = 0.3918$). We observe that the profile, and hence the full likelihood regions, have two identifiable subsets, one for each MLE mode.

There is an interesting observation, which must be noted here. At $Conf_2(c) = 0.7415$ and $n = 100$ the shapes of two constructed modal regions appear to bear less resemblance to two ellipses, even though they are star-shaped. In this case using the Wald sets to describe the likelihood sets for the parameter of interest is likely to be a poor approximation.

4.5.2 Mixture likelihood region simulation and visualisation

To show the performance of the modal simulation strategy in reconstructing the mixture profile likelihood confidence sets, we construct the modal simulation-based profile plots for (π_1, ξ_1) in Figure 4.3. The left plot of Figure 4.3 shows the modal simulation-based profile plot for (π_1, ξ_1) at the elevation c with $Conf_2(c) = 0.95$ when $n = 500$. For the modal simulation we used 3000 rays and found no star-shaped region outlier.

We simulated points on the boundaries of the targeted likelihood region for just one MLE mode $\hat{\boldsymbol{\theta}}$, setting the first component to be $(\pi_1, \xi_1) = (0.588, 0.948)$ in $\hat{\boldsymbol{\theta}}$. Thus, there is just one nearly elliptical, and locally identifiable, 95 % profile confidence region, corresponding the upper mode in the left plot of Figure 4.2. Note that one can visualize the second profile modal region by simply plotting the simulated parameter values from the second component. We observe that the simulated samples successfully describe the boundary of the elliptical-shaped likelihood contour.

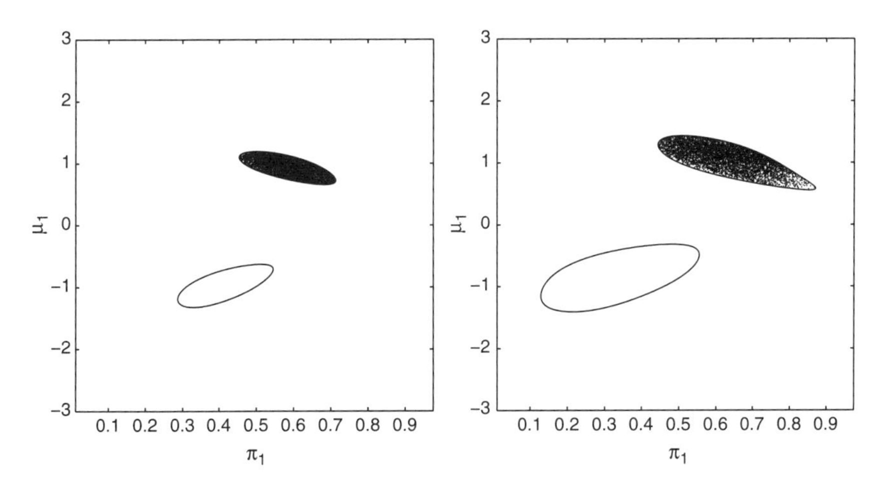

Figure 4.3 Modal simulation-based profile plots (black dots) for (π_1, ξ_1): $n = 500$ with $Conf_2(c) = 0.95$ (left) and $n = 100$ with $Conf_2(c) = 0.7415$ (right).

Table 4.2 Two simulated examples: volumetric analysis and worst case analysis.

	$Conf_1(c)$	Volumetric analysis					Worst case analysis	
		$\hat{V}(I)$	$VE+$	$VE-$	$\frac{\hat{V}(I)}{\hat{V}(W)}$	$\frac{\hat{V}(I)}{\hat{V}(L)}$	S_HL	L_HL
$n = 500$	0.70	0.935	0.062	0.219	0.810	0.938	0.725	1.151
	0.90	5.870	0.808	1.450	0.802	0.879	0.689	1.249
	0.95	11.761	2.265	2.996	0.797	0.839	0.671	1.324
	0.99	34.471	12.562	9.551	0.783	0.733	0.641	1.532
$n = 100$	0.70	1.011	0.366	0.143	0.876	0.735	0.777	1.482
	0.90	6.038	6.386	1.282	0.825	0.486	0.705	2.271

The nonexistence of star-shaped region outliers suggests that both modal regions are reasonably shaped and well described by the simulation.

To see the performance of a simulation method at the $n = 100$ case, we construct the simulation-based profile plot for (π_1, ξ_1) at the elevation c with $Conf_2(c) = 0.7415$ (see the right plot of Figure 4.3). Here we used 3000 rays in the modal simulation and there was no star-shaped region outlier. We see that the sampled values for (π_1, ξ_1) successfully captured the numerical boundary of the star-shaped region for $(\pi_1, \xi_1) = (0.623, 1.094)$.

4.5.3 Adequacy of using the Wald confidence region

In order to assess the differences between the likelihood confidence region and the Wald confidence region for both simulated datasets ($n = 500$ and $n = 100$), we apply both diagnostics introduced in Section 4.4. These diagnostics need outputs from the modal simulation at the confidence levels that generate locally identifiable confidence regions. Thus, we used 3000 rays and the following confidence levels: $Conf_4(c) = (0.1017, 0.3918, 0.5721, 0.8435)$ for $n = 500$ and $Conf_4(c) = (0.1017, 0.3918)$ for $n = 100$. Note that for these two datasets one can construct locally identifiable confidence sets for any confidence levels below the level for degeneracy, $Conf_4(\psi_{dgnt}) = 0.9856$ for $n = 500$ and 0.5050 for $n = 100$. We found that there was no star-shaped region outlier for the confidence levels used for both datasets. The corresponding values of $Conf_1(c)$ are $(0.7, 0.9, 0.95, 0.99)$ for $n = 500$ and $(0.7, 0.9)$ for $n = 100$, and the corresponding values of $Conf_1(\psi_{dgnt})$ for $n = 500$ and $n = 100$ are 0.9996 and 0.9344, respectively.

Table 4.2 shows the results from the volumetric analysis and worst case analysis for both simulated data sets. In the volumetric analysis of the $n = 500$ case we observe that the discrepancy between the two sets for the confidence levels, 0.70 and 0.90, results mainly from $VE-$, an error in the regions where the Wald set includes points outside the likelihood set. However, when $Conf_1(c)$ was closer to the level of degeneracy, both volumetric errors are increased, relative to the agreement set $\hat{V}(I)$. At $Conf_1(c) = 0.99$ a volumetric error, $VE+$, is much bigger than the counterpart, $VE-$, so that the likelihood regions are much larger.

For the second diagnostic, the worst case analysis at $n = 500$ shows similar results. For the small/moderate confidence level the likelihood half-interval was shorter than the Wald set by about 30 % in the worst case. When the confidence level approaches the level of degeneracy, say $Conf_1(c) = 0.99$, the likelihood half-interval was longer than the Wald one by more than 50 % in the worst case, but could also be as much as 35 % shorter. In general, we must conclude that the likelihood sets, and the corresponding profiles, are really very different from the Wald sets in this example.

The volumetric analysis at sample size 100 shows that $VE+$ was large compared to $VE-$, so that the likelihood set was much bigger than the Wald set, especially for the confidence level close to $Conf_1(\psi_{dgnt}) = 0.9344$. In the worst case analysis the length of the likelihood half-interval was much longer (shorter) than that of the Wald set by more than 200 % (30 %) when $Conf_1(c) = 0.90$. In this example there is even a stronger difference between the two types of region than in the example with $n = 500$.

Although statistical users should draw their own conclusions, we are not confident in using the Wald confidence sets (and the Fisher information for standard errors) when either $n = 100$ or the confidence levels are close to the level of degeneracy at $n = 500$, say 0.99.

4.6 Data analysis

In this section we provide two real examples to show application of the two diagnostics proposed in Section 4.4 as tools for comparing the likelihood and the Wald sets in a finite mixture model. In the modal simulation we use the chi-square distribution as the distribution of the likelihood-ratio statistic for calculating the corresponding elevations for the confidence levels of interest.

Example 1: SLC data

The first example concerns red blood cell sodium–lithium countertransport (SLC) data from 190 individuals that are characterised by a single type of three genotypes A_1A_1, A_1A_2 and A_2A_2; see the left-hand plot of Figure 4.4 for a histogram. Roeder (1994) fitted a three-component normal mixture model with equal variances to the SLC data and the computed MLE for $\boldsymbol{\theta} = [(\pi_1, \xi_1, \omega)', (\pi_2, \xi_2, \omega)', (\pi_3, \xi_3, \omega)']$ was $\hat{\boldsymbol{\theta}} = [(0.777, 0.224, 0.003)', (0.200, 0.380, 0.003)', (0.023, 0.579, 0.003)']$. Since ξ_j is univariate and we have only the MLE modal group, one can construct identifiable confidence regions at any confidence level below $Conf_6(\psi_{dgnt}) = 0.86$ ($Conf_1(\psi_{dgnt}) = 0.998$), which is the confidence level corresponding to the elevation of the MLE for a class of degenerate parameters.

We first obtained a set of samples describing the targeted likelihood sets from a modal simulation. Here we used the four confidence levels, $Conf_6(c) = (0.017, 0.155, 0.302, 0.644)$ (corresponding to

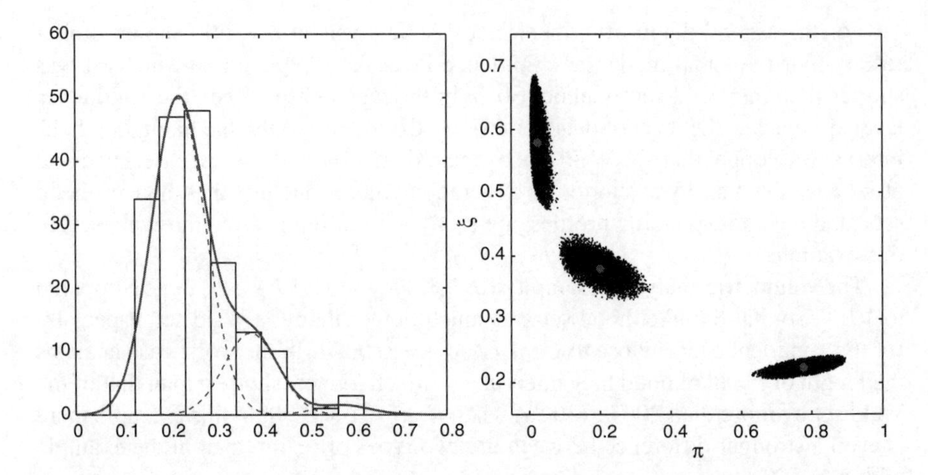

Figure 4.4 SLC data: a histogram and a fitted mixture density (solid line) (left); a simulation-based profile plot for (π, ξ) with $Conf_1(c) = 0.99$ (circles are the MLEs for π and ξ) (right).

$Conf_1(c) = (0.70, 0.90, 0.95, 0.99))$, and 4000 rays for each level. Note that there was no star-shaped region outlier.

Table 4.3 shows the volume-based analysis and worst case analysis based on the simulation outputs. We can observe that an error in the regions where the likelihood region excludes points from the Wald set, VE_-, is responsible for the dissimilarity between the two sets, especially when the confidence levels are 0.70 and 0.90. Both volumetric errors became larger relative to the agreement set $\hat{V}(I)$ as the confidence level approaches the level of degeneracy. If we carry out the worst case analysis we find that the likelihood half-interval was shorter than the Wald one by at least 35 % in the worst case, even for small/moderate $Conf_1(c)$. When $Conf_1(c)$ is equal to or larger than 0.95, the likelihood half-interval was longer (shorter) than the Wald one by more than 35 % in the worst case. Thus our conclusions are very similar to the simulated examples. The Wald set is a poor approximation to the likelihood set.

The right-hand plot of Figure 4.4 is a simulation-based profile plot for (π, ξ) with $Conf_1(c) = 0.99$, which shows three separated sets.

Table 4.3 SLC data: volumetric analysis and worst case analysis.

$Conf_1(c)$	Volumetric analysis					Worst case analysis	
	$\hat{V}(I)$	$VE+$	$VE-$	$\frac{\hat{V}(I)}{\hat{V}(W)}$	$\frac{\hat{V}(I)}{\hat{V}(L)}$	S_HL	L_HL
0.70	0.802	0.081	0.438	0.647	0.908	0.654	1.175
0.90	12.464	2.701	7.340	0.629	0.822	0.633	1.314
0.95	35.217	10.847	21.470	0.621	0.765	0.621	1.399
0.99	176.284	106.153	115.797	0.604	0.624	0.575	1.703

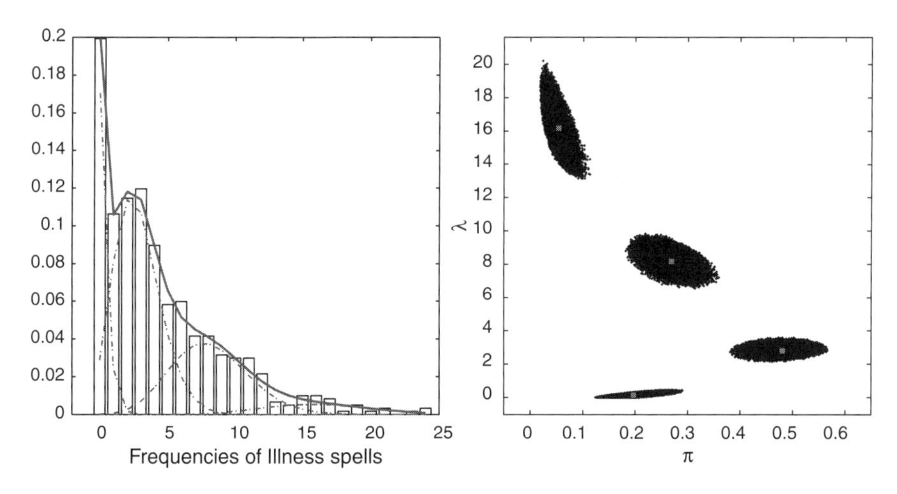

Figure 4.5 Morbidity data: a histogram and a fitted mixture density (solid lines) (left); a simulation-based profile plot for $(\boldsymbol{\pi}, \boldsymbol{\xi})$ with $Conf_1(c) = 0.99$ (circles are the MLEs for $\boldsymbol{\pi}$ and $\boldsymbol{\xi}$) (right).

Example 2: Morbidity data

Our second example concerns the health status of 602 preschool children checked every two weeks from June 1982 until September 1985 (Schelp *et al.*, 1990). Each child was examined if she/he showed symptoms of fever, cough, running nose or these symptoms together. The data were the frequencies of these illness spells during the study period; see the left plot of Figure 4.5 for a histogram.

Böhning *et al.*, (1992) and Schlattmann (2005) fitted a four-component Poisson mixture model to these data. The MLE for $\boldsymbol{\theta} = [(\pi_1, \xi_1)', (\pi_2, \xi_2)', (\pi_3, \xi_3)', (\pi_4, \xi_4)']$ was $\hat{\boldsymbol{\theta}} = [(0.197, 0.143)', (0.48, 2.817)', (0.269, 8.164)', (0.054, 16.156)']$ with $\ell(\hat{\boldsymbol{\theta}}) = -1553.81$. The log-likelihood at the MLE for the class of degenerate parameters (i.e. a three-component Poisson mixture model) was -1568.28 and then $Conf_7(\psi_{dgnt})$ was larger than 0.99. Since ξ_j was univariate and there was no secondary mode, $Conf_7(\psi_{dgnt})$ was the exact upper bound on the confidence levels for a locally identifiable likelihood confidence region at these data. Therefore, the $24(= 4!)$ modal regions in the 99 % full-likelihood confidence region were identifiable subsets.

To assess similarity between the likelihood and Wald sets at given data, we first applied a modal simulation using 6000 rays and three confidence levels, $Conf_7(c) = (0.089, 0.202, 0.532)$ (corresponding to $Conf_1(c) = (0.90, 0.95, 0.99)$). There was no star-shaped region outlier. Table 4.4 gives results of the volume-based analysis and worst case analysis based on modal simulation outputs. It appears from the volume-based analysis that the VE_- is relatively larger than VE_+ so that discrepancy between the two sets occurs from the regions where the the likelihood sets exclude points from the Wald sets. The size of estimated volumes in the two sets for $Conf_1(c) = 0.99$ seem to be about seven times those in the two sets for

Table 4.4 Morbidity data: volumetric analysis and worst case analysis .

$Conf_1(c)$	Volumetric analysis					Worst case analysis	
	$\hat{V}(I)$	$VE+$	$VE-$	$\frac{\hat{V}(I)}{\hat{V}(W)}$	$\frac{\hat{V}(I)}{\hat{V}(L)}$	S_HL	L_HL
0.90	26.960	2.034	5.614	0.828	0.930	0.805	1.190
0.95	91.010	9.851	20.090	0.819	0.902	0.778	1.243
0.99	605.600	115.230	146.750	0.805	0.840	0.727	1.371

$Conf_1(c) = 0.95$. The worst case analysis shows that the likelihood half-interval was longer (shorter) than the Wald one by less than 25 % in the worst case when the confidence levels are less than 0.99. On the whole, we find that the Wald sets are still quite dissimilar to the likelihood sets.

The right plot of Figure 4.5 is a simulation-based profile plot for (π, ξ) with $Conf_1(c) = 0.99$. We see four confidence sets that are separated by order constraints on ξ, not π.

4.7 Discussion

In this chapter we have proposed two diagnostics to evaluate the difference between the two commonly used confidence sets, the likelihood set and the Wald confidence sets, and showed how one can apply them to a locally identifiable confidence set in a finite mixture model. One important conclusion from our mixture examples is that the Wald sets (in the parameterisations we used) give intervals and regions for parameters of interest that were strikingly different from the likelihood regions. The disagreements between the regions were in their basic shape, not their size, and so could not be repaired in any simple fashion, such as adjusting the Wald critical value. We think that this provides evidence of the value of our simulation method. We would argue that, instead of reporting the Wald covariance matrix for the ML estimates, one could provide a pairwise scatterplot matrix based on the simulation outputs to see the correlative relationships between all the various mixture estimates. This would have the advantage of showing the asymmetries of the confidence sets relative to the ML estimates.

Our proposed diagnostics used the outputs from the modal simulation method (Kim and Lindsay, 2011) that efficiently reconstructs the boundaries of the targeted likelihood confidence sets. Since the modal simulation generates a set of samples uniformly on the sphere in the information-transformed coordinate system, one can easily estimate the volumes of the two sets and calculate volumetric errors in the region of sampled points that are in one set but not the other set. A set of sampled rays in a new coordinate system also helps us evaluate how much shorter or longer the likelihood half-interval would be than the Wald one in the worst case.

Before checking the adequacy of using the Wald confidence sets for the mixture parameters at given data, one needs to know if the confidence level of interest can generate a well-defined likelihood region that is topologically identifiable in the full-dimensional parameter space. As we showed in our examples, where the component specific parameter is univariate, this can be done by using a simple rule that provides the exact upper bound on the confidence levels where one can construct a topologically identifiable likelihood region.

We restricted ourselves to univariate examples in this chapter because, when the component-specific parameter is multivariate, it is difficult to find the upper bound for the confidence levels that generate identifiable likelihood confidence regions (Kim and Lindsay, 2010). We will be investigating this point further in the near future.

Finally, the accuracy of our proposed methods depends on the modal simulation that provides a random boundary to the targeted confidence set. It would be valuable to investigate the ability of the modal simulation to generate a sample of points from the targeted confidence set with a reasonable number of rays in higher dimensions.

References

Agresti, A. (2002) *Categorical Data Analysis*, 2nd edn. John Wiley & Sons, Ltd.

Basford, K. E., Greenway, D. R., McLachlan, G. J. and Peel, D. (1997) Standard errors of fitted means under normal mixture models. *Computational Statistics*, **12**, 1–17.

Böhning, D., Schlattmann, P. and Lindsay, B. G. (1992) C.A.MAN-computer assisted analysis of mixtures: statistical algorithms. *Biometrics*, **48**, 283–303.

Cox, D. R. and Hinkley, D. V. (1974) *Theoretical Statistics*. Chapman & Hall.

Crawford, S. L. (1994) An application of the Laplace method to finite mixture distributions. *Journal of the American Statistical Association*, **89**, 259–267.

Davison, A. C. and Hinkley, D. V. (2006) *Bootstrap Methods and Their Application*, 8th edn. Cambridge University Press.

Dempster, A. P., Laird, N. M., and Rubin, D. B. (1977) Maximum likelihood from incomplete data via the EM algorithm. *Journal of the Royal Statistical Society, Series B*, **39**, 1–38.

Efron, B. and Tibshirani, R. J. (1993) *An Introduction to the Bootstrap*. Chapman & Hall.

Hall, P. (1992) *The Bootstrap and Edgeworth Expansion*. Springer-Verlag.

Kalbfleisch, J. D. and Prentice, R. L. (2002) *The Statistical Analysis of Failure Time Data*, 2nd edn. John Wiley & Sons Ltd.

Kim, D. and Lindsay, B. G. (2010) Topology of mixture likelihood regions (submitted for publication).

Kim, D. and Lindsay, B. G. (2011) Using confidence distribution sampling to visualize confidence sets. *Statistica Sinica* (to appear).

Klemelä, J. (2009) *Smoothing of Multivariate Data: Density Estimation and Visualization*. John Wiley & Sons, Ltd.

Lang, J. B. (2008) Score and profile likelihood confidence intervals for contingency table parameters. *Statistics in Medicine Science*, **27**, 5975–5990.

Lindsay, B. G. (1995) *Mixture Models: Theory, Geometry, and Applications*. NSF-CBMS Regional Conference Series in Probability and Statistics, Vol. 5. Institute of Mathematical Statistics. Hayward, California.

McLachlan, G. J. (1987) On bootstrapping the likelihood-ratio test statistic for the number of components in a normal mixture. *Applied Statistics*, **36**, 318–324.

McLachlan, G. J. and Peel, D. (2000) *Finite Mixture Models*. John Wiley & Sons, Ltd.

Matsumoto, Y. (2002) *An Introduction to Morse Theory*. Translations of Mathematical Monographs, Vol. 208. American Mathematical Society.

Meeker, W. Q. and Escobar, L. A. (1995) Teaching about approximate confidence regions based on maximum likelihood estimation. *The American Statistician*, **49**, 48–53.

Moore, D. S. and McCabe, G. P. (1989) *Introduction to the Practice of Statistics*. W.H. Freeman.

Owen, A. B. (2001) *Empirical Likelihood*. Chapman & Hall.

Redner, R. A. and Walker, H. F. (1984) Mixture densities, maximum likelihood and the EM algorithm. *SIAM Review*, **26**, 195–239.

Roeder, K. (1994) A graphical technique for determining the number of components in a mixture of normals. *Journal of the American Statistical Association*, **89**, 487–495.

Schelp, F. P., Vivatanasept, P., Sitaputra, P., Sornmani, S., Pongpaew, P., Vudhivai, N., Egormaiphol, S. and Böhning, D. (1990) Relationship of the morbidity of under-fives to anthropometric measurements and community health intervention. *Tropical Medicine and Parasitology*, **41**, 121–126.

Schlattmann, P. (2005) On bootstrapping the number of components in finite mixtures of Poisson distributions. *Statistics and Computing*, **15**, 179–188.

5

Mixture of experts modelling with social science applications

Isobel Claire Gormley and Thomas Brendan Murphy

5.1 Introduction

Clustering methods are used to group observations into homogeneous subgroups. Clustering methods are usually either algorithmically based (e.g. k-means or hierarchical clustering) (see Hartigan, 1975) or based on statistical models (e.g. Fraley and Raftery, 2002; McLachlan and Basford, 1988).

Clustering methods have been widely used in the social sciences. Examples of clustering applications include market research (Punj and Stewart, 1983), archaeology (Hall, 2004), education (Aitkin *et al.*, 1981; Gormley and Murphy, 2006) and sociology (Lee *et al.*, 2005). In Section 5.2, we outline two applications of clustering in the social sciences: studying voting blocs in elections (Section 5.2.1) and exploring organisational structure in a corporation (Section 5.2.2).

In any cluster analysis application, it is common that clustering is implemented on outcome variables of interest without reference to concomitant covariate information on the objects being clustered. Once a clustering of objects has been produced, the user must probe the clusters to investigate their structure. Interpretations of the clusters can be produced with reference to values of the outcome variables within each cluster and/or with reference to the concomitant covariate information that was not used in the construction of the clusters.

The use of a model-based approach to clustering allows for any uncertainty to be accounted for in a probabilistic framework. Mixture models are the basis

Mixtures: Estimation and Applications, First Edition. Edited by Kerrie L. Mengersen, Christian P. Robert and D. Michael Titterington.

of many model-based clustering methods. In Section 5.3, we briefly describe the use of mixture models for clustering. The mixture of experts model (Jacobs *et al.*, 1991) provides a framework for extending the mixture model to allow the model parameters to depend on concomitant covariate information; these models are reviewed in Section 5.4.

Examples of mixture of experts models and their application are motivated in Section 5.2 and implemented for the study of voting blocs in Section 5.5 and for studying organisational structure in Section 5.6.

We conclude, in Section 5.7, by discussing mixture of experts models and their interpretation in statistical applications.

5.2 Motivating examples

5.2.1 Voting blocs

In any election, members of the electorate exhibit different voting behaviours by choosing to vote for different candidates. Differences in voting behaviour may be due to allegiance to a political party or faction, choosing familiar candidates, choosing geographically local candidates or one of many other reasons. Such different voting behaviours lead to a collection of votes from a heterogeneous population.

The discovery and characterisation of voting blocs (i.e. groups of voters with similar preferences) is of considerable interest. For example, Tam (1995) studies Asian voting behaviour within the American political arena via a multinomial logistic regression model and concludes that Asians should not be treated as a monolithic group. Holloway (1995) examines the differences between voting blocs when analysing United Nations roll call data using a multidimensional scaling technique. Stern (1993), Murphy and Martin (2003) and Busse *et al.* (2007) use mixtures of distance-based models to characterise voting blocs in the American Psychological Association presidential election of 1980. Gormley and Murphy (2008a) use a mixture of Plackett-Luce (Plackett, 1975) and Benter (1994) models to characterize voting blocs in the electorate for Irish governmental and Irish presidential elections. Spirling and Quinn (2010) use a Dirichlet process mixture model to study voting blocs in the UK House of Commons.

Many of the above studies investigate the existence of voting blocs by clustering voting data and then subsequently investigating the resulting clusters by examining the cluster parameters and by exploring concomitant voter covariates (when available) for members of each cluster. Such explorations can assist in determining what factors influence voting bloc membership and as a result voting behaviour.

A more principled approach to investigating which factors influence voting behaviour is to incorporate voter covariates in the modelling process that is used to construct the clusters. The mixture of experts framework provides a modelling structure to do this. In Section 5.5, we outline the use of mixture of experts models to characterise intended voting behaviour in the 1997 Irish presidential election.

5.2.2 Social and organisational structure

The study of social mechanisms which underlie cooperation among peers within an organisation is an important area of study within sociology and within organisations in general (Lazega, 2001). Social network analysis (e.g. Wasserman and Faust, 1994) is a highly active research area which provides an approach to examining structures within an organisation or a network of 'actors'. The study of such networks has recently attracted attention from a broad spectrum of research communities including sociology, statistics, mathematics, physics and computer science.

Specifically, social network data record the interactions (relationships) between a group of actors or social entities. For example, a social network dataset may detail the friendship links among a group of colleagues or it may detail the level of international trade between countries. Network data may be binary, indicating the presence/absence of a link between two actors, or it may be nonbinary, indicating the level of interaction between two actors. The aim of social network analysis is to explore the structure within the network, to aid understanding of underlying phenomena and the relations that may or may not exist within the network.

Many statistical approaches to modelling the interactions between actors in a network are available (Goldenberg *et al.*, 2009; Kolaczyk, 2009; Snijders *et al.*, 2006; Wasserman and Faust, 1994); many recent modelling advances tend to employ the idea of locating actors in a latent social space. In particular, Hoff *et al.* (2002) develop the idea of a latent social space and define the probability of a link between two actors as a function of their separation in the latent social space; this idea has been developed in various directions in Handcock *et al.* (2007), Krivitsky and Handcock (2008) and Krivitsky *et al.* (2009) in order to accommodate clusters (or communities) of highly connected actors in the network and other network effects. Airoldi *et al.* (2008) develop an alternative latent variable model for social network data where a soft clustering of network actors is achieved; this has been further extended by Xing *et al.* (2010) to model dynamic networks. More recently, Mariadassou *et al.* (2010) and Latouche *et al.* (2011) developed novel latent variable models for finding clusters of actors (or nodes) in network data.

In many social network modelling applications, concomitant covariate information on each actor is not used in the clustering of actors in the network. The clusters discovered in the network are explored and explained by examining the actor attributes that were not used in the clustering process. We endorse the use of mixture of experts models to provide a principled framework for clustering actors in a social network when concomitant covariate information is available for the actors.

In Section 5.6, a mixture of experts model for social network data is employed to explore the organisational structure within a northeastern USA corporate law firm.

5.3 Mixture models

Let $y_1, y_2, \ldots, y_N$ be an independent and identically distributed (iid) sample of outcome variables from a population that is modelled by a probability density

$p(\cdot)$. The mixture model assumes that the population consists of G components or subpopulations. The probability of component g occurring is τ_g and each component is modelled using a probability density $p(y_i|\theta_g)$, for $g = 1, 2, \ldots, G$. Hence, the overall model for a member of the population is of the form

$$p(y_i) = \sum_{g=1}^{G} \tau_g p(y_i|\theta_g).$$

In many mixture-modelling contexts, an augmented form of the mixture model, which includes the unknown subpopulation membership vectors l_i, for $i = 1, \ldots, N$, greatly assists computations. The augmented model is

$$p(y_i, l_i) = \prod_{g=1}^{G} \left[\tau_g p(y_i|\theta_g)\right]^{l_{ig}},$$

where $l_{ig} = 1$ if observation i comes from subpopulation g and $l_{ig} = 0$ otherwise.

Inference for the mixture model is usually implemented by maximum likelihood using the EM algorithm (Dempster *et al.*, 1977) or in a Bayesian framework using Markov chain Monte Carlo (Diebolt and Robert, 1994). The clustering of observations is based on the posterior probability of component membership for each observation,

$$\mathbf{P}(l_{ig} = 1|y_i) = \mathbf{E}(l_{ig}|y_i) = \frac{\tau_g p(y_i|\theta_g)}{\sum_{g'=1}^{G} \tau_{g'} p(y_i|\theta_{g'})}.$$

The maximum a posteriori estimate of cluster membership assigns each observation to its most probable group, thus achieving a clustering of the observations.

Among the most commonly studied and applied mixture models are the Gaussian mixture model (e.g. Fraley and Raftery, 2002) and the latent class analysis model, which is a mixture of products of independent Bernoulli models (Lazarsfeld and Henry, 1968). Extensive reviews of mixture models and their application are given in Everitt and Hand (1981), Titterington *et al.* (1985), McLachlan and Basford (1988), McLachlan and Peel (2000), Fraley and Raftery (1998, 2002) and Melnykov and Maitra (2010).

Software for fitting mixture models in R (R Development Core Team, 2009) include mclust (Fraley and Raftery, 2006), mixtools (Benaglia *et al.*, 2009) and flexmix (Leisch, 2004) among others. Other software for mixture modelling includes MIXMOD (Biernacki *et al.*, 2006) and EMMIX (McLachlan *et al.*, 1999).

5.4 Mixture of experts models

The mixture of experts model (Jacobs *et al.*, 1991) extends the mixture model by allowing the parameters of the model to be functions of an observation's concomitant covariates w_i:

$$p(y_i|w_i) = \sum_{g=1}^{G} \tau_g(w_i)p(y_i|\theta_g(w_i)). \tag{5.1}$$

Bishop (2006,Chapter 14.5) refers to the mixture of experts model as a conditional mixture model, because for a given set of concomitant covariates w_i the distribution of y_i is a mixture model.

The terminology used in the mixture of experts model literature calls the $p(y_i|\theta_g(w_i))$ densities 'experts' and the $\tau_g(w_i)$ probabilities 'gating networks'. In its original formulation in Jacobs *et al.* (1991), the model for $\tau_1(w_i), \tau_2(w_i), \ldots, \tau_G(w_i)$ is a multinomial logistic regression model and $p(y_i|\theta_g(w_i))$ is a general linear model.

Figure 5.1 illustrates a graphical model representation of the mixture of experts model. This representation aids the interpretation of the full mixture of experts model (in which all model parameters are functions of covariates (Figure 5.1(d))) and the special cases where some of the model parameters do not depend on the covariates (Figures 5.1(a) to 5.1(c)). The four models detailed in Figure 5.1 have the following interpretations.

(a) In the mixture model, the outcome variable distribution depends on the latent cluster membership variable l_i and the model is independent of the covariates w_i.
(b) In the expert-network mixture of experts model, the outcome variable distribution depends on both the covariates w_i and the latent cluster membership variable l_i; the distribution of the latent variable is independent of the covariates.
(c) In the gating-network mixture of experts model, the outcome variable distribution depends on the latent cluster membership variable l_i and the distribution of the latent variable depends on w_i.
(d) In the full mixture of experts model, the outcome variable distribution depends on both the covariates w_i and on the latent cluster membership variable l_i. Additionally, the distribution of the latent variable l_i depends on the covariates w_i.

Mixture of experts models have been employed in a wide range of modelling settings; Peng *et al.* (1996) use a mixture of experts model and a hierarchical mixture of experts model in speech recognition applications. Thompson *et al.* (1998) use a mixture of experts model for studying the diagnosis of diabetic patients. Rosen and Tanner (1999) develop a mixture of experts proportional hazards model and analyse a multiple myeloma dataset. Hurn *et al.* (2003) use MCMC to fit a mixture of regressions model, which is a special case of the mixture of experts model, but where the mixing proportions do not depend on the covariates w_i. Carvalho and Tanner (2007) use a mixture of experts model for nonlinear time-series modelling. Geweke and Keane (2007) use a model similar to the mixture of experts model, but where the gating network has a probit structure, in a number of econometric applications.

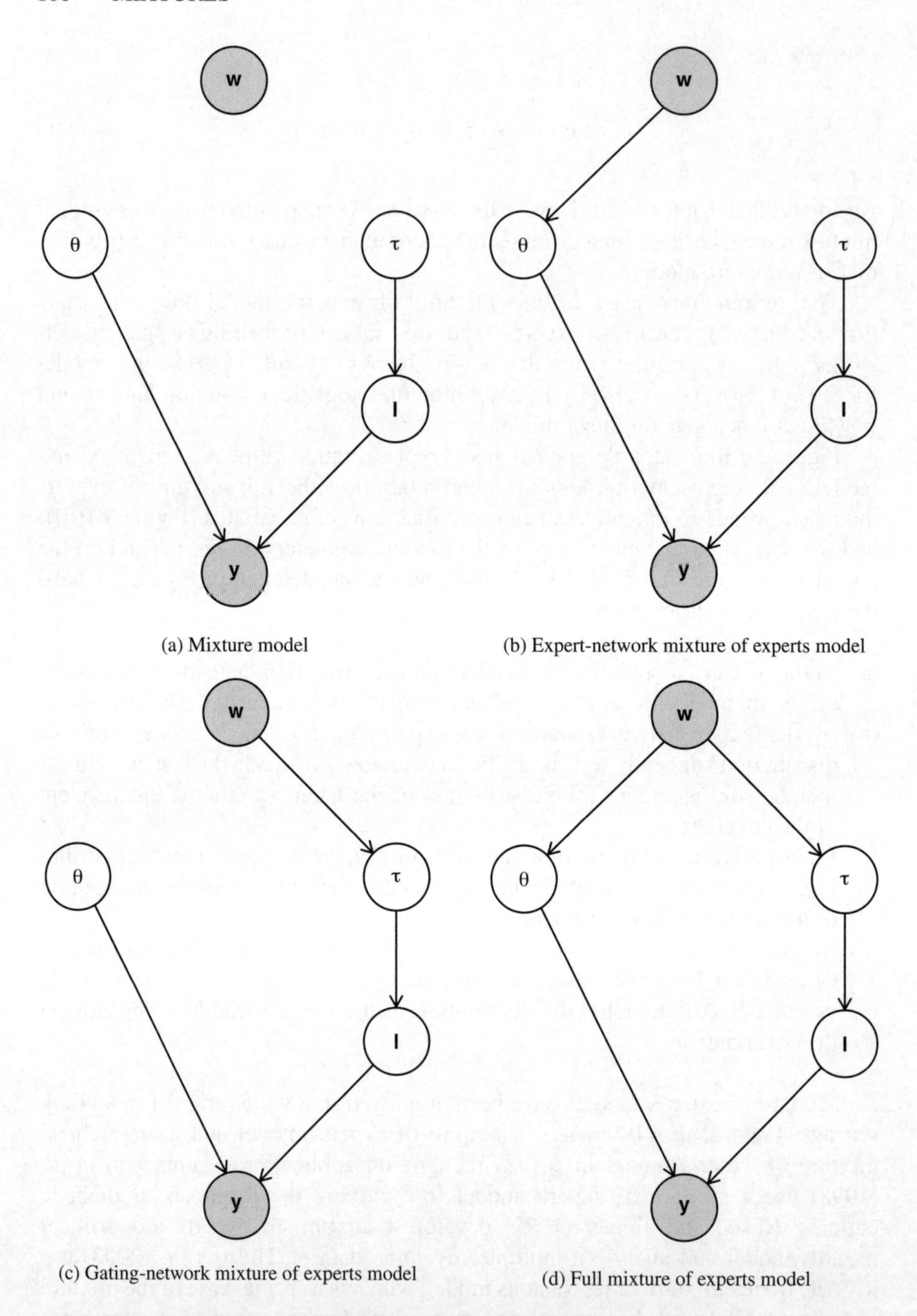

Figure 5.1 The graphical model representation of the mixture of experts model. The differences between the four special cases are due to the presence or absence of edges between the covariates w and the parameters τ and θ.

Further details on mixture of experts models are given in McLachlan and Peel (2000, Chapter 5.13), Tanner and Jacobs (2001) and Bishop (2006, Chapter 14.5), where extensions including the hierarchical mixture of experts model (Jordan and Jacobs, 1994) are discussed. Software for fitting mixture of experts models in the R programming environment (R Development Core Team, 2009) include hme (Evers, 2007), mixtools (Benaglia *et al.*, 2009) and integrativeME (Cao, 2010).

5.5 A mixture of experts model for ranked preference data

The current President of Ireland, Mary McAleese, was first elected in 1997 under the Single Transferable Vote electoral system. Under this system voters rank, in order of their preference, some or all of the electoral candidates. The vote counting system, which results in the elimination of candidates and the subsequent election of the President, is an intricate process involving the transfer of votes between candidates as specified by the voters' ballots. Details of the electoral system, the counting process and the 1997 Irish presidential election are given in Coakley and Gallagher (2004), Sinnott (1995, 1999) and Marsh (1999).

The 1997 presidential election race involved five candidates: Mary Banotti, Mary McAleese, Derek Nally, Adi Roche and Rosemary Scallon. Derek Nally and Rosemary Scallon were independent candidates while Mary Banotti and Adi Roche were endorsed by the then current opposition parties Fine Gael and Labour respectively. Mary McAleese was endorsed by the Fianna Fáil party who were in power at that time. In terms of candidate type, McAleese and Scallon were deemed to be conservative candidates with the other candidates regarded as liberal. Gormley and Murphy (2008a, 2008b, 2010a) provide further details on the 1997 presidential election and on the candidates.

One month prior to election day a survey was conducted by Irish Marketing Surveys on 1083 respondents. Respondents were asked to list some or all of the candidates in order of preference, as if they were voting on the day of the poll. In addition, pollsters gathered data on attributes of the respondents as detailed in Table 5.1.

Table 5.1 Covariates recorded for each respondent in the Irish Marketing Surveys poll.

Age	Area	Gender	Government satisfaction	Marital status	Social class
–	City	Housewife	Satisfied	Married	AB
	Town	Nonhousewife	Dissatisfied	Single	C1
	Rural	Male	No opinion	Widowed	C2
					DE
					F50+
					F50–

Interest lies in determining if groups of voters with similar preferences (i.e. voting blocs) exist within the electorate. If such voting blocs do exist, the influence that the recorded socioeconomic variables may have on the clustering structure and/or on the preferences that characterise a voting bloc is also of interest. Jointly modelling the rank preference votes and the covariates through a mixture of experts model for rank preference data when clustering the electorate provides this insight.

Given the rank nature of the outcome variables or votes y_i ($i = 1, \ldots, N = 1083$) the probability density $p(\cdot)$ in the mixture of experts model (5.1) must have an appropriate form. The Plackett-Luce model (Plackett, 1975) (or exploded logit model) for rank data provides a suitable model; Benter's model (Benter, 1994) provides another alternative. Let $y_i = [c(i, 1), \ldots, c(i, m_i)]$ denote the ranked ballot of voter i where $c(i, j)$ denotes the candidate ranked in the jth position by voter i and m_i is the number of candidates ranked by voter i. Under the Plackett-Luce model, given that voter i is a member of voting bloc g and given the 'support parameter' $p_g = (p_{g1}, \ldots, p_{gM})$, the probability of voter i's ballot is

$$\mathbb{P}(y_i | p_g) = \frac{p_{gc(i,1)}}{\sum_{s=1}^{M} p_{gc(i,s)}} \cdot \frac{p_{gc(i,2)}}{\sum_{s=2}^{M} p_{gc(i,s)}} \cdots \frac{p_{gc(i,m_i)}}{\sum_{s=m_i}^{M} p_{gc(i,s)}},$$

where $M = 5$ denotes the number of candidates in the electoral race. The support parameter p_{gj} (typically restricted such that $\sum_{j=1}^{M} p_{gj} = 1$) can be interpreted as the probability of ranking candidate j first, out of the currently available choice set. Hence, the Plackett-Luce model models the ranking of candidates by a voter as a set of independent choices by the voter, conditional on the cardinality of the choice set being reduced by one after each choice is made.

In the full mixture of experts model, the parameters of the group densities are modelled as functions of covariates. Here the support parameters are modelled as a logistic function of the covariates:

$$\log \left[\frac{p_{gj}(w_i)}{p_{g1}(w_i)} \right] = \gamma_{gj0} + \gamma_{gj1} w_{i1} + \cdots + \gamma_{gjL} w_{iL},$$

where $w_i = (w_{i1}, \ldots, w_{iL})$ is the set of L covariates associated with voter i. Note that for identifiability reasons candidate 1 is used as the baseline choice and $\gamma_{g1} = (0, \ldots, 0)$ for all $g = 1, \ldots, G$. The intuition behind this version of the model is that a voter's covariates may influence their support for each candidate beyond what is explained by their membership of a voting bloc.

In the full mixture of experts model, the gating networks (or mixing proportions) are also modelled as functions of covariates. In a similar vein to the support parameters, the mixing proportions are modelled via a multinomial logistic regression model,

$$\log \left[\frac{\tau_g(w_i)}{\tau_1(w_i)} \right] = \beta_{g0} + \beta_{g1} w_{i1} + \cdots + \beta_{gL} w_{iL},$$

Table 5.2 The model with the smallest BIC within each type of the mixture of experts model for ranked preference data applied to the 1997 Irish presidential election data. In the 'government satisfaction' variable the 'no opinion' level was used as the baseline category.

	BIC	G	Covariates
The gating network MoE model	8491	4	τ_g: age, government satisfaction
The full MoE model	8512	3	τ_g: age, government satisfaction p_g: age
The mixture model	8513	3	–
The expert network MoE model	8528	1	p_g: government satisfaction

where voting bloc 1 is used as the baseline voting bloc. Here, the motivation for this model term is that a voter's covariates may influence their voting bloc membership.

Modelling the group parameters and/or the mixing proportions as functions of covariates, or as constant with respect to covariates, results in the four types of mixture of experts models, as illustrated in Figure 5.1. Each model can be fitted in a maximum-likelihood framework using an EM algorithm (Dempster *et al.*, 1977). Model fitting details for each model are outlined in Gormley and Murphy (2008a, 2008b, 2010a).

Each of the four mixture of experts models for rank preference data illustrated in Figure 5.1 was fitted to the data from the electorate in the Irish presidential election poll. A range of groups $G = 1, \ldots, 5$ was considered and a forwards-selection method was employed to select influential covariates. The Bayesian Information Criterion (BIC) (Kass and Raftery, 1995; Schwartz, 1978) was used to select the optimal model; this criterion is a penalised likelihood criterion which rewards model fit while penalising nonparsimonious models. Small BIC values indicate a preferable model. Table 5.2 details the optimal models for each type of mixture of experts model fitted.

Based on the BIC values, the optimal model is a gating-network MoE model with four groups where 'age' and 'government satisfaction' are important covariates for determining group or 'voting bloc' membership. Under this gating-network MoE model, the covariates are not informative within voting blocs, but only in determining voting bloc membership. The maximum-likelihood estimates of the model parameters are reported in Figure 5.2 and in Table 5.3.

The support parameter estimates illustrated in Figure 5.2 have an interpretation in the context of the 1997 Irish presidential election. Voting bloc 1 could be characterised as the 'conservative voting bloc' due to its large support parameters for McAleese and Scallon. Voting bloc 2 has large support for the liberal candidate Adi Roche. Voting bloc 3 is the largest voting bloc in terms of marginal mixing proportions and intuitively has larger support parameters for the high-profile candidates McAleese and Banotti. These candidates were endorsed by the two largest political parties in the country at that time. Voters belonging to voting bloc 4 favour Banotti and have more uniform levels of support for the other candidates. A detailed discussion of this optimal model is also given in Gormley and Murphy (2008b).

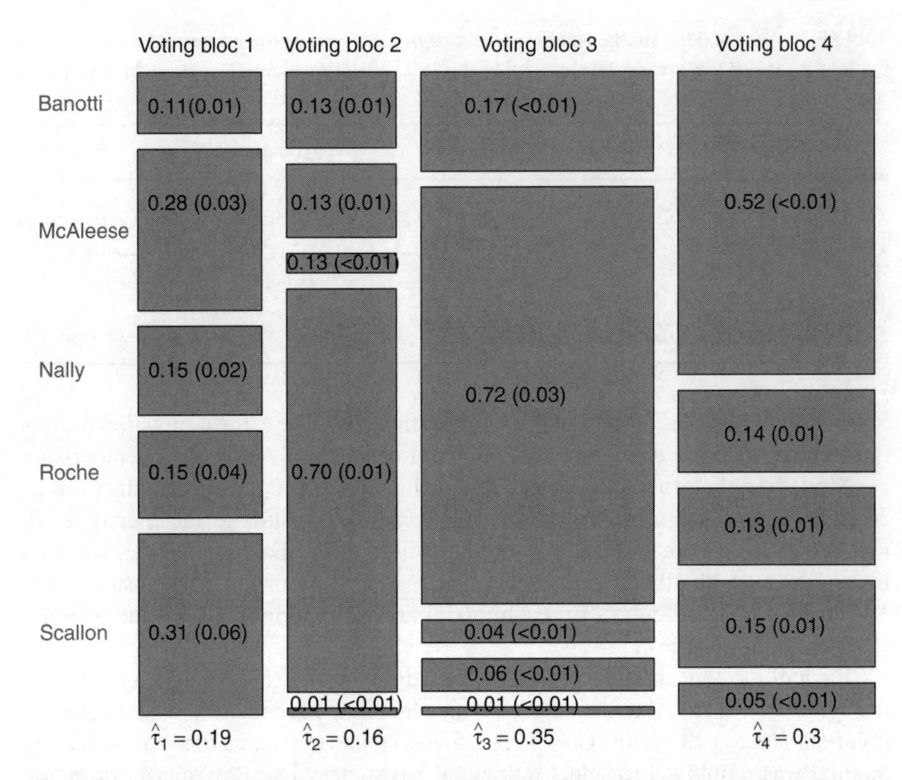

Figure 5.2 A mosaic-plot representation of the parameters of the group densities of the gating-network mixture of experts model for rank preference data. The width of each block is proportional to the marginal probability of belonging to that group and the blocks are divided in proportion to the Plackett-Luce support parameters. Standard errors are given in parentheses.

Table 5.3 details the odds ratios computed for the mixing proportion (or gating network) parameters $\beta = (\beta_1, \ldots, \beta_G)$. In the model, voting bloc 1 (the conservative voting bloc) is the baseline voting bloc. Two covariates were selected as influential: age and government satisfaction levels. In the 'government satisfaction' covariate, the baseline was chosen to be 'no opinion'.

Table 5.3 Odds ratios for the mixing proportion parameters in the gating-network MoE model for rank preference data. The covariates 'age' and 'government satisfaction' were selected as influential.

	Age		Satisfied		Not satisfied	
	Odds ratio	95 % CI	Odds ratio	95 % CI	Odds ratio	95 % CI
Voting bloc 2	0.01	[0.00, 0.05]	1.14	[0.42, 3.11]	2.80	[0.77, 10.15]
Voting bloc 3	0.95	[0.32, 2.81]	3.12	[0.94, 10.31]	3.81	[0.90, 16.13]
Voting bloc 4	1.56	[0.35, 6.91]	0.35	[0.12, 0.98]	3.50	[1.07, 11.43]

Interpreting the odds ratios provides insight into the type of voter that characterises each voting bloc. For example, older (and generally more conservative) voters are much less likely to belong to the liberal voting bloc 2 than to the conservative voting bloc 1 ($\beta_{21} = 0.01$). Also, voters with some interest in government are more likely to belong to voting bloc 3 ($\beta_{32} = 3.12$ and $\beta_{33} = 3.81$), the bloc favouring candidates backed by large government parties, than to belong to the conservative voting bloc 1. Voting bloc 1 had high levels of support for the independent candidate Scallon. The mixing proportions parameter estimates further indicate that voters dissatisfied with the current government are more likely to belong to voting bloc 4 than to voting bloc 1 ($\beta_{43} = 3.50$). This is again intuitively plausible as voting bloc 4 favours Mary Banotti who was backed by the main government opposition party, while voting bloc 1 favours the government backed Mary McAleese. Further interpretation of the mixing proportion parameters are given in Gormley and Murphy (2008b).

5.5.1 Examining the clustering structure

It is important that the clusters found by the mixture of experts model correspond to distinct voting blocs. Baudry *et al.* (2010) propose a method to check if mixture components are really modelling distinct clusters or whether multiple mixture components are being used to model each cluster because the component density in the mixture model is too restrictive. Hennig (2010) proposes an alternative approach to this problem specifically for normal mixture models.

The method developed by Baudry *et al.* (2010) uses the estimated a posteriori cluster membership probabilities,

$$\hat{l}_{ig} = \frac{\hat{\tau}_g(w_i) f(y_i|\hat{\theta}_g(w_i))}{\sum_{g'=1}^{G} \hat{\tau}_{g'}(w_i) f(y_i|\hat{\theta}_{g'}(w_i))}.$$

In particular, suppose the mixture components $\{1, 2, \ldots, G\}$ are partitioned into sets $\{\rho_1, \rho_2, \ldots, \rho_K\}$, where ρ_k are the mixture components used to model distinct cluster k. Further, let $t_{ik} = \sum_{g \in \rho_k} \hat{l}_{ig}$ be the estimated a posteriori probability of membership in cluster k. Then the entropy of a particular clustering is given by

$$\mathcal{E}_K = -\sum_{i=1}^{N} \sum_{k=1}^{K} t_{ik} \log t_{ik}.$$

A greedy algorithm is used to combine mixture components to reduce K from G to $G - 1$, from $G - 1$ to $G - 2$ and so on, until $K = 1$. A plot of $\mathcal{E}_K$ versus K gives an indication of the number of clusters in the population, where a large drop in $\mathcal{E}_K$ when K is decreased indicates that multiple components are modelling a cluster in the population. The results of applying the component-merging algorithm to the mixture of experts model are shown in Figure 5.3.

These results suggest that the entropy does not decrease substantially when the two closest components (i.e. 1 and 3) are combined to form a single cluster. The two components are therefore distinct from each other. Hence, it appears from this

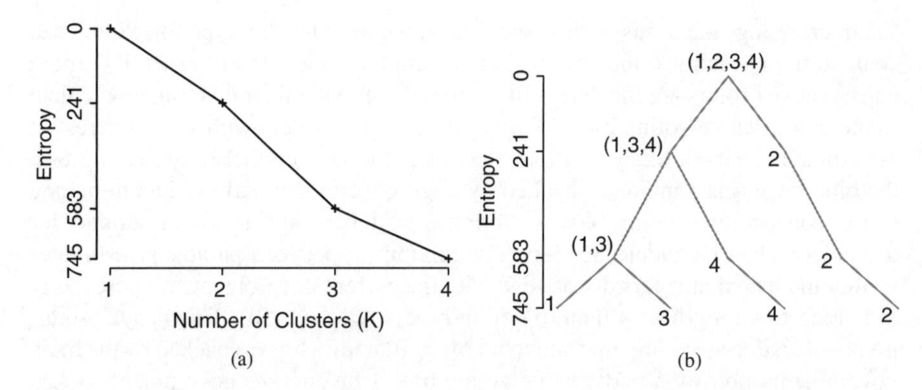

Figure 5.3 (a) The value of $\mathcal{E}_K$ plotted as a function of K. (b) A dendrogram representation of the combination of mixture of experts components when clustered into K clusters. Note that the vertical scale on each plot is inverted.

analysis that the four components are modelling distinct clusters in the data. Dean and Nugent (2010) develop a mixture component tree to visualize the similarity of mixture components. Following their work, we use the dendrogram in Figure 5.3(b) to give a visualization of the connections between the four voting blocs found in this analysis.

5.6 A mixture of experts latent position cluster model

The latent position cluster model (LPCM) (Handcock *et al.*, 2007) develops the idea of the latent social space model (Hoff *et al.*, 2002) by extending the model to accommodate clusters of actors in the latent space. Under the latent position cluster model, the latent location of each actor is assumed to be drawn from a finite normal mixture model, each component of which represents a cluster of actors. In contrast, the model outlined in Hoff *et al.* (2002) assumed that the latent positions were normally distributed. Thus, the latent position cluster model offers a more flexible version of the latent space model for modelling heterogeneous social networks.

The latent position cluster model provides a framework in which actor covariates may be explicitly included in the model; the probability of a link between two actors may be modelled as a function of both their separation in the latent space and of their relative covariates. However, the covariates may contribute more to the structure of the network than solely through the link probabilities; the covariates may influence both the cluster membership of an actor and their link probabilities. A latent position cluster model in which the cluster membership of an actor is modelled as a function of their covariates lies within the mixture of experts framework.

Specifically, social network data take the form of a set of relations $\{y_{i,j}\}$ between a group of $i, j = 1, \ldots, N$ actors, represented by an $N \times N$ sociomatrix $\mathbf{Y}$. Here it is assumed that the relation $y_{i,j}$ between actors i and j is a binary relation, indicating the presence or absence of a link between the two actors; the mixture of experts latent position cluster model is easily extended to other forms of relation (such as

count data). Covariate data $w_i = (w_{i1}, \ldots, w_{iL})$ associated with actor i is assumed to be available, where L denotes the number of observed covariates.

Each actor i is assumed to have a location $z_i = (z_{i1}, \ldots, z_{id})$ in the d-dimensional latent social space. The probability of a link between any two actors is assumed to be independent of all other links in the network, given the latent locations of the actors. Let $x_{i,j} = (x_{ij1}, \ldots, x_{ijL})$ denote an L-vector of dyadic-specific covariates where $x_{ijk} = d(w_{ik}, w_{jk})$ is a measure of the similarity in the value of the kth covariate for actors i and j. Given the link probabilities parameter vector β the likelihood function is then

$$\mathbb{P}(\mathbf{Y}|\mathbf{Z}, \mathbf{X}, \beta) = \prod_{i=1}^{N} \prod_{j \neq i} \mathbb{P}(y_{i,j}|z_i, z_j, x_{i,j}, \beta),$$

where $\mathbf{Z}$ is the $N \times d$ matrix of latent locations and $\mathbf{X}$ is the matrix of dyadic-specific covariates. The probability of a link between actors i and j is then modelled using a logistic regression model where both dyadic specific covariates and Euclidean distance in the latent space are dependent variables:

$$\log \left[\frac{\mathbb{P}(y_{i,j} = 1)}{\mathbb{P}(y_{i,j} = 0)} \right] = \beta_0 + \beta_1 x_{ij1} + \cdots + \beta_L x_{ijL} - ||z_i - z_j||.$$

To account for clustering of actor locations in the latent space, it is assumed that the latent locations z_i are drawn from a finite mixture model. Moreover, in the mixture of experts latent position cluster model, the latent locations are assumed drawn from a finite mixture model in which actor covariates may influence the mixing proportions:

$$z_i \sim \sum_{g=1}^{G} \tau_g(w_i) \mathrm{MVN}(\mu_g, \sigma_g^2 \mathbf{I}),$$

where

$$\tau_g(w_i) = \frac{\exp(\tau_{g0} + \tau_{g1} w_{i1} + \cdots + \tau_{gL} w_{iL})}{\sum_{g'=1}^{G} \exp(\tau_{g'0} + \tau_{g'1} w_{i1} + \cdots + \tau_{g'L} w_{iL})}$$

and $\tau_1 = (0, \ldots, 0)$. This model has an intuitive motivation: the covariates of an actor may influence their cluster membership, their cluster membership influences their latent location and in turn their latent location determines their link probabilities.

The mixture of experts latent position cluster model can be fitted within the Bayesian paradigm; a Metropolis-within-Gibbs sampler can be employed to draw samples from the posterior distribution of interest. As is standard in Bayesian estimation of mixture models (Diebolt and Robert, 1994; Hurn *et al.*, 2003), the problem is greatly simplified by augmenting the observed data with an indicator variable K_i for each actor i where $K_i = g$ if actor i belongs to cluster g. The indicator variable K_i therefore has a multinomial distribution with a single trial and probabilities

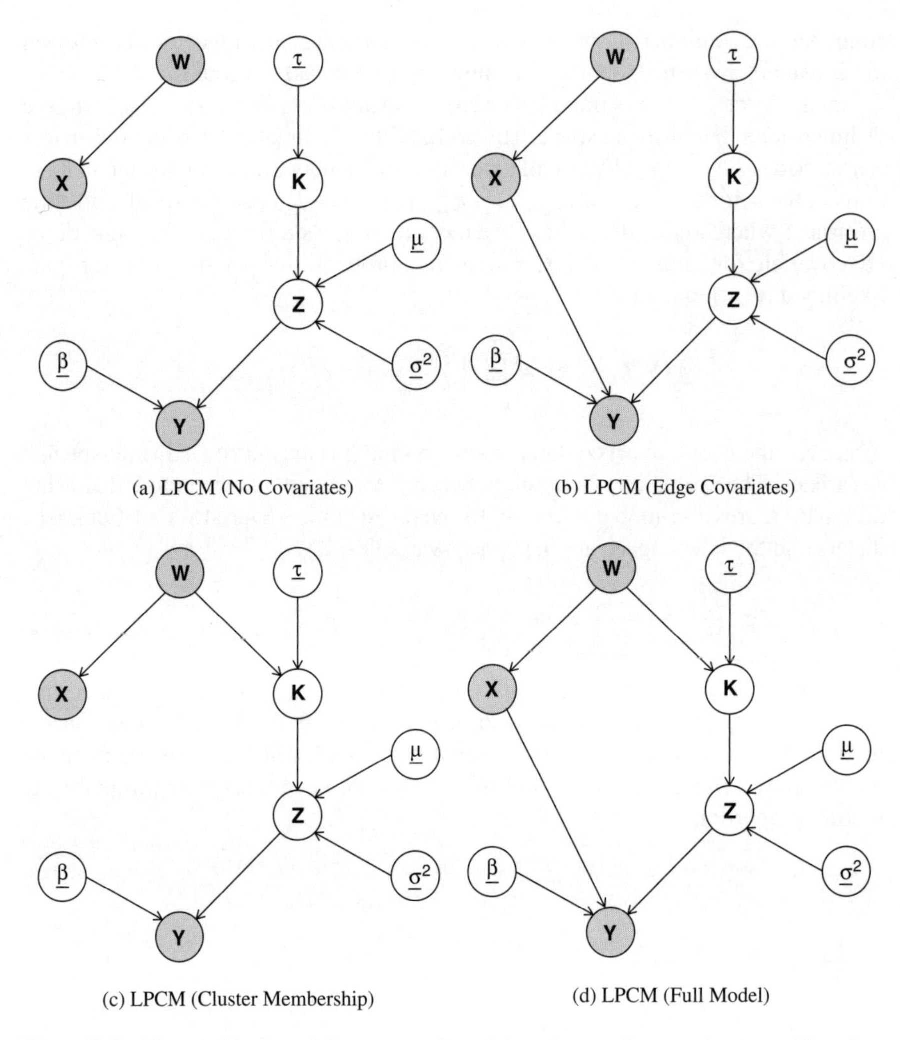

Figure 5.4 The graphical-model representation of the mixture of experts latent position cluster model. The differences between the four special cases are due to the presence or absence of edges between the covariates X and W and the data Y and K.

equal to $\tau_g(w_i)$ for $g = 1, \ldots, G$. Model issues such as likelihood invariance to distance-preserving transformations of the latent space and label switching must be considered during the model-fitting process; an approach to dealing with such model identifiabilty and full model-fitting details are available in Gormley and Murphy (2010b).

Figure 5.4 illustrates a graphical-model representation of the mixture of experts latent position cluster model. Similarly to Figure 5.1, four different models are available by allowing or disallowing covariates to influence the mixing proportions and/or the link probabilities.

Table 5.4 Covariates associated with the 71 lawyers in the US corporate law firm. The last category in each categorical covariate is treated as the baseline category in all subsequent analyses.

Covariate	Levels
Seniority	1 = partner
	2 = associate
Gender	1 = male
	2 = female
Office	1= Boston
	2 = Hartford
	3 = Providence
Practice	1 = litigation
	2 = corporate
Law school	1 = Harvard or Yale
	2 = University of Connecticut
	3 = other
Years with the firm	–
Age	–

An illustrative example of the mixture of experts latent position cluster model methodology is provided through the analysis of a network dataset detailing interactions between a set of 71 lawyers in a corporate law firm in the USA (Lazega, 2001). The data include measurements of the co-worker network, an advice network and a friendship network. Covariates associated with each lawyer in the firm are also included and are detailed in Table 5.4. Interest lies in identifying social processes within the firm such as knowledge sharing and organisational structures.

The four mixture of experts latent position cluster models illustrated in Figure 5.4 were fitted to the advice network; data in this network detail links between lawyers who sought basic professional advice from each other over the previous 12 months. Gormley and Murphy (2010b) explore the co-workers network dataset and the friendship network dataset using similar methodology. Figure 5.5 illustrates the resulting latent space locations of the lawyers under each fitted model with $(G, d) = (2, 2)$. These parameter values were selected using BIC after a range of latent position cluster models (with no covariates) were fitted to the network data only (Handcock *et al.*, 2007). Table 5.5 details the resulting regression parameter estimates and their associated uncertainty for the four fitted models.

The results of the analysis show some interesting patterns. The model with the highest AICM (Raftery *et al.*, 2007) value is the model that has covariates in the link probabilities and in the cluster membership probabilities.

The coefficients of the covariates in the link probabilities are very similar in models (b) and (d) in Table 5.5. These coefficients indicate that a number of factors have a positive or negative effect on whether or not a lawyer asks another for advice.

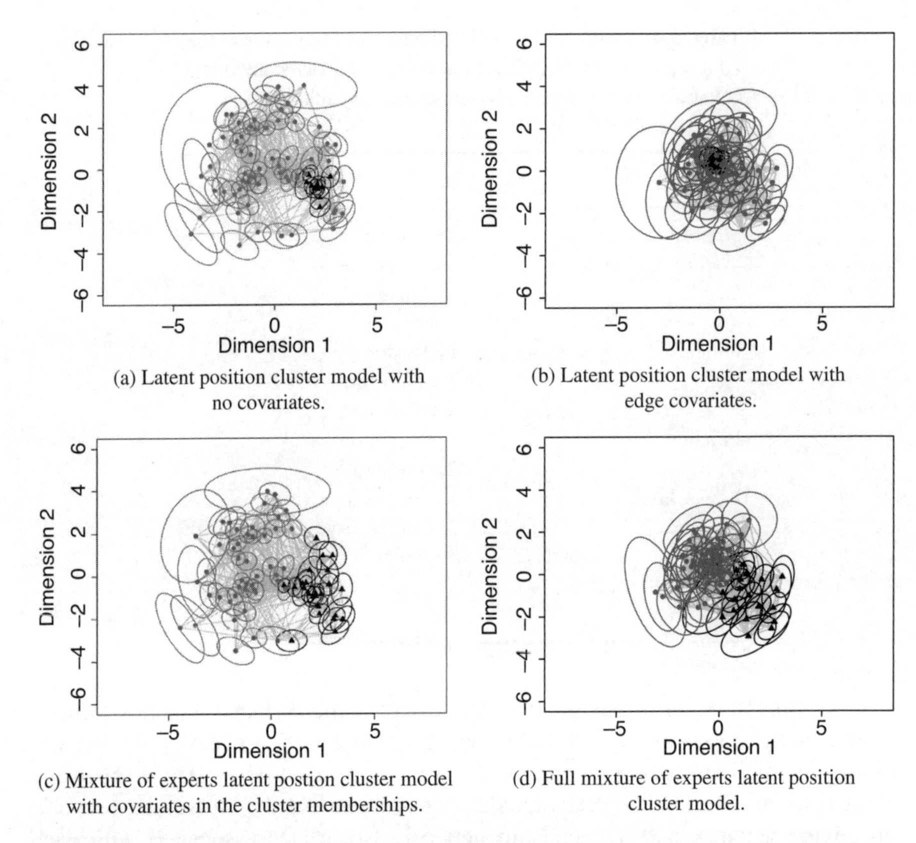

(a) Latent position cluster model with no covariates.

(b) Latent position cluster model with edge covariates.

(c) Mixture of experts latent postion cluster model with covariates in the cluster memberships.

(d) Full mixture of experts latent position cluster model.

Figure 5.5 Estimates of clusters and latent positions of the lawyers from the advice network data. The ellipses are 50 % posterior sets illustrating the uncertainty in the latent locations. Lawyers who are members of the same cluster are shown using the same type of darkness and symbol. Observed links between lawyers are also illustrated.

In summary, lawyers who are similar in seniority, gender, office location and practice type are more likely to ask each other for advice. The effects of years and age seem to have a negative effect, but these variables are correlated with seniority and with each other, so their marginal effects are more difficult to interpret.

Importantly, the latent positions are very similar in models (a) and (c) which do not have covariates in the link probabilities and models (b) and (d) which do have covariates in the link probabilities. This can be explained because of the different role that the latent space plays in the models with covariates in the link probabilities and in those that do not have such covariates. When the covariates are in the link probabilities, the latent space models the part of the network structure that could not be explained by the link covariates, whereas in the other case the latent space models much more of the network structure.

Interestingly, in the model with the highest AICM value, there are covariates in the cluster membership probabilities as well as in the link probabilities. This means

Table 5.5 Posterior mean parameter estimates for the four mixture of experts models fitted to the lawyers advice data as detailed in Figure 5.5. Standard deviations are given in parentheses. Note that cluster 1 was used as the baseline cluster in the case of the cluster membership parameters. Baseline categories for the covariates are detailed in Table 5.4.

	(a)	(b)	(c)	(d)
Link probabilities				
Intercept	1.26 (0.10)	-2.87 (0.17)	1.23 (0.10)	-2.65 (0.17)
Seniority		0.89 (0.11)		0.81 (0.11)
Gender		0.60 (0.09)		0.62 (0.09)
Office		2.02 (0.10)		1.97 (0.10)
Practice		1.63 (0.10)		1.57 (0.10)
Years		-0.04 (0.005)		-0.04 (0.005)
Age		-0.02 (0.004)		-0.02 (0.004)
Cluster memberships				
Intercept	-1.05 (1.75)	0.94 (0.79)	-0.62 (1.23)	1.27 (1.29)
Office (=1)			1.94 (1.02)	2.40 (1.14)
Office (=2)			-2.08 (1.09)	-0.97 (1.19)
Practice			3.18 (0.85)	2.14 (1.08)
Age			-0.09 (0.04)	-0.14 (0.06)
Latent space model				
Cluster 1 mean	- 0.50 (0.52)	0.09 (0.19)	-1.09 (0.31)	-0.54 (0.21)
	0.21 (0.58)	-0.09 (0.26)	0.40 (0.28)	0.40 (0.20)
Cluster 1 variance	3.35 (1.29)	2.12 (0.77)	3.19 (0.58)	1.25 (0.34)
Cluster 2 mean	1.66 (0.92)	-0.24 (0.20)	2.10 (0.30)	1.32 (0.51)
	-0.67 (0.58)	0.35 (0.23)	-0.77 (0.30)	-0.98 (0.47)
Cluster 2 variance	1.29 (1.58)	0.27 (0.68)	1.16 (0.40)	1.63 (0.69)
AICM	-3644.24	-3346.87	-3682.71	-3325.95

that the structure in the latent space, which is modelling what could not be explained directly in the link probabilities, has a structure that can be further explained using the covariates. The office location, practice and age of the lawyers retain explanatory power in explaining the clustering found in the latent social space.

The difference in the cluster membership coefficients in models (c) and (d) is due to the different interpretation of the latent space in these models. However, it is interesting to note that the signs of the coefficients are identical. This is because the cluster memberships shown for these models in Figures 5.5(c) and 5.5(d) are similar; this phenomenon does not hold generally (see Gormley and Murphy, 2010b, Section 5.3).

The results of this analysis offer a cautionary message in automatically selecting the type of mixture of experts latent position cluster model for analysing social network data. The role of the latent space in the model is very different depending on how the covariates enter the model. If the latent space is to be interpreted as a social space that explains network structure, then the covariates should not directly enter the link probabilities. However, if the latent space is being used to find interesting or anomalous structure in the network that cannot be explained by the covariates, then one should consider allowing the covariates to enter the link probabilities and cluster membership probabilities.

5.7 Discussion

This chapter has illustrated the utility of mixture of experts models in two social science clustering applications where concomitant covariate information is available. The mixture of experts framework provides a systematic method for describing and exploring the clustering found in the outcome variables.

This inclusion of covariates through the mixture of experts model can give different clustering results from when a two-stage process of clustering followed by cluster interrogation is taken. This is because both the outcome variables and the concomitant covariates provide information that is relevant in defining the clustering. The result of the use of both sources of information is often a clearer clustering structure.

When the mixture of experts model is used, it is important to consider how the covariates enter the model. The interpretation of the latent structure (clustering or other latent variables) in the mixture of experts model depends heavily on how the covariates enter the model. Therefore, this choice needs to be directed by the interpretation of the latent structure in the context of the application.

Acknowledgements

We would like to thank the participants of the ICMS Workshop on Mixture Estimation and Applications for their insightful feedback on this work. This work has been supported by Science Foundation Ireland Research Frontiers grants (06/RFP/M040 and 09/RFP/MTH2367) and Clique, a Science Foundation Ireland Strategic Research Cluster grant (08/SRC/I1407).

References

Airoldi, E. M., Blei, D. M., Fienberg, S. E. and Xing, E. (2008) Mixed-membership stochastic blockmodels. *Journal of Machine Learning Research*, **9**, 1981–2014.

Aitkin, M., Anderson, D. and Hinde, J. (1981) Statistical modelling of data on teaching styles (with discussion). *Journal of the Royal Statistical Society Series* A, **144**, 419–461.

Baudry, J. P., Raftery, A. E., Celeux, G., Lo, K. and Gottardo, R. G. (2010) Combining mixture components for clustering. *Journal of Computational and Graphical Statistics*, **19**, 332–353.

Benaglia, T., Chauveau, D., Hunter, D. R. and Young, D. (2009) mixtools: An R package for analyzing finite mixture models. *Journal of Statistical Software* , **32**, 1–29.

Benter, W. (1994) Computer-based horse race handicapping and wagering systems: A report. In *Efficiency of Racetrack Betting Markets* (eds W. T., Ziemba, V. S. Lo and D. B. Haush), pp. 183–198. Academic Press.

Biernacki, C., Celeux, G., Govaert, G. and Langrognet, F. (2006) Model-based cluster and discriminant analysis with the MIXMOD software. *Computational Statistics and Data Analysis*, **51**, 587–600.

Bishop, C. M. (2006) *Pattern Recognition and Machine Learning*. Springer-Verlag.

Busse, L. M., Orbanz, P. and Buhmann, J. M. (2007) Cluster analysis of heterogeneous rank data. In *Proceedings of the 24th International Conference on Machine Learning* (ed. Z. Ghahramani), Vol. 227, pp. 113–120. ACM International Conference Proceeding Series.

Cao, K. A. L. (2010) *Integrative ME: Integrative Mixture of Experts*. R package version 1.2.

Carvalho, A. X. and Tanner, M. A. (2007) Modelling nonlinear count time series with local mixtures of Poisson autoregressions. *Computational Statistics and Data Analysis*, **51**, 5266–5294.

Coakley, J. and Gallagher, M. (2004) *Politics in the Republic of Ireland*, 4th edn. Routledge in association with PSAI Press, London.

Dean, N. and Nugent, R. (2010) Mixture Model Component Trees: Visualizing the Hierarchical Structure of Complex Groups. *Technical Report*.

Dempster, A. P., Laird, N. M. and Rubin, D. B. (1977) Maximum likelihood from incomplete data via the EM algorithm (with discussion). *Journal of the Royal Statistical Society, Series B*, **39**, 1–38.

Diebolt, J. and Robert, C. P. (1994) Estimation of finite mixture distributions through Bayesian sampling. *Journal of the Royal Statisical Society, Series B*, **56**, 363–375.

Everitt, B. S. and Hand, D. J. (1981) *Finite Mixture Distributions*. Chapman & Hall.

Evers, L. (2007) *hme: Methods for Fitting Hierarchical Mixtures of Experts (HMEs)*. R package version 0.1-0.

Fraley, C. and Raftery, A. E. (1998) How many clusters? Which clustering method? – Answers via model-based cluster analysis. *Computer Journal*, **41**, 578–588.

Fraley, C. and Raftery, A. E. (2002) Model-based clustering, discriminant analysis, and density estimation. *Journal of the American Statistical Association*, **97**, 611–631.

Fraley, C. and Raftery, A. E. (2006) MCLUST version 3 for R: normal mixture modelling and model-based clustering. Technical Report 504, Department of Statistics, University of Washington.

Geweke, J. and Keane, M. (2007) Smoothly mixing regressions. *Journal of Econometrics*, **136**, 252–290.

Goldenberg, A., Zheng, A. X., Fienberg, S. E. and Airoldi, E. M. (2009) A survey of statistical network models. *Foundations and Trends in Machine Learning*, **2**, 129–233.

Gormley, I. C. and Murphy, T. B. (2006) Analysis of Irish third-level college applications data. *Journal of the Royal Statistical Society, Series A*, **169**, 361–379.

Gormley, I. C. and Murphy, T. B. (2008a) Exploring voting blocs within the Irish electorate: a mixture modelling approach. *Journal of the American Statistical Association*, **103**, 1014–1027.

Gormley, I. C. and Murphy, T. B. (2008b) A mixture of experts model for rank data with applications in election studies. *Annals of Applied Statistics*, **2**, 1452–1477.

Gormley, I. C. and Murphy, T. B. (2010a) Clustering ranked preference data using sociodemographic covariates In *Choice Modelling: The State-of-the-Art and the State-of-Practice* (eds S. Hess and A. Daly), pp. 543–569. Emerald United Kingdom.

Gormley, I. C. and Murphy, T. B. (2010b) A mixture of experts latent position cluster model for social network data. *Statistical Methodology*, **7**, 385–405.

Hall, M. E. (2004) Pottery production during the late Jomon period: insights from the chemical analyses of Kasori B pottery. *Journal of Archaeological Science*, **31**, 1439–1450.

Handcock, M., Raftery, A. and Tantrum, J. M. (2007) Model-based clustering for social networks. *Journal of the Royal Statistical Society, Series A*, **170**, 301–354.

Hartigan, J. A. (1975) *Clustering Algorithms*. John Wiley & Sons, Ltd.

Hennig, C. (2010) Methods for merging Gaussian mixture components. *Advances in Data Analysis and Classification*, **4**, 3–34.

Hoff, P. D., Raftery, A. E. and Handcock, M.S. (2002) Latent space approaches to social network analysis. *Journal of the American Statistical Association*, **97**, 1090–1098.

Holloway, S. (1995) Forty years of United Nations General Assembly voting. *Canadian Journal of Political Science*, **17**, 223–249.

Hurn, M., Justel, A. and Robert, C. P. (2003) Estimating mixtures of regressions. *Journal of Computational and Graphical Statistics*, **12**, 55–79.

Jacobs, R. A., Jordan, M. I., Nowlan, S. J. and Hinton, G. E. (1991) Adaptive mixture of local experts. *Neural Computation*, **3**, 79–87.

Jordan, M. I. and Jacobs, R. A. (1994) Hierarchical mixtures of experts and the EM algorithm. *Neural Computation*, **6**, 181–214.

Kass, R. E. and Raftery, A. E. (1995) Bayes factors. *Journal of the American Statistical Association*, **90**, 773–795.

Kolaczyk, E. D. (2009) *Statistical Analysis of Network Data: Methods and Models*. Springer-Verlag.

Krivitsky, P. N. and Handcock, M. S. (2008) Fitting position latent cluster models for social networks with latentnet. *Journal of Statistical Software*, **24**, 1–23.

Krivitsky, P. N., Handcock, M. S., Raftery, A. E. and Hoff, P. D. (2009) Representing degree distributions, clustering, and homophily in social networks with latent cluster random effects models. *Social Networks*, **31**, 204–213.

Latouche, P., Birmelé, E. and Ambroise, C. (2011) Overlapping Stochastic Block Models with Application to the French Political Blogosphere. *Annals of Applied Statistics* (in press).

Lazarsfeld, P. F. and Henry, N. W. (1968) *Latent Structure Analysis*. Houghton Mifflin.

Lazega, E. (2001) *The Collegial Phenomenon: The Social Mechanisms of Cooperation Among Peers in a Corporate Law Partnership*. Oxford University Press.

Lee, C. K., Lee, Y. K., Bernhard, B. J. and Yoon, Y. S. (2005) Segmenting casino gamblers by motivation: a cluster analysis of Korean gamblers. *Tourism Management*, **27**, 856–866.

Leisch, F. (2004) FlexMix: a general framework for finite mixture models and latent class regression in R. *Journal of Statistical Software*, **11**, 1–18.

McLachlan, G. J. and Basford, K. E. (1988) *Mixture Models: Inference and Applications to Clustering*. Marcel Dekker.

McLachlan, G. J. and Peel, D. (2000) *Finite Mixture Models*. John Wiley & Sons, Ltd.

McLachlan, G., Peel, D., Basford, K. and Adams, P. (1999) The EMMIX software for the fitting of mixtures of normal and t-components. *Journal of Statistical Software*.

Mariadassou, M., Robin, S. and Vacher, C. (2010) Uncovering latent structure in valued graphs: a variational approach. *Annals of Applied Statistics*, **4**, 715–742.

Marsh, M. (1999) The making of the eighth President. In *How Ireland Voted 1997* (eds M. Marsh and P. Mitchell), pp. 215–242. Westview and PSAI Press. Boulder, Colorado.

Melnykov, V. and Maitra, R. (2010) Finite mixture models and model-based clustering. *Statistics Surveys*, **4**, 80–116.

Murphy, T. B. and Martin, D. (2003) Mixtures of distance-based models for ranking data. *Computational Statistics and Data Analysis*, **41**, 645–655.

Peng, F., Jacobs, R. A. and Tanner, M. A. (1996) Bayesian inference in Mixtures-of-Experts and Hierarchical Mixtures-of-Experts models with an application to speech recognition. *Journal of the American Statististical Association*, **91**, 953–960.

Plackett, R. L. (1975) The analysis of permutations. *Applied Statistics* , **24**, 193–202.

Punj, G. and Stewart, D. W. (1983) Cluster analysis in marketing research: review and suggestions for application. *Journal of Marketing Research*, **20**, 134–148.

R Development Core Team (2009) *R: A Language and Environment for Statistical Computing*. R Foundation for Statistical Computing, Vienna, Austria.

Raftery, A. E., Newton, M. A., Satagopan, J. M. and Krivitsky, P. N. (2007) Estimating the integrated likelihood via posterior simulation using the harmonic mean identity (with discussion). In *Bayesian Statistics 8* (eds M. J. Bernardo, M. J. Bayarri, J. O. Berger, A. P. Dawid, D. Heckermen A. F. M. Smith and M. West), pp. 371–415. Oxford University Press.

Rosen, O. and Tanner, M. (1999) Mixtures of proportional hazards regression models. *Statistics in Medicine*, **18**, 1119–1131.

Schwartz, G. (1978) Estimating the dimension of a model. *Annals of Statistics*, **6**, 461–464.

Sinnott, R. (1995) *Irish Voters Decide: Voting Behaviour in Elections and Referendums since 1918*. Manchester University Press.

Sinnott, R. (1999) The electoral system. In *Politics in the Republic of Ireland* (eds J. Coakley and M. Gallagher) 3rd edn, pp. 99–126. Routledge & PSAI Press.

Snijders, T., Pattison, P., Robins, G. and Handcock, M. (2006) New specifications for exponential random graph models. *Sociological Methodology*, 99–153.

Spirling, A. and Quinn, K. (2010) Identifying intraparty voting blocs in the U.K. House of Commons. *Journal of the American Statistical Association*, **105**, 447–457.

Stern, H. S. (1993) Probability models on rankings and the electoral process. In *Probability Models and Statistical Analyses For Ranking Data* (eds M. A. Fligner and J. S. Verducci), pp. 173–195. Springer-Verlag.

Tam, W. K. (1995) Asians – a monolithic voting bloc? *Political Behaviour*, **17**, 223–249.

Tanner, M. A. and Jacobs, R. A. (2001) Neural networks and related statistical latent variable models In *International Encyclopedia of the Social and Behavioral Sciences* (eds N. J. Smelser and P. B. Baltes), pp. 10526–10534. Elsevier.

Thompson, T. J., Smith, P. J. and Boyle, J. P. (1998) Finite mixture models with concomitant information: assessing diagnostic criteria for diabetes. *Journal of the Royal Statistical Socity, Series C*, **47**, 393–404.

Titterington, D. M., Smith, A. F. M and Makov, U. E. (1985) *Statistical Analysis of Finite Mixture Distributions*. John Wiley & Sons, Ltd.

Wasserman, S. and Faust, K. (1994) *Social Network Analysis: Methods and Applications*. Cambridge University Press.

Xing, E., Fu, W. and Song, L. (2010) A state-space mixed membership blockmodel for dynamic network tomography. *Annals of Applied Statistics*, **4**, 535–566.

6

Modelling conditional densities using finite smooth mixtures

Feng Li, Mattias Villani and Robert Kohn

6.1 Introduction

Finite smooth mixtures, or *mixtures of experts* (ME) as they are known in the machine learning literature, have become increasingly popular in the statistical literature since their introduction by Jacobs *et al.* (1991). A smooth mixture is a mixture of regression models where the mixing probabilities are functions of the covariates, leading to a partitioned covariate space with stochastic (soft) boundaries. The first applications of smooth mixtures focused on flexible modelling of the mean function $E(y|x)$, but more recent works explore their potential for nonparametric modelling of conditional densities $p(y|x)$. A smooth mixture models $p(y|x)$ nonparametrically for any given x, but is also flexible across different covariate values.

Smooth mixtures are capable of approximating a large class of conditional distributions. For example, Jiang and Tanner (1999a, 1999b) show that smooth mixtures with sufficiently many (generalised) linear regression mixture components can approximate any density in the exponential family with arbitrary smooth mean function. More recently, Norets (2010) proves results for a mixture of Gaussian components under fairly general regularity conditions. See also Zeevi and Meir (1997) for additional results along these lines.

Like any mixture model, a smooth mixture may have a fairly complex multimodal likelihood surface. The choice of estimation method is therefore a key ingredient for successfully implementing smooth mixture models. Jordan and Jacobs (1994) employ the expectation maximisation (EM) algorithm

---

Mixtures: Estimation and Applications, First Edition. Edited by Kerrie L. Mengersen, Christian P. Robert and D. Michael Titterington.

for the ME model, and similar optimisation algorithms are popular in the machine learning field. Some recent approaches to smooth mixtures are Bayesian, with the computation implemented by Markov chain Monte Carlo (MCMC) methods. The first Bayesian paper on smooth mixtures is Peng *et al.* (1996) who use the random walk Metropolis algorithm to sample from the posterior. More sophisticated algorithms are proposed by Wood *et al.* (2002), Geweke and Keane (2007) and Villani *et al.* (2009).

The initial work on smooth mixtures in the machine learning literature advocated what may be called a *simple-and-many* approach with very simple mixture components (constants or linear homoscedastic regressions), but many of them. This practice is partly because estimating complicated component models was somewhat difficult in the time before and early days of MCMC, but probably also reflects an underlying divide-and-conquer philosophy in the machine learning literature. More recent implementations of smooth mixtures with access to MCMC technology successively introduce more flexibility within the components. This *complex-and-few* strategy tries to model nonlinearities and non-Gaussian features within the components and relies less on the mixture to generate the required flexibility, i.e. mixtures are used only when needed. For example, Wood *et al.* (2002) and Geweke and Keane (2007) use basis expansion methods (splines and polynomials) to allow for nonparametric component regressions. Further progress is made in Villani *et al.* (2009) who propose the smooth adaptive Gaussian mixture (SAGM) model as a flexible model for regression density estimation. Their model is a finite mixture of Gaussian densities with the mixing probabilities, the component means and component variances modelled as (spline) functions of the covariates. Li *et al.* (2010) extend this model to asymmetric Student's t components with the location, scale, skewness and degrees of freedom all modelled as functions of covariates. Villani *et al.* (2009) and Li *et al.* (2010) show that a single complex component can often give a better and numerically more stable fit in substantially less computing time than a model with many simpler components. As an example, simulations and real applications in Villani *et al.* (2009) show that a mixture of homoscedastic regressions can fail to fit heteroscedastic data, even with a very large number of components. Having heteroscedastic components in the mixture is therefore crucial for accurately modelling heteroscedastic data. The empirical stock returns example in Li *et al.* (2010) shows that including heavy-tailed components in the mixture can improve on the SAGM model when heteroscedastic heavy-tailed distributions are modelled. This finding is backed up by the theoretical results in Norets (2010).

This chapter further explores the simple-and-many versus complex-and-few issue by modelling regression data with a skewed response variable. A simulation study shows that it may be difficult to model a skewed conditional density by a smooth mixture of heteroscedastic Gaussian components (like SAGM). Introducing skewness within the components can improve the fit substantially.

We use the efficient Markov chain Monte Carlo (MCMC) method in Villani *et al.* (2009) to simulate draws from the posterior distribution in smooth mixture models; see Section 6.3.1. This algorithm allows for Bayesian variable selection in all parameters of the density, and in the mixture weights. Variable selection mitigates problems with overfitting, which is particularly important in models with complex mixture components. The automatic pruning effect achieved by variable selection in a mixture context is illustrated in Section 6.4.2 on the LIDAR data. Reducing the number of effective parameters by variable selection also helps the MCMC algorithm to converge faster and mix better.

Section 6.4.3 uses smooth mixtures of Gaussians and split-t components to model the electricity expenditure of households. To take into account that expenditures are positive, and more generally to handle positive dependent variables, we also introduce two smooth mixtures for strictly positively valued data: a smooth mixture of gamma densities and smooth mixture of log-normal densities. In both cases we use an interpretable reparameterised density where the mean and the (log) variance are modelled as functions of the covariates.

6.2 The model and prior

6.2.1 Smooth mixtures

Our model for the conditional density $p(y|x)$ is a finite mixture density with weights that are smooth functions of the covariates,

$$p(y|x) = \sum_{k=1}^{K} \omega_k(x) p_k(y|x), \tag{6.1}$$

where $p_k(y|x)$ is the kth component density with weight $\omega_k(x)$. The next subsection discusses specific component densities $p_k(y|x)$. The weights are modelled by a multinomial logit function,

$$\omega_k(x) = \frac{\exp(x'\gamma_k)}{\sum_{r=1}^{K} \exp(x'\gamma_r)}, \tag{6.2}$$

with $\gamma_1 = 0$ for identification. The covariates in the components can in general be different from the covariates in the mixture weights.

To simplify the MCMC simulation, we express the mixture model in terms of latent variables as in Diebolt and Robert (1994) and Escobar and West (1995). Let $s_1, \ldots, s_n$ be unobserved indicator variables for the observations in the sample such that $s_i = k$ means that the ith observation belongs to the kth component, $p_k(y|x)$. The model in (6.1) and (6.2) can then be written as

$$\Pr(s_i = k|x_i, \gamma) = \omega_k(x_i),$$
$$y_i|(s_i = k, x_i) \sim p_k(y_i|x_i).$$

Conditional on $s = (s_1, \ldots, s_n)'$, the mixture model decomposes into K separate component models $p_1(y|x), \ldots, p_K(y|x)$, with each data observation being allocated to one and only one component.

6.2.2 The component models

The component densities in SAGM (Villani *et al.*, 2009) are Gaussian with both the mean and variance functions of covariates,

$$y|x, s = k \sim N\left[\mu_k(x), \sigma_k^2(x)\right],$$

where

$$\mu_k(x) = \beta_{\mu_0,k} + x'\beta_{\mu,k}, \; \log \sigma_k^2(x) = \beta_{\sigma_0^2,k} + x'\beta_{\sigma^2,k}. \tag{6.3}$$

Note that each mixture component has its own set of parameters. We will suppress the component subscript k in the remainder of this section, but, unless stated otherwise, all parameters are component-specific. SAGM uses a linear link function for the mean and a log link for the variance, but any smooth link function can equally well be used in our MCMC methodology. Additional flexibility can be obtained by letting a subset of the covariates be a nonlinear basis expansion, e.g. additive splines or splines surfaces (Ruppert *et al.*, 2003) as in Villani *et al.* (2009); see also the LIDAR example in Section 6.4.2.

SAGM is in principle capable of capturing heavy-tailed and skewed data. In line with the complex-and-few approach it may be better, however, to use mixture components that allow for skewness and excess kurtosis. Li *et al.* (2010) extend the SAGM model to components that are split-t densities according to the following definition.

Definition 1: Split-t distribution

The random variable y follows a split-t distribution with $\nu > 0$ degrees of freedom, if its density function is of the form

$$p(y; \mu, \phi, \lambda, \nu) = c \cdot \kappa(y; \mu, \phi, \nu)\mathbf{1}_{y\leq\mu} + c \cdot \kappa(y; \mu, \lambda\phi, \nu)\mathbf{1}_{y>\mu},$$

where

$$\kappa(y; \mu, \phi, \nu) = \left[1 + \left(\frac{y-\nu}{\phi}\right)^2 \nu^{-1}\right]^{-(\nu+1)/2},$$

is the kernel of a Student's t density with variance $\phi^2\nu/(\nu-2)$ and $c = 2[(1+\lambda)\phi\sqrt{\nu}Beta(\nu/2, 1/2)]^{-1}$ is the normalisation constant.

The location parameter μ is the mode, $\phi > 0$ is the scale parameter and $\lambda > 0$ is the skewness parameter. When $\lambda < 1$ the distribution is skewed to the left, when $\lambda > 1$ it is skewed to the right and when $\lambda = 1$ it reduces to the usual symmetric Student's t density. The split-t distribution reduces to the split-normal distribution in Gibbons and Mylroie (1973) and John (1982) as $\nu \to \infty$. Any other asymmetric t density can equally well be used in our MCMC methodology; see Section 6.3.1.

Each of the four parameters, μ, ϕ, λ and ν, is connected to covariates as

$$\begin{aligned} \mu &= \beta_{\mu_0} + x'\beta_\mu, \quad \log \phi = \beta_{\phi_0} + x'\beta_\phi, \\ \log \nu &= \beta_{\nu_0} + x'\beta_\nu, \quad \log \lambda = \beta_{\lambda_0} + x'\beta_\lambda, \end{aligned} \tag{6.4}$$

but, as mentioned above, any smooth link function can equally well be used in the MCMC methodology.

Section 6.4.3 applies smooth mixtures in a situation where the response is non-negative. Natural mixture components are then gamma and log-normal densities. The gamma components are of the form

$$y|s, x \sim \text{Gamma}\left(\frac{\mu^2}{\sigma^2}, \frac{\sigma^2}{\mu}\right),$$

where

$$\log \mu(x) = \beta_{\mu_0} + x'\beta_\mu, \quad \log \sigma^2(x) = \beta_{\sigma_0^2} + x'\beta_{\sigma^2}, \tag{6.5}$$

where we have again suppressed the component labels. Note that we use an interpretable parameterisation of the gamma distribution where μ and σ^2 are the mean and variance, respectively.

Similarly, the log-normal components are of the form

$$y|s, x \sim \text{Log N}\left[\log \mu - \frac{1}{2}\log\left(1 + \frac{\sigma^2}{\mu^2}\right), \sqrt{\log\left(1 + \frac{\sigma^2}{\mu^2}\right)}\right],$$

where

$$\log \mu(x) = \beta_{\mu_0} + x'\beta_\mu, \quad \log \sigma^2(x) = \beta_{\sigma_0^2} + x'\beta_{\sigma^2}. \tag{6.6}$$

Again, the two parameters, μ and σ^2, are the mean and variance.

A smooth mixture of complex densities is a model with a large number of parameters, however, and is therefore likely to overfit the data unless model complexity is controlled effectively. We use Bayesian variable selection on all the component's parameters and in the mixing function. This can lead to important simplifications of the mixture components. Not only does this control complexity for a given number of components but it also simplifies the existing components if an additional component is added to the model; the LIDAR example in Section 6.4.2 illustrates this well. Increasing the number of components can therefore in principle even reduce the number of effective parameters in the model. It may nevertheless be useful to put additional structure on the mixture components before estimation. One particularly important restriction is that one or more component parameters are common to all components. A component parameter (e.g. ν in the split-t model in (6.4)) is said to be *common* to the components when only the intercepts in (6.4) are allowed to be different across components. The unrestricted model is said to have *separate* components.

The regression coefficient vectors, such as β_μ, β_ϕ, β_ν and β_λ in the split-t model, are all treated in a unified way in the MCMC algorithm. Whenever we refer to a regression coefficient vector without subscript, β, the argument applies to any of the regression coefficient vectors of the split-t parameters in (6.4).

6.2.3 The prior

We now describe an easily specified prior for smooth mixtures, proposed by Villani *et al.* (2010) that builds on Ntzoufras *et al.* (2003) and depends only on a few hyperparameters. Since there can be a large number of covariates in the model, the strategy in Villani *et al.* (2010) is to incorporate available prior information via the intercepts, and to use a unit-information prior that automatically takes the model geometry and link function into account.

We standardise the covariates to have zero mean and unit variance, and assume prior independence between the intercept and the remaining regression coefficients. The intercepts then have the interpretation of being the (possibly transformed) density parameters at the mean of the original covariates. The strategy in Villani *et al.* (2010) is to specify priors directly on the parameters of the mixture component, e.g. the degrees of freedom ν in the split-t components, and then back out the implied prior on the intercept β_{ν_0}. For example, a normal prior for a parameter with identity link (e.g. μ in the split-t model) trivially implies a normal prior on $\beta_{\mu 0}$; a log-normal prior with mean m^* and variance s^{*2} for a parameter with log link (e.g. ϕ in the split-t model) implies a normal prior $N(m_0, s_0^2)$ for β_{ϕ_0}, where

$$m_0 = \log m^* - \frac{1}{2}\log\left[\left(\frac{s^*}{m^*}\right)^2 + 1\right] \quad \text{and} \quad s_0^2 = \log\left[\left(\frac{s^*}{m^*}\right)^2 + 1\right].$$

The regression coefficient vectors are assumed to be independent a priori. We allow for Bayesian variable selection by augmenting each parameter vector β by a vector of binary covariate selection indicators $\mathcal{I} = (i_1, ..., i_p)$ such that $\beta_j = 0$ if $i_j = 0$. Let $\beta_{\mathcal{I}}$ denote the subset of β selected by $\mathcal{I}$. In a Gaussian linear regression one can use a g-prior (Zellner, 1986) $\beta \sim N[0, \tau_\beta^2(X'X)^{-1}]$ on the full β and then condition on the restrictions imposed by $\mathcal{I}$. Setting $\tau^2 = n$, where n is the number of observations, gives the unit-information prior, i.e. a prior that carries information equivalent to a single observation from the model. More generally, the unit information prior is $\beta \sim N[0, \tau_\beta^2 \mathcal{I}^{-1}(\beta)]$, where

$$\mathcal{I}(\beta) = -E\left[\left.\frac{\partial^2 \log p(\beta|y)}{\partial\beta\partial\beta'}\right|_{\beta=\bar{\beta}}\right]$$

and $\bar{\beta} = (\beta_0, 0, ..., 0)'$ is the prior mean of β. When the analytical form of the expected Hessian matrix is not available in closed form, we simulate replicated datasets from a model with parameter vector β_0 and approximate the expected Hessian by the average Hessian over the simulated datasets.

The variable selection indicators are assumed to be independent Bernoulli variables with probability π_β a priori, but more complicated distributions are easily accommodated; see, for example, the extension in Villani *et al.* (2009) for splines in a mixture context, or a prior which is uniform on the variable selection indicators for a given model size in Denison *et al.* (2002). It is also possible to estimate π_β as proposed in Kohn *et al.* (2001) with an extra Gibbs sampling step. Note also that π_β may be different for each parameter in the mixture components. Our default prior has $\pi_\beta = 0.5$.

The prior on the mixing function decomposes as

$$p(\gamma, \mathcal{Z}, s) = p(s|\gamma, \mathcal{Z})p(\gamma|\mathcal{Z})p(\mathcal{Z}),$$

where $\mathcal{Z}$ is the $p \times (K-1)$ matrix with variable selection indicators for the p covariates in the mixing function; recall that $\gamma_1 = 0$ for identification. The variable indicators in $\mathcal{Z}$ are assumed to be independent and identically distributed (iid) Bernoulli (ω_γ). Let $\gamma_\mathcal{Z}$ be the prior on $\gamma = (\gamma_2', \ldots, \gamma_m')'$ of the form

$$\gamma_\mathcal{Z}|\mathcal{Z} \sim N(0, \tau_\gamma^2 I),$$

and $\gamma_{\mathcal{Z}^c} = 0$ with probability one. We use $\tau_\gamma^2 = 10$ as the default value. Finally, $p(s|\gamma, \mathcal{Z})$ is given by the multinomial logit model in (6.2). To reduce the number of parameters and to speed up the MCMC algorithm we restrict the columns of $\mathcal{Z}$ to be identical, i.e. we make the assumption that a covariate is either present in the mixing function in all components or does not appear at all, but the extension to general $\mathcal{Z}$ is straightforward; see Villani *et al.* (2009).

6.3 Inference methodology

6.3.1 The general MCMC scheme

We use MCMC methods to sample from the joint posterior distribution, and draw the parameters and variable selection indicators in blocks. The algorithm below is the preferred algorithm from the experiments in Villani *et al.* (2009). The number of components is determined by a Bayesian version of cross-validation discussed in Section 6.3.3.

The MCMC algorithm is very general, but for conciseness we describe it for the smooth mixture of split-*t* components. The algorithm is a Metropolis-within-Gibbs sampler that draws parameters using the following six blocks:

1. $\{(\beta_\mu^{(k)}, \mathcal{I}_\mu^{(k)})\}_{k=1,\ldots,K}$;
2. $\{(\beta_\phi^{(k)}, \mathcal{I}_\phi^{(k)})\}_{k=1,\ldots,K}$;
3. $\{(\beta_\lambda^{(k)}, \mathcal{I}_\lambda^{(k)})\}_{k=1,\ldots,K}$;
4. $\{(\beta_\nu^{(k)}, \mathcal{I}_\nu^{(k)})\}_{k=1,\ldots,K}$;
5. $s = (s_1, \ldots, s_n)$;
6. γ and $\mathcal{I}_Z$.

The parameters in the different components are independent conditional on s. This means that each of the first four blocks splits up into K independent updating steps. Each updating step in the first four blocks is sampled using highly efficient tailored MH proposals following a general approach described in the next subsection. The latent component indicators in s are independent conditional on the model parameters and are drawn jointly from their full conditional posterior. Conditional on s, Step 6 is a multinomial logistic regression with variable selection and γ and $\mathcal{I}_Z$ are drawn jointly using a generalisation of the method used to draw blocks 1 to 4; see Villani *et al.* (2009) for details.

It is well known that the likelihood function in mixture models is invariant with respect to permutations of the components; see, for example, Celeux *et al.* (2000), Jasra *et al.* (2005) and Frühwirth-Schnatter (2006). The aim here is to estimate the predictive density, so label switching is neither a numerical nor a conceptual problem (Geweke, 2007). If an interpretation of the mixture components is required, then it is necessary to impose some identification restrictions on some of the model parameters, e.g. an ordering constraint (Jasra *et al.*, 2005). Restricting some parameters to be common across components is clearly also helpful for identification.

6.3.2 Updating β and $\mathcal{I}$ using variable-dimension finite-step Newton proposals

Nott and Leonte (2004) extend the method which was introduced by Gamerman (1997) for generating MH proposals in a generalised linear model (GLM) to the variable selection case. Villani *et al.* (2009) extend the algorithm to a general setting not restricted to the exponential family. We first treat the problem without variable selection. The algorithm in Villani *et al.* (2009) only requires that the posterior density can be written as

$$p(\beta|y) \propto p(y|\beta)p(\beta) = \prod_{i=1}^{n} p(y_i|\varphi_i)p(\beta), \tag{6.7}$$

where $\varphi_i = x_i'\beta$ and x_i is a covariate vector for the ith observation. Note that $p(\beta|y)$ may be a conditional posterior density and the algorithm can then be used as a step in a Metropolis-within-Gibbs algorithm. The full conditional posteriors for blocks 1 to 4 in Section 6.3.1 are clearly all of the form in (6.7). Newton's method can be used to iterate R steps from the current point β_c in the MCMC sampling towards the mode of $p(\beta|y)$, to obtain $\hat{\beta}$ and the Hessian at $\hat{\beta}$. Note that $\hat{\beta}$ may not be the mode but is typically close to it already after a few Newton iterations, so setting $R = 1, 2$ or 3 is usually sufficient. This makes the algorithm fast, especially when the gradient and Hessian are available in closed form, which is the case here; see the Appendix.

Once we have obtained good approximations of the posterior mode and covariance matrix from the Newton iterations, the proposal β_p is now drawn from the multivariate t-distribution with $g > 2$ degrees of freedom:

$$\beta_p|\beta_c \sim t\left[\hat{\beta}, -\left(\frac{\partial^2 \log p(\beta|y)}{\partial\beta\partial\beta'}\right)^{-1}\Bigg|_{\beta=\hat{\beta}}, g\right],$$

where the second argument of the density is the covariance matrix.

In the variable selection case we propose β and $\mathcal{I}$ simultaneously using the decomposition

$$g(\beta_p, \mathcal{I}_p|\beta_c, \mathcal{I}_c) = g_1(\beta_p|\mathcal{I}_p, \beta_c)g_2(\mathcal{I}_p|\beta_c, \mathcal{I}_c),$$

where g_2 is the proposal distribution for $\mathcal{I}$ and g_1 is the proposal density for β conditional on $\mathcal{I}_p$. The Metropolis–Hastings acceptance probability is

$$\begin{aligned} &a[(\beta_c, \mathcal{I}_c) \to (\beta_p, \mathcal{I}_p)] \\ &\quad = \min\left[1, \frac{p(y|\beta_p, \mathcal{I}_p)p(\beta_p|\mathcal{I}_p)p(\mathcal{I}_p)g_1(\beta_c|\mathcal{I}_c, \beta_p)g_2(\mathcal{I}_c|\beta_p, \mathcal{I}_p)}{p(y|\beta_c, \mathcal{I}_c)p(\beta_c|\mathcal{I}_c)p(\mathcal{I}_c)g_1(\beta_p|\mathcal{I}_p, \beta_c)g_2(\mathcal{I}_p|\beta_c, \mathcal{I}_c)}\right]. \end{aligned}$$

The proposal density at the current point $g_1(\beta_c|\mathcal{I}_c, \beta_p)$ is a multivariate t-density with mode $\widetilde{\beta}$ and covariance matrix equal to the negative inverse Hessian evaluated at $\widetilde{\beta}$, where $\widetilde{\beta}$ is the point obtained by iterating R steps of the Newton algorithm, this time starting from β_p. A simple way of proposing $\mathcal{I}_p$ is to select randomly a small subset of $\mathcal{I}_c$ and then always propose a change of the selected indicators. It is important to note that β_c and β_p may now be of different dimensions, so the original Newton iterations no longer apply. We will instead generate β_p using the following generalisation of Newton's method. The idea is that, when the parameter vector β changes dimensions, the dimension of the functionals $\varphi_c = x'\beta_c$ and $\varphi_p = x'\beta_p$ stay the same, and the two functionals are expected to be quite close. A generalised Newton update is

$$\beta_{r+1} = A_r^{-1}(B_r\beta_r - s_r), \qquad r = 0, ..., R-1, \tag{6.8}$$

where $\beta_0 = \beta_c$, the dimension of β_{r+1} equals the dimension of β_p and

$$\begin{aligned} s_r &= X'_{r+1}d + \frac{\partial \log p(\beta)}{\partial\beta}, \\ A_r &= X'_{r+1}DX_{r+1} + \frac{\partial^2 \log p(\beta)}{\partial\beta\partial\beta'}, \\ B_r &= X'_{r+1}DX_r + \frac{\partial^2 \log p(\beta)}{\partial\beta\partial\beta'}, \end{aligned} \tag{6.9}$$

where d is an n-dimensional vector with gradients $\partial \log p(y_i|\varphi_i)/\partial\varphi_i$ for each observation currently allocated to the component being updated. Similarly, D is a diagonal matrix with Hessian elements

$$\frac{\partial^2 \log p(y_i|\varphi_i)}{\partial\varphi_i \partial\varphi_i'},$$

X_r is the matrix with the covariates that have nonzero coefficients in β_r and all expressions are evaluated at $\beta \doteq \beta_r$. For the prior gradient this means that $\partial \log p(\beta)/\partial\beta$ is evaluated at β_r, including all zero parameters, and that the sub-vector conformable with β_{r+1} is extracted from the result. The same applies to the prior Hessian (which does not depend on β, however, if the prior is Gaussian). Note that we only need to compute the scalar derivatives $\partial \log p(y_i|\phi_i)/\partial\phi_i$ and $\partial^2 \log p(y_i|\phi_i)/\partial\phi_i^2$.

6.3.3 Model comparison

The number of components is assumed known in our MCMC scheme above. A Bayesian analysis via mixture models with an unknown number of components is possible using, for example, Dirichlet process mixtures (Escobar and West, 1995), reversible jump MCMC (Richardson and Green, 1997) or birth-and-death MCMC (Stephens, 2000). The fundamental quantity determining the posterior distribution of the number of components is the marginal likelihood of the models with a different number of components. It is well-known, however, that the marginal likelihood is sensitive to the choice of prior, and this is especially true when the prior is not very informative; see, for example, Kass (1993) for a general discussion and Richardson and Green (1997) in the context of density estimation.

Following Geweke and Keane (2007) and Villani *et al.* (2009), we therefore compare and select models based on the out-of-sample log predictive density score (LPDS). By sacrificing a subset of the observations to update/train the vague prior, we remove much of the dependence on the prior and obtain a better assessment of the predictive performance that can be expected for future observations. To deal with the arbitrary choice of which observations to use for estimation and model evaluation, we use B-fold cross-validation of the log predictive density score (LPDS):

$$\frac{1}{B}\sum_{b=1}^{B} \log p(\tilde{y}_b|\tilde{y}_{-b}, x),$$

where $\tilde{y}_b$ is an n_b-dimensional vector containing the n_b observations in the bth test sample and $\tilde{y}_{-b}$ denotes the remaining observations used for estimation. If we assume that the observations are independent conditional on θ, then

$$p(\tilde{y}_b|\tilde{y}_{-b}, x) = \int \prod_{i\in\mathcal{T}_b} p(y_i|\theta, x_i)p(\theta|\tilde{y}_{-b})\mathrm{d}\theta,$$

where $\mathcal{T}_b$ is the index set for the observations in $\tilde{y}_b$ and the LPDS is easily computed by averaging $\prod_{i\in\mathcal{T}_b} p(y_i|\theta, x_i)$ over the posterior draws from $p(\theta|\tilde{y}_{-b})$. This requires sampling from each of the B posteriors $p(\theta|\tilde{y}_{-b})$, for $b = 1, ..., B$, but these MCMC runs can all be run in isolation from each other and are therefore ideal for straightforward parallel computing on widely available multicore processors. Cross-validation is less appealing in a time-series setting since it is typically false that the observations are independent conditional on the model parameters for time-series data. A more natural approach is to use the most recent observations in a single test sample; see Villani *et al.* (2009).

6.4 Applications

6.4.1 A small simulation study

The simulation study in Villani *et al.* (2009) explores the out-of-sample performance of a smooth mixture of homoscedastic Gaussian components for heteroscedastic data. The study shows that a smooth mixture of heteroscedastic regressions is likely to be a much more effective way of modelling heteroscedastic data. This section uses simulations to explore how different smooth mixture models cope with skewed and heavy-tailed data. We generate data from the following models.

1. A one-component normal with mean $\mu = 0$ and variance $\phi^2 = 1$ at $x = \bar{x}$.
2. A split-normal with mean $\mu = 0$, variance $\phi^2 = 0.5^2$ and skewness parameter $\lambda = 5$ at $x = \bar{x}$.
3. A Student-t with mean $\mu = 0$, variance $\phi^2 = 1$ and $\nu = 5$ degrees of freedom at $x = \bar{x}$.
4. A split-t with mean $\mu = 0$, variance $\phi^2 = 1$, $\nu = 5$ degrees of freedom and skewness parameter $\lambda = 5$ at $x = \bar{x}$.

Each of the parameters μ, ϕ, ν and λ are connected to four covariates (drawn independently from the $N(0, 1)$ distribution) as in (6.4). Two of the covariates have nonzero coefficients in the data-generating process and the other two have zero coefficients. The number of observations in each simulated dataset is 1000. We generate 30 datasets for each model and analyse them with both SAGM and a smooth mixture of split-t components using 1 to 5 mixture components. The priors for the parameters in the estimated models are set as in Table 6.1.

We analyse the relative performance of SAGM and split-t by comparing the estimated conditional densities $q(y|x)$ with the true data-generating densities $p(y|x)$ using estimates of both the Kullback–Leibler divergence and the L_2 distance,

Table 6.1 Priors in the simulation study.

	μ	ϕ	ν	λ
Mean	0	1	10	1
Std	10	1	7	0.8

defined respectively as

$$D_{\mathrm{KL}}(p,q)=\sum_{i=1}^{n}p(y_i|x_i)\log\frac{p(y_i|x_i)}{q(y_i|x_i)},$$

$$D_{L2}(p,q)=100\left(\sum_{i=1}^{n}\left[q(y_i|x_i)-p(y_i|x_i)\right]^2\right)^{1/2},$$

where $\{y_i,x_i\}_{i=1}^{n}$ is the estimation data.

Table 6.2 shows that, when the true data are normal (DGP 1), both SAGM and split-t do well with a single component. The extra coefficients in the degrees of freedom and skewness in the split-t are effectively removed by variable selection. SAGM improves a bit when components are added, while the split-t gets slightly worse.

When the DGP also exhibits skewness (DGP 2), SAGM(1) performs much worse than split-t(1). SAGM clearly improves with more components, but the fit of SAGM(5) is still much worse than the one-component split-t. Note how variable selection makes the performance of the split-t deteriorate only very slowly as we add unnecessary components.

The same story as in the skewed-data situation holds when the data are heavy-tailed (DGP 3) and when the data are both skewed and heavy-tailed (DGP 4).

In conclusion, smooth mixtures with a few complex components can greatly outperform smooth mixtures with many simpler components. Moreover, variable selection is effective in down-weighting unnecessary aspects of the components and makes the results robust to mis-specification of the number of components, even when the components are complicated.

Table 6.2 Kullback–Leibler and L_2 distance between estimated models and the true DGPs.

	Split-t					SAGM				
K	1	2	3	4	5	1	2	3	4	5
					DGP 1 – Normal					
D_{KL}	1.06	1.40	1.54	1.79	2.19	1.31	1.03	0.90	0.95	1.05
D_{L2}	1.73	2.64	3.18	6.11	8.33	2.21	1.52	1.34	1.46	1.71
					DGP 2 – Split-normal					
D_{KL}	3.67	3.67	4.76	4.74	5.57	51.05	14.16	7.30	7.33	8.01
D_{L2}	6.05	6.82	9.51	9.55	13.11	106.13	31.49	16.46	16.20	17.59
					DGP 3 – Student-t					
D_{KL}	1.12	1.72	1.79	2.05	2.20	13.30	1.94	1.78	2.16	2.65
D_{L2}	2.14	4.82	4.70	5.72	5.42	35.79	4.33	3.91	4.70	6.61
					DGP 4 – Split-t					
D_{KL}	3.99	3.24	4.24	4.66	5.67	75.80	21.02	8.89	7.35	7.36
D_{L2}	9.02	8.22	11.78	13.13	16.90	199.99	59.54	27.06	22.43	22.63

6.4.2 LIDAR data

Our first real dataset comes from a technique that uses laser-emitted light to detect chemical compounds in the atmosphere (LIDAR, or light detection and ranging). The response variable (`logratio`) consists of 221 observations on the log ratio of received light from two laser sources, one at the resonance frequency of the target compound and the other from a frequency different from this target frequency. The predictor is the distance travelled before the light is reflected back to its source (`range`). The original data come from Holst *et al.* (1996) and has been analysed by, for example, Ruppert *et al.* (2003) and Leslie *et al.* (2007). Our aim is to model the predictive density $p(\texttt{logratio} \mid \texttt{range})$.

Leslie *et al.* (2007) show that a Gaussian model with nonparametric mean and variance can capture this dataset quite well. We will initially use the SAGM model in Villani *et al.* (2009) with the mean, variance and mixing functions all modelled nonparametrically by thin plate splines (Green and Silverman, 1994). Ten equidistant knots in each component are used for each of these three aspects of the model. We use a version of SAGM where the variance functions of the components are proportional to each other i.e. only the intercepts in the variance functions are allowed to be different across components. The more general model with completely separate variance functions gives essentially the same LPDS, and the posterior distributions of the component variance functions, identified by order restrictions, are largely overlapping. We use the variable selection prior in Villani *et al.* (2009) where the variable selection indicator for a knot κ in the kth mixture component is distributed as *Bernoulli* $\left[\pi_{\beta}\omega_k(\kappa)\right]$. This has the desirable effect of down-weighting knots in regions where the corresponding mixture component has small probability. We compare our results to the smoothly mixing regression (SMR) in Geweke and Keane (2007), which is a special case of SAGM where the components' variance functions are independent of the covariates and any heteroscedasticity is generated solely by the mixture. We use a prior with $m^* = 0$ and $s^{*2} = 10$ in the mean function and $m^* = 1$ and $s^{*2} = 1$ in the variance function; see Section 6.2.3. Given the scale of the data, these priors are fairly noninformative. As documented in Villani *et al.* (2009) and Li *et al.* (2010), the estimated conditional density and the LPDS are robust to variations in the prior.

Table 6.3 displays the fivefold cross-validated LPDS for the SMR and SAGM models, both when the components are linear in covariates and when they are

Table 6.3 Log predictive density score (LPDS) over the five cross-validation samples for the LIDAR data.

	Linear components			Thin-plate components		
	$K = 1$	$K = 2$	$K = 3$	$K = 1$	$K = 2$	$K = 3$
SMR	26.564	59.137	63.162	48.399	61.571	62.985
SAGM	30.719	61.217	64.223	64.267	64.311	64.313

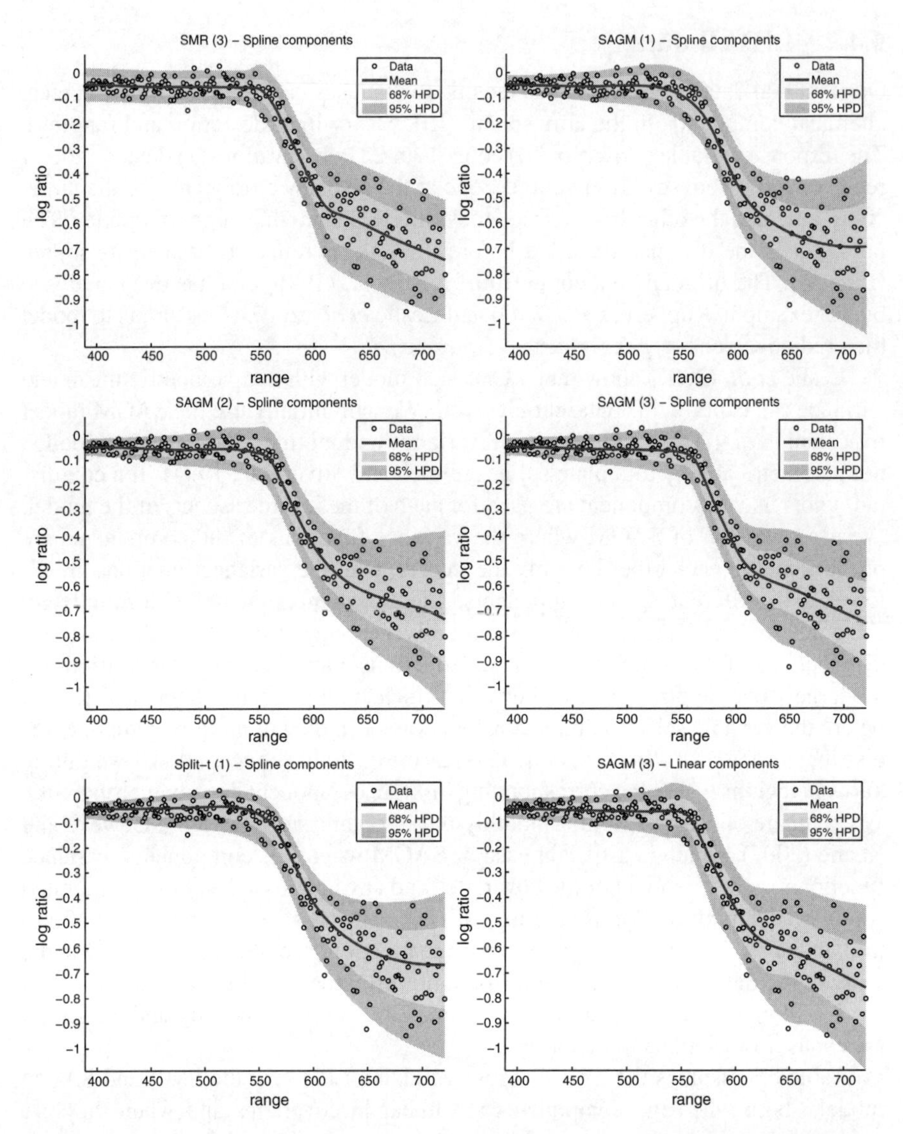

Figure 6.1 Assessing the in-sample fit of the smooth mixture models for the LIDAR data. The figure displays the actual data overlaid on HPD predictive regions. The solid line is the predictive mean.

modelled by thin plate splines. The three SAGM models with splines have roughly the same LPDS. The SMR model needs three components to come close to the LPDS of the SAGM(1) model with splines, and even then does not quite reach it. All the knots in the variance function of the SAGM models have posterior inclusion probabilities smaller than 0.1, suggesting strongly that the (log) variance function is linear in `range`. Figure 6.1 plots the LIDAR data and the 68 % and 95 %

highest posterior density (HPD) regions in the predictive distribution $p(\texttt{logratio} \mid \texttt{range})$ from the SMR(3) and the SAGM models with 1, 2 and 3 components. Perhaps the most interesting result in Table 6.3 and Figure 6.1 is that SAGM models with more than one component do not seem to overfit. This is quite remarkable since the one-component model fits the data well, and additional components should therefore be a source of overfitting. This is due to the self-adjusting mechanism provided by the variable/knot selection prior where the already present components automatically become simpler (more linear) as more components are added to the model. The estimation results for the SAGM(3) model with spline components (not shown) reveals that the SAGM(3) model with spline components is in fact reduced to essentially a model with linear components. Figure 6.1 also shows that the fit of the SAGM(3) models with linear components (bottom row, second column) and spline components (second row, second column) are strikingly similar. The same holds for the LPDS in Table 6.3. Finally, Figure 6.1 also displays the fit of the split-t model with one component. The estimation results for this model show that only two knots are really active in the mean function; all of the knots in the scale, degrees of freedom and skewness have posterior probabilities smaller than 0.3. The number of degrees of freedom is roughly 43 for the smallest values of range and then decreases smoothly towards 7 when range is 720. The skewness parameter λ is roughly 0.5 for all values of range, a sizeable skewness, which is also visible in Figure 6.1. The LPDS of the one-component split-t model is 64.014, which is only slightly worse than SAGM(1).

6.4.3 Electricity expenditure data

Our second example uses a dataset with electricity expenditures in 1602 households from South Australia (Bartels *et al.*, 1996). Leslie *et al.* (2007) analyse this dataset and conclude that a heteroscedastic regression with errors following a Dirichlet process mixture fits the data well. They also document that the response variable is quite skewed. We consider both in-sample and out-of-sample performance of smooth mixture models, using the dataset in Leslie *et al.* (2007) without interactions. The 13 covariates used in our application are defined in Table 6.4. Following Leslie *et al.* (2007), we mean-correct the covariates but keep their original scale.

The prior means of μ and ϕ are set equal to the median and the standard deviation of the response variable, respectively. This data-snooping is innocent as we set the standard deviation of μ and ϕ to 100, so the prior is largely noninformative. The prior mean and standard deviation of the skewness parameter, λ, are both set to unity. This means that we are centring the prior on a symmetric model, but allowing for substantial skewness a priori. The prior mean of the degrees of freedom is set to 10 with a standard deviation of 7, which is wide enough to include both the Cauchy and essentially the Gaussian distributions. Since the data sample is fairly large and we base model choice on the LPDS, the results are insensitive to the exact choice of priors.

We first explore the out-of-sample performance of several smooth mixture models using fivefold cross-validation of the LPDS. The five subsamples are chosen

Table 6.4 The electricity bill regressors (subsets).

Variable name	Description
log(rooms)	log of the number of rooms in the house
log(income)	log of the annual pre-tax household income in Australian dollars
log(people)	log of the number of usual residents in the house
mhtgel	indicator for electric main heating
sheonly	indicator for electric secondary heating only
whtgel	indicator for peak electric water heating
cookel	indicator for electric cooking only
poolfilt	indicator for pool filter
airrev	indicator for reverse cycle air conditioning
aircond	indicator for air conditioning
mwave	indicator for microwave
dish	indicator for dishwasher
dryer	indicator for dryer

by sampling systematically from the dataset. Table 6.5 displays the results for a handful of models. Every model is estimated both under the assumption of separate parameters and when all parameters except the intercepts are common across components; see Section 6.2.2.

If we look first at the LPDS of the one-component models, it is clear that data are skewed (the skewed models are all doing better than SAGM), but the nature of

Table 6.5 Log predictive density score (LPDS) from fivefold cross-validation of the electricity expenditure data.

Model		$K = 1$	$K = 2$	$K = 3$
SMR	*separate*	−2,086	−2,027	−2,020
	common	–	−2,030	−2,020
SAGM	*separate*	−2,040	−2,022	−2,024
	common	–	−2,022	−2,017
Split-normal	*separate*	−2,014	−2,012	−2,015
	common	–	−2,064	−2,251
Student's *t*	*separate*	−2,025	−2,014	−2,014
	common	–	−2,029	−2,022
Split-*t*	*separate*	−2,034	−2,006	−1,996
	common	–	−2,073	−2,041
Gamma	*separate*	−2,007	−2,002	−2,003
	common	–	−2,008	−2,009
Log-normal	*separate*	−2,011	−2,006	−2,009
	common	–	−2,007	−2,010

The numerical standard errors of the LPDS are smaller than one for all models.

the skewness is clearly important (gamma is doing a lot better than split-normal). The best one-component model is the gamma model.

The best model overall is the split-t model with three separate components, closely followed by the log-normal model, with two separate components. It seems that this particular dataset has a combination of skewness and heavy-tailedness, which is better modelled by a mixture than by a single skewed and heavy-tailed component.

One way of checking the in-sample fit of the models on the full dataset is to look at the normalised residuals. We define the normalised residual as $\Phi^{-1}[F(y_i)]$, where $F(\cdot)$ is the distribution function from the model. If the model is correctly specified, the normalised residuals should be an iid $N(0, 1)$ sample. Figure 6.2 displays QQ

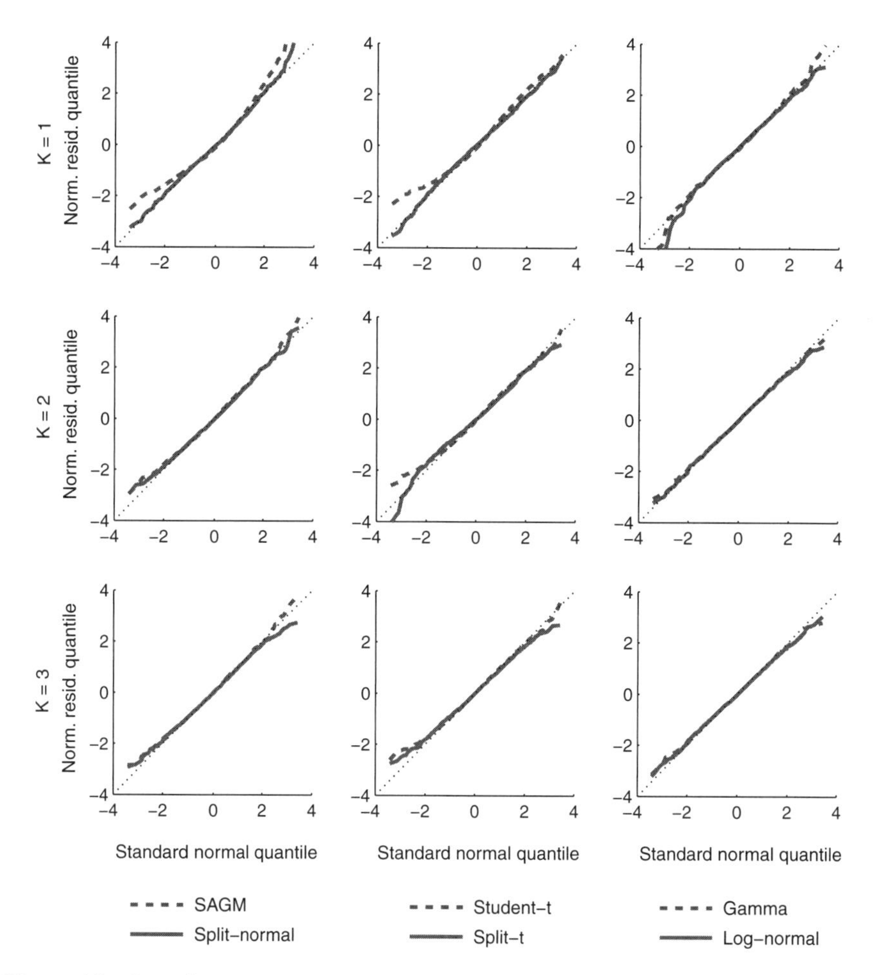

Figure 6.2 Quantile–quantile plots of the normalised residuals resulting from SAGM and split-normal (first column); Student-t and split-t (second column); gamma and log-normal (third column) with one to three separate components respectively. If the model is correct, the normalised residuals should lie on the dotted reference line.

Table 6.6 Posterior means and inclusion probabilities in the one-component split-t model for the electricity expenditure data.

Variable	β_μ	$\mathcal{I}_\mu$	β_ϕ	$\mathcal{I}_\phi$	β_ν	$\mathcal{I}_\nu$	β_λ	$\mathcal{I}_\lambda$
Intercept	256.62	–	3.82	–	2.83	–	1.34	–
log(rooms)	**49.47**	0.90	−0.65	0.43	−0.05	0.04	**0.97**	1.00
log(income)	2.71	0.48	**−0.36**	1.00	−0.05	0.02	**0.55**	1.00
log(people)	**40.62**	1.00	−0.20	0.22	0.06	0.03	**0.34**	1.00
mhtgel	**27.28**	1.00	0.07	0.12	−0.18	0.03	0.13	0.15
sheonly	10.11	0.72	0.01	0.04	**2.10**	0.99	0.04	0.05
whtgel	17.74	0.68	−0.23	0.18	0.33	0.04	**0.82**	0.99
cookel	**27.80**	0.99	−0.19	0.14	0.01	0.04	**0.39**	1.00
poolfilt	−6.50	0.50	−0.11	0.23	1.62	0.07	0.32	0.76
airrev	**14.06**	0.91	0.06	0.07	−0.03	0.03	0.12	0.16
aircond	5.58	0.46	0.03	0.11	0.01	0.03	**0.29**	0.96
mwave	8.08	0.75	−0.38	0.49	−0.39	0.05	0.43	0.49
dish	12.96	0.66	0.08	0.05	1.16	0.04	0.11	0.07
dryer	**19.64**	0.99	0.06	0.12	−0.29	0.05	**0.20**	0.90

plots for the models with one to three components. The QQ plots should be close to the 45 degree line if the model is correctly specified. It is clear from the first row of Figure 6.2 that a model with one component has to be skewed in order to fit the data. As expected, most of the models provide a better fit as we add components, the main exception being the split-t, which deteriorates as we go from one to two components. This may be due to the MCMC algorithm getting stuck in a local mode, but several MCMC runs gave very similar results.

Table 6.6 presents estimation results from the best one-component model, the split-t model. We choose to present results for this model as it is easy to interpret and requires no additional identifying restriction. Table 6.6 shows that many of the covariates, including `log(room)` and `log(people)`, are important in the mean function. The variable `log(income)` gives a relatively low posterior inclusion probability in the mean function, but is an important covariate in the scale, ϕ. The covariate `sheonly` is the only important variable in the degrees of freedom function, but at least seven covariates are very important determinants of the skewness parameter.

Figure 6.3 depicts the conditional predictive densities $p(y|x)$ from three of the models: split-t(1) (the most feature-rich one-component model), Student-t(2) (the best mixture of symmetric densities model with a minimal number of components) and gamma(1) (the most efficient model with a minimum number of potential parameters). The predictive densities are displayed for three different conditioning values of the most important covariates: `log(rooms)`, `log(income)`, `sheonly` and `whtgel`. All other covariates except the one indicated below the horizontal axis are fixed at their sample means. It is clear from Figure 6.3 that the predictive densities are very skewed, but also that the different models tend to produce very different types of skewness. The predictive densities from the

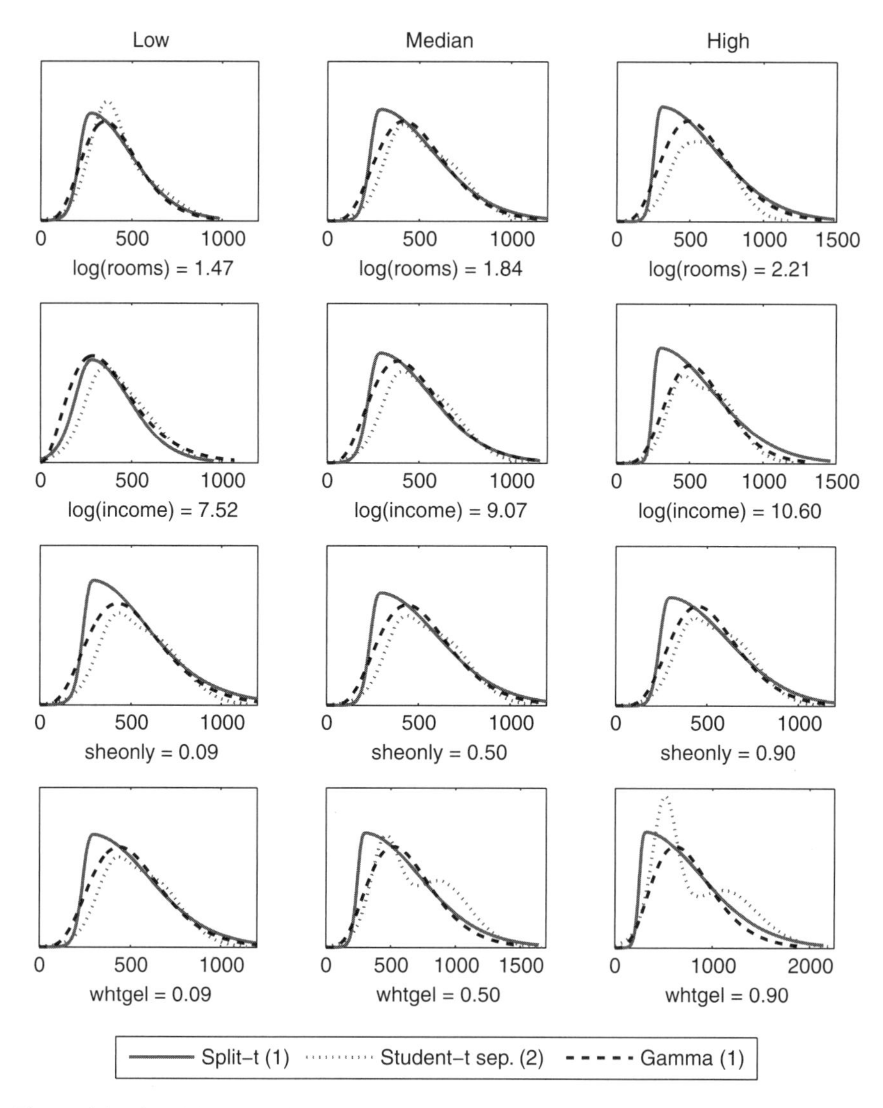

Figure 6.3 Conditional predictive densities for different values of the most important covariates. All other covariates are held fixed at their mean.

two-component Student-t model are unimodal except for median and high values of `whtgel`, where the two components are clearly visible.

6.5 Conclusions

We have presented a general model class for estimating the distribution of a continuous variable conditional on a set of covariates. The models are finite smooth mixtures of component densities where the mixture weights and all component parameters are functions of covariates. The inference methodology is a fully unified Bayesian approach based on a general and efficient MCMC algorithm. Easily specified priors are used and Bayesian variable selection is carried out to obtain model parsimony and guard against overfitting. We use the log predictive density score to determine the number of mixture components. Simulation and real examples show that using fairly complex components in the mixture is a wise strategy and that variable selection is an efficient approach to guarding against overfitting.

Acknowledgements

We would like to thank Denzil Fiebig for the use of the electricity expenditure data. Robert Kohn's research was partially supported by ARC Discovery grant DP0988579.

Appendix: Implementation details for the gamma and log-normal models

The general MCMC algorithm documented in Section 6.3 only requires the gradient and Hessian matrix of the conditional posteriors for each of the parameters in the component densities. The gradient and Hessian for the split-t model are documented in Li *et al.* (2010). We now present the gradient and Hessian for the gamma model and log-normal model for completeness.

1. Gradient and Hessian with respect to μ and ϕ for the gamma density:

$$\frac{\partial \log p(y|\mu,\phi)}{\partial \mu} = \frac{1}{\phi}\left[\mu + 2\mu \log\left(\frac{y\mu}{\phi}\right) - 2\mu\psi\left(\frac{\mu^2}{\phi}\right) - y\right],$$

$$\frac{\partial \log p(y|\mu,\phi)}{\partial \phi} = \frac{\mu}{\phi^2}\left[y - \mu - \mu\log\left(\frac{y\mu}{\phi}\right) + \mu\psi\left(\frac{\mu^2}{\phi}\right)\right],$$

$$\frac{\partial^2 \log p(y|\mu,\phi)}{\partial \mu^2} = \frac{1}{\phi}\left[3 + 2\log\left(\frac{y\mu}{\phi}\right)\right] - \frac{2}{\phi}\psi\left(\frac{\mu^2}{\phi}\right) - \frac{\mu^2}{\phi^2}\psi_1\left(\frac{\mu^2}{\phi}\right),$$

$$\begin{aligned}\frac{\partial^2 \log p(y|\mu,\phi)}{\partial \phi^2} &= -\frac{\mu}{\phi^3}\left[2y - 3\mu - 2\mu\log\left(\frac{y\mu}{\phi}\right)\right] \\ &\quad -\frac{2\mu^2}{\phi^3}\psi\left(\frac{\mu^2}{\phi}\right) - \frac{\mu^4}{\phi^4}\psi_1\left(\frac{\mu^2}{\phi}\right),\end{aligned}$$

where $\psi(\cdot)$ and $\psi_1(\cdot)$ are the digamma function, and trigamma function, respectively.

2. Gradient and Hessian with respect to μ and ϕ for the log-normal density. It is convenient to define $h = \log(y/\mu)$ and $l = \log\left(1 + \phi^2/\mu^2\right)$:

$$\frac{\partial \log p(y|\mu,\phi)}{\partial \mu} = \frac{\phi^2\left(3l^2 - 4h^2 + 4hl + 4l\right) + 2\mu^2\left(l^2 + 2hl\right)}{4\mu\left(\mu^2+\phi^2\right)l^2},$$

$$\frac{\partial \log p(y|\mu,\phi)}{\partial \phi} = \frac{\phi\left(4h^2 - l^2 - 4l\right)}{4\left(\mu^2+\phi^2\right)l^2},$$

$$\begin{aligned}\frac{\partial^2 \log p(y|\mu,\phi)}{\partial \mu^2} &= -\frac{4\phi^4 h^2}{\left(\mu^2+\phi^2\right)^2\mu^2 l^3} + \frac{2\phi^4 + 4\left(\mu^2+\phi^2\right)\phi^2 h + \left(3\mu^2+\phi^2\right)\phi^2 h^2}{\left(\mu^2+\phi^2\right)^2\mu^2 l^2}\\ &\quad - \frac{2\phi^4 + \left(\mu^2+5\phi^2\right)\mu^2 + \left(\mu^2+\phi^2\right)^2 h}{\left(\mu^2+\phi^2\right)^2\mu^2 l} - \frac{\left(2\mu^2+\phi^2\right)\left(\mu^2+3\phi^2\right)}{4\left(\mu^2+\phi^2\right)^2\mu^2},\end{aligned}$$

$$\frac{\partial^2 \log p(y|\mu,\phi)}{\partial \phi^2} = -\frac{4\phi^2 h^2}{\left(\mu^2+\phi^2\right)^2 l^3} + \frac{2\phi^2 + (\mu^2-\phi^2)\left(h^2 + l^2/4 - l\right)}{\left(\mu^2+\phi^2\right)^2 l^2}.$$

References

Bartels, R., Fiebig, D. and Plumb, M. (1996) Gas or electricity, which is cheaper? An econometric approach with application to Australian expenditure data. *The Energy Journal*, **17**, 33–58.

Celeux, G., Hurn, M. and Robert, C. (2000) Computational and inferential difficulties with mixture posterior distributions. *Journal of the American Statistical Association*, **95**, 957–970.

Denison, D., Holmes, C., Mallick, B. and Smith, A. (2002) *Bayesian Methods for Nonlinear Classification and Regression*. John Wiley & Sons, Ltd.

Diebolt, J. and Robert, C. (1994) Estimation of finite mixture distributions through Bayesian sampling. *Journal of the Royal Statistical Society, Series B*, **56**, 363–375.

Escobar, M. and West, M. (1995) Bayesian density estimation and inference using mixtures. *Journal of the American Statistical Association*, **90**, 577–588.

Frühwirth-Schnatter, S. (2006) *Finite Mixture and Markov Switching Models*. Springer.

Gamerman, D. (1997) Sampling from the posterior distribution in generalized linear mixed models. *Statistics and Computing*, **7**, 57–68.

Geweke, J. (2007) Interpretation and inference in mixture models: simple MCMC works. *Computational Statistics and Data Analysis*, **51**, 3529–3550.

Geweke, J. and Keane, M. (2007) Smoothly mixing regressions. *Journal of Econometrics*, **138**, 252–290.

Gibbons, J. F. and Mylroie, S. (1973) Estimation of impurity profiles in ion-implanted amorphous targets using joined half-Gaussian distributions. *Applied Physics Letters*, **22**, 568–569.

Green, P. J. and Silverman, B. W. (1994) *Nonparametric Regression and Generalized Linear Models*. Chapman & Hall.

Holst, U., Hössjer, O., Björklund, C., Ragnarson, P. and Edner, H. (1996) Locally weighted least squares kernel regression and statistical evaluation of lidar measurements. *Environmetrics*, **7**, 401–416.

Jacobs, R., Jordan, M., Nowlan, S. and Hinton, G. (1991) Adaptive mixtures of local experts. *Neural Computation*, **3**, 79–87.

Jasra, A., Holmes, C. and Stephens, D. (2005) Markov chain Monte Carlo methods and the label switching problem in Bayesian mixture modeling. *Statistical Science*, **20**, 50–67.

Jiang, W. and Tanner, M. (1999a) Hierarchical mixtures-of-experts for exponential family regression models: approximation and maximum likelihood estimation. *Annals of Statistics*, **27**, 987–1011.

Jiang, W. and Tanner, M. (1999b) On the approximation rate of hierarchical mixtures-of-experts for generalized linear models. *Neural Computation*, **11**, 1183–1198.

John, S. (1982) The three-parameter two-piece normal family of distributions and its fitting. *Communications in Statistics – Theory and Methods*, **11**, 879–885.

Jordan, M. and Jacobs, R. (1994) Hierarchical mixtures of experts and the EM algorithm. *Neural Computation*, **6**, 181–214.

Kass, R. (1993) Bayes factors in practice. *The Statistician*, **42**, 551–560.

Kohn, R., Smith, M. and Chan, D. (2001) Nonparametric regression using linear combinations of basis functions. *Statistics and Computing*, **11**, 313–322.

Leslie, D., Kohn, R. and Nott, D. (2007) A general approach to heteroscedastic linear regression. *Statistics and Computing*, **17**, 131–146.

Li, F., Villani, M. and Kohn, R. (2010) Flexible modeling of conditional distributions using smooth mixtures of asymmetric Student *t* densities. *Journal of Statistical Planning and Inference*, **140**, 3638–3654.

Norets, A. (2010) Approximation of conditional densities by smooth mixtures of regressions. *Annals of Statistics*, **38**, 1733–1766.

Nott, D. and Leonte, D. (2004) Sampling schemes for Bayesian variable selection in generalized linear models. *Journal of Computational and Graphical Statistics*, **13**, 362–382.

Ntzoufras, I., Dellaportas, P. and Forster, J. (2003) Bayesian variable and link determination for generalised linear models. *Journal of Statistical Planning and Inference*, **111**, 165–180.

Peng, F., Jacobs, R. A. and Tanner, M. A. (1996) Bayesian inference in mixtures-of-experts and hierarchical mixtures-of-experts models with an application to speech recognition. *Journal of the American Statistical Association*, **91**, 953–960.

Richardson, S. and Green, P. (1997) On Bayesian analysis of mixtures with an unknown number of components (with discussion). *Journal of the Royal Statistical Society, Series B*, **59**, 731–792.

Ruppert, D., Wand, M. and Carroll, R. (2003) *Semiparametric Regression*. Cambridge University Press.

Stephens, M. (2000) Bayesian analysis of mixture models with an unknown number of components – an alternative to reversible jump methods. *Annals of Statistics*, **28**, 40–74.

Villani, M., Kohn, R. and Giordani, P. (2009) Regression density estimation using smooth adaptive Gaussian mixtures. *Journal of Econometrics*, **153**, 155–173.

Villani, M., Kohn, R. and Nott, D. (2010) Generalized smooth finite mixtures. Manuscript.

Wood, S., Jiang, W. and Tanner, M. (2002) Bayesian mixture of splines for spatially adaptive nonparametric regression. *Biometrika*, **89**, 513–528.

Zeevi, A. and Meir, R. (1997) Density estimation through convex combinations of densities: approximation and estimation bounds. *Neural Networks*, **10**, 99–109.

Zellner, A. (1986) On assessing prior distributions and Bayesian regression analysis with g-prior distributions. *Bayesian Inference and Decision Techniques: Essays in Honor of Bruno de Finetti*, **6**, 233–243.

7

Nonparametric mixed membership modelling using the IBP compound Dirichlet process

Sinead Williamson, Chong Wang, Katherine A. Heller and David M. Blei

7.1 Introduction

Often the assumptions of mixture modelling, namely that each data point belongs to one of a finite or countable number of distributions, are overly restrictive. In many real-life datasets, individual data points exhibit features associated with multiple clusters: a movie may contain elements of both romance and comedy or an individual member of a population may exhibit traits from multiple subpopulations.

Mixed membership models are a hierarchical variant of mixture models used for modelling grouped data, where each individual data point consists of a collection of observations. Rather than being assigned to a single component, each data point is associated with a distribution over components, allowing us to capture more complicated relationships between data points than is possible with a simple mixture model. One example of a dataset where a mixed membership assumption is appropriate is a corpus of text documents: each document is a data point and consists of a collection of words. In such an application, each component of the

Mixtures: Estimation and Applications, First Edition. Edited by Kerrie L. Mengersen, Christian P. Robert and D. Michael Titterington.

mixture model is a distribution over words, each document is associated with a distribution over these components and each word is associated with a single component. This framework is often referred to as 'topic modelling', since we typically find that the posterior components (called 'topics') reflect the semantic themes of the documents.

The hierarchical Dirichlet process (HDP Teh *et al.*, 2006, Section 7.2.2) has proved a useful prior for mixed membership models where the number of components is not known a priori. However, as we will see in Section 7.3, the HDP makes certain assumptions that are often unrealistic. Specifically, the HDP assumes that if a component contributes to few data points in our dataset it will make up a low proportion of the observations for those data points and if a component is present in most data points it will contribute a high proportion of the observations for each such data point.

To see why this might not be desirable, consider modelling a text corpus consisting of news articles. We know that most articles in this corpus will not be about baseball. However, those articles that *are* about baseball tend to be dominated by that topic. This is in direct disagreement with the assumptions made by the HDP, which state that if a topic occurs rarely in a corpus, it will make up a low proportion of those documents where it does appear.

The IBP compound Dirichlet process (ICD) (see Williamson *et al.*, 2010b) is a recently developed nonparametric prior for mixed membership models that avoids unwanted correlation between across-dataset prevalence and within-data-point proportion of components. The ICD uses a random binary matrix drawn from the Indian buffet process (IBP) (see Griffiths and Ghahramani, 2006) to select *which* components are used in each data point (across-data prevalence) and an infinite series of gamma random variables to model *how much* they are used (within-data proportions), thus decoupling across-data prevalence and within-data proportion.

We will begin by reviewing mixed membership models in Section 7.2 before describing in more detail the source of the correlation between across-data prevalence and within-data proportion in Section 7.3. The ICD is presented in Section 7.4.2 and a specific topic-modelling application of the ICD is described in Section 7.4.3. After presenting some experimental results in Section 7.6 we conclude by discussing how the framework described herein could be extended to new models and applications.

7.2 Mixed membership models

Mixed membership models (Erosheva *et al.*, 2004) provide a framework for modelling datasets where each data point is a collection of observations. Examples include documents, which are a collection of words (Blei *et al.*, 2003), individuals in a social network, who are represented by the collection of people they interact with (Airoldi *et al.*, 2008), and images, which are a collection of subimages (Fei-Fei and Perona, 2005).

Under the mixed membership framework, each data point is associated with a (typically unobserved) distribution over components, describing the degree of membership of that data point in each of the components. Each observation associated with the data point, for example each word in a document, is associated with a single component. These component assignments are conditionally independent and identically distributed (i.i.d.) given the data-point-specific distribution over components.

7.2.1 Latent Dirichlet allocation

Latent Dirichlet allocation (LDA) (see Blei *et al.*, 2003), one of the earliest probabilistic mixed membership models, was introduced as a model for text corpora. LDA describes a generative model for documents where the word ordering is ignored (the so-called 'bag-of-words' representation), as follows.

1. For each component $k = 1, \ldots, K$, sample a distribution over a vocabulary of words, $\boldsymbol{\beta}_k \sim Dirichlet(\boldsymbol{\eta})$.
2. For each document $m = 1, \ldots, M$,
   (a) sample a distribution over components, $\boldsymbol{\theta}_m \sim Dirichlet(\boldsymbol{\alpha})$,
   (b) for each word in the document, $n = 1, \ldots, N_m$,
      i. sample a component $z_n \sim Discrete(\boldsymbol{\theta})$,
      ii. sample a word from the corresponding distribution over words, $w_n \sim Discrete(\boldsymbol{\beta}_{z_n})$.

We find that the posterior distributions over words for each component assign high probability to semantically coherent groups of words. For this reason, mixed membership models for text corpora are often referred to as 'topic models' and the components $\boldsymbol{\beta}_k$ as 'topics'.

7.2.2 Nonparametric mixed membership models

Finite mixed membership models such as LDA require an a priori specification of the number of components used to model the data. Establishing an appropriate number is often difficult, requiring costly model comparisons. Moreover, the number of components required may increase with the number of data points observed, causing problems if we wish to evaluate previously unseen data.

A similar problem is faced in finite mixture models. The Dirichlet process mixture model (DPMM) (see Antoniak, 1974; Lo, 1984) uses a nonparametric prior, the Dirichlet process, to avoid the need to specify the number of components a priori.

The Dirichlet process (DP) (see Ferguson, 1973; Ghosal, 2010) is a distribution over discrete probability measures with a countably infinite number of atoms. It is parameterised by a base measure, H, which governs the atom locations and can be thought of as a mean, and a concentration parameter, ζ, which controls how close the random distribution is to the base distribution. Samples $G \sim DP(\zeta, H)$ can be represented as

$$G = \sum_{k=1}^{\infty} p_k \delta_{\beta_k},$$

where $\sum_k p_k = 1$ and $\beta_k \sim H$, i.i.d. for each k. In the DPMM, the atom locations β_k parameterise component distributions in a nonparametric mixture model, allowing an unbounded number of components to be represented in the posterior distribution.

The hierarchical Dirichlet process (HDP) (see Teh *et al.*, 2006) employs a hierarchy of DPs to construct a prior appropriate for Bayesian nonparametric mixed membership modelling. In an HDP mixed membership model, each data point is modelled using a DPMM, with the mixing components, $\boldsymbol{\beta}$, shared across the dataset. The sharing of component locations is ensured by the use of a discrete base measure, which is itself drawn from a DP. More precisely, the generative process for the per-data-point distributions G_m is given by

$$\begin{aligned} G_0 &\sim DP(\zeta, H), \\ G_m &\sim DP(\tau, G_0), \qquad \text{for each } m. \end{aligned}$$

Each atom in the probability measures G_0 and G_m represents a component and is described by a location, a weight in each G_m and a weight in G_0. The location of an atom, which is identical in both G_0 and G_m due to the discreteness of G_0, gives the parameters associated with the component, e.g. a Gaussian mean or a distribution over terms. The weight in G_m gives the proportion for that component in the mth data point, and is drawn from a distribution centred around the corresponding weight in G_0. The distributions G_m can be used as the distributions over components in a nonparametric version of LDA.

7.3 Motivation

The HDP has proved a useful prior for mixed membership models where the number of components is not known a priori, with applications including haplotype modelling (Xing *et al.*, 2006), category learning (Griffiths *et al.*, 2007), topic modelling (Teh *et al.*, 2006) and natural language processing (Liang *et al.*, 2007). However, as we will show in this section, the HDP makes certain assumptions that do not always reflect our beliefs about data.

The observations constituting the mth data point in an HDP mixed membership model are sampled from the multinomial distribution with component probabilities G_m. A component with low weight in G_m is likely, therefore, to contribute few or indeed no observations to the mth data point. Conversely, a component with high weight in G_m is very likely to appear in the mth data point, and moreover is likely to be responsible for a high proportion of the observations therein. Since the weight of an atom in G_0 determines the distribution over the weight of that atom in all of the G_m, this leads to a global coupling between the probability of a component contributing to a document and the proportion of the document attributed to that component. If a component has low weight in G_0 then it will also have low weight in most G_m, meaning that the component will appear in few data points and, when it does appear, its contribution will be small.

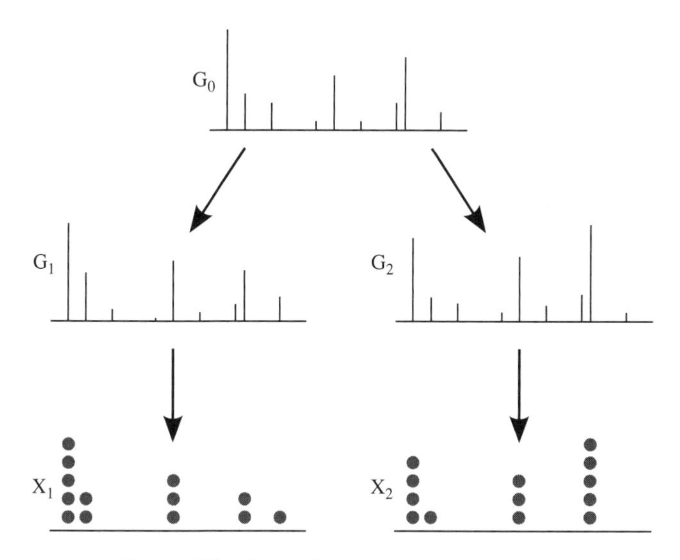

Figure 7.1 Generating samples from the HDP.

This inherent correlation between the prevalence of a component throughout the dataset and the proportion of data points attributed to that component is illustrated in Figure 7.1. The top level of the diagram shows the shared base measure G_0; the second level shows the distributions $G_1, G_2 \sim DP(G_0, \tau)$ corresponding to two data points; and the bottom level shows the collections of observations belonging to these two data points, $X_1 \sim Multinomial(G_1, N)$ and $X_2 \sim Multinomial(G_2, N)$. In the HDP, the common base measure G_0 determines both whether a component contributes to a data point and the proportion of the data point it contributes. Thus, components that are frequently present in the dataset will tend to appear with high proportion in those data points where they occur, and vice versa. It is clear that this correlation is a result of the fact that a single distribution, G_0, controls both the *prevalence* of a given component across the dataset and the *proportion* of that component in the data points where it occurs.

The assumptions made by the HDP are flawed. It is, in general, unreasonable to assume that a component that appears in few data points will necessarily contribute only a small proportion of the observations associated with those data points. For example, a topic that appears in few documents within a corpus may dominate those documents where it does occur. Conversely, a component that contributes to every data point will not necessarily contribute a large number of observations in each case; if we were to model an email network we may find that everyone in a company receives emails from the payroll department, but that these emails make up a very small fraction of their email activity.

7.4 Decorrelating prevalence and proportion

In this section we describe the *IBP compound Dirichlet process* (ICD), a Bayesian nonparametric prior that avoids unwanted correlation between component prevalence and proportion. Rather than control these two properties via a single variable, as is the case in the HDP, the ICD uses two separate variables to control the distribution over components for each data point – one to determine which components occur and another to determine their relative proportions.

We begin this section by reviewing the Indian buffet process (IBP) (see Griffiths and Ghahramani, 2006), which provides a means of selecting a subset of components for each data point. We describe the ICD in Section 7.4.2, before presenting an application of the ICD to document analysis in Section 7.4.3 and briefly discussing inference in Section 7.4.4.

7.4.1 Indian buffet process

The ICD uses the Indian buffet process (IBP) (see Griffiths and Ghahramani, 2006) to control component occurrence separately from component proportion. The IBP defines a distribution over binary matrices with an infinite number of columns, only a finite number of which contain nonzero entries. It can be derived by taking the limit as $K \rightarrow \infty$ of a finite $M \times K$ binary matrix $\mathbf{B}$, with elements b_{mk} distributed according to

$$\begin{aligned}\pi_k &\sim \mathit{Beta}(\alpha/K, 1),\\ b_{mk} &\sim \mathit{Bernoulli}(\pi_k).\end{aligned}$$

The mth row of $\mathbf{B}$ is $\mathbf{b}_m$, the kth cell of $\mathbf{b}_m$ is b_{mk} and π_k is the probability of observing a nonzero value in column k. As K tends to infinity, we can obtain a strictly decreasing ordering of the latent probabilities π_k by starting with a 'stick' of unit length and recursively breaking it at a point $\mathit{Beta}(\alpha, 1)$ along its length, discarding the excess (Teh *et al.*, 2007). For $k = 1, 2, \ldots$ and $m = 1, \ldots, M$,

$$\begin{aligned}\mu_k &\sim \mathit{Beta}(\alpha, 1),\\ \pi_k &= \textstyle\prod_{j=1}^{k} \mu_j,\\ b_{mk} &\sim \mathit{Bernoulli}(\pi_k).\end{aligned} \tag{7.1}$$

Figure 7.2 shows a series of 'stick lengths' $\boldsymbol{\pi}$ and a binary matrix $\mathbf{B}$ generated according to the above procedure.

7.4.2 The IBP compound Dirichlet process

We now incorporate the IBP in a hierarchical Bayesian nonparametric prior. Rather than assigning positive probability mass to all components for every data point, as in the HDP, our model assigns positive probability to only a subset of components, selected independently of their masses.

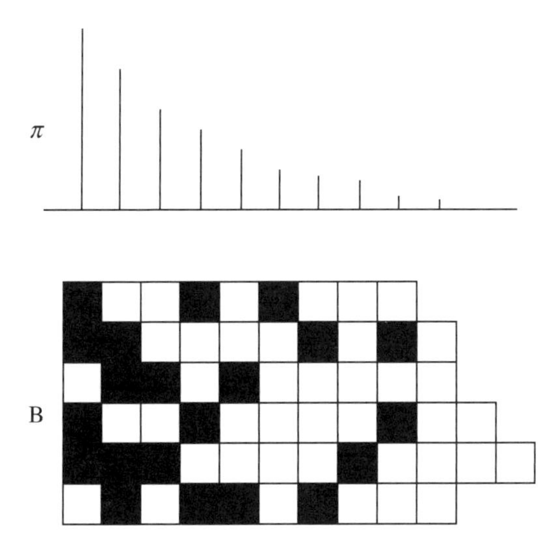

Figure 7.2 Component probabilities $\boldsymbol{\pi}$ and binary matrix **B** sampled from the IBP.

The IBP provides a method for selecting subsets from a countably infinite set of components. In our new model, we associate each data point with a row of a matrix sampled from the IBP, and associate each column of the matrix with a mixture component. Thus, the total number of components assigned probability mass in the dataset is unbounded, and each data point assigns probability mass to a finite subset of these components determined by the nonzero elements of the IBP distributed matrix.

We associate each column $k = 1, 2, \ldots$ of the IBP distributed matrix with a gamma random variable ϕ_k and with a mixture component β_k. We can now associate each data point with a discrete measure over the space of mixture components. This distribution has a finite number of atoms – their locations are given by the β_k and their masses are given by the ϕ_k. The locations and masses are selected from a countably infinite set of locations and masses that is shared between all data points.

Finally, we use this measure as the base measure of a Dirichlet process, generating samples θ_m that are finite discrete probability measures over the space of mixture components. This allows variation in the relative proportions of the components, so that two data points exhibiting the same subset of components do not have the same distribution over these components. The subset of atoms assigned positive mass in θ_m is determined by the IBP distributed matrix, and the masses assigned to these atoms are controlled by the ϕ_k.

The model assumes the following generative process.

1. For $k = 1, 2, \ldots,$
   (a) sample the stick length π_k according to Equation (7.1),
   (b) sample the relative mass parameter $\phi_k \sim Gamma(\gamma, 1)$,
   (c) sample the atom location $\beta_k \sim H$.

2. For $m = 1, \ldots, M$,
   (a) sample a binary vector $\mathbf{b}_m$ according to Equation (7.1),
   (b) sample the lower level DP, $G_m \sim DP(\sum_k b_{mk}\phi_k, \sum_k b_{mk}\phi_k\delta_{\beta_k}/\sum_k b_{mk}\phi_k)$.

In sampling the lower-level DP, masses are assigned to the atoms δ_{β_k} independent of their locations. Since the number of locations selected by the binary vector $\mathbf{b}_m$ is finite almost surely, these masses can be sampled from a Dirichlet distribution defined over the selected ϕ_k:

$$\begin{aligned} \boldsymbol{\theta}_m &\sim Dirichlet(\mathbf{b} \cdot \boldsymbol{\phi}), \\ G_m &= \textstyle\sum_k \theta_{mk}\delta_{\beta_k}, \end{aligned}$$

where $\mathbf{b} \cdot \boldsymbol{\phi}$ is the vector composed of those ϕ_k for which the corresponding $b_k = 1$. If we marginalise out the sparse binary matrix $\mathbf{B}$ and the gamma random variables ϕ_k, the atom masses are distributed according to a mixture of Dirichlet distributions governed by the IBP:

$$p(\{\boldsymbol{\theta}_m\}_{m=1}^M|\gamma, \alpha) = \int \nu_\alpha(\mathrm{d}\mathbf{B}) \int \mathrm{d}\boldsymbol{\phi} \prod_{m=1}^M p(\boldsymbol{\theta}_m|\mathbf{B}, \boldsymbol{\phi})p(\boldsymbol{\phi}|\gamma), \tag{7.2}$$

where $\nu_\alpha(\cdot)$ is the IBP measure with parameter α, $\phi_k \sim Gamma(\gamma, 1)$ for each $k = 1, 2, \ldots$ and $\boldsymbol{\theta}_m \sim Dirichlet(\mathbf{b}_m \cdot \boldsymbol{\phi})$.

We call this model the *IBP compound Dirichlet process* (ICD), since the IBP provides the mixing measure for a mixture of Dirichlet distributions. Figure 7.3 shows the generative process for the ICD. The top level shows the binary matrix $\mathbf{B}$ and the collection of gamma random variables $\boldsymbol{\phi}$, which are combined to

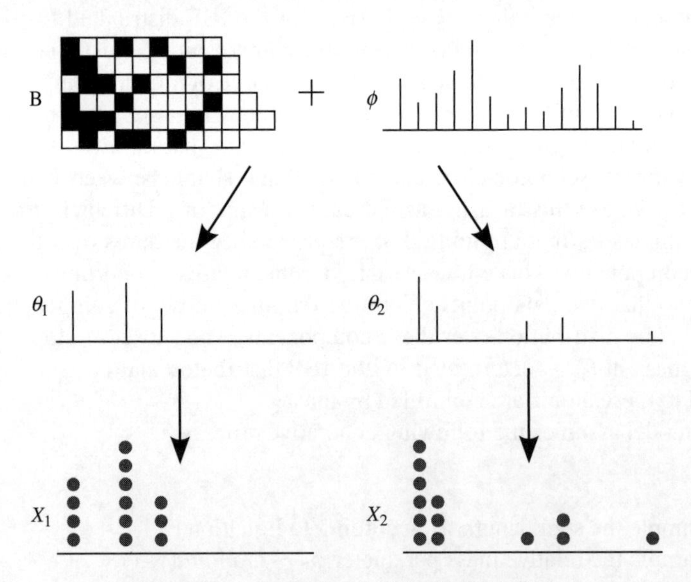

Figure 7.3 Generating samples from the ICD.

form the base measures for the individual data points. The second level shows the distributions $\boldsymbol{\theta}_1 \sim Dirichlet(\mathbf{b}_1 \cdot \boldsymbol{\phi})$ and $\boldsymbol{\theta}_2 \sim Dirichlet(\mathbf{b}_2 \cdot \boldsymbol{\phi})$ corresponding to two data points. The bottom level shows the collections of observations belonging to these two data points, $X_1 \sim Multinomial(\boldsymbol{\theta}_1, N)$ and $X_2 \sim Multinomial(\boldsymbol{\theta}_2, N)$. The ICD decides which components are present in a data point separately from their proportions. Thus, components that appear in few data points can contribute heavily to those data points where they do appear, and vice versa.

We note that the ICD can be interpreted as an infinite spike and slab model (Ishwaran and Rao, 2005). Spike and slab models describe a mixture model between a continuous distribution (the 'slab') and the measure degenerate at zero. (In its original form, the slab was a uniform distribution. However, the concept and terminology have also been employed in models where the slab is not the uniform distribution (see, for example, Ishwaran and Rao, 2005; it is in this more general sense that we use the term.) A 'spike' distribution determines which variables are drawn from the slab and which are zero. In the model above, the spikes are provided by the IBP and the slab is provided by the gamma random variables.

7.4.3 An application of the ICD: focused topic models

Topic modelling is a common application of mixed membership models such as the HDP. If H parameterises distributions over words, the ICD defines a generative topic model, where each data point, or document, is generated from a distribution over a subset of an infinite number of shared components, or topics. As with LDA (Section 7.2.1), these topics are drawn from a Dirichlet distribution over a vocabulary of words.

We note that, since rows of a matrix sampled from the IBP can contain no nonzero element, the ICD assigns finite probability to the measure degenerate at zero. We avoid complications that might arise from this by sampling the number of words for each document from a negative binomial distribution, $n_{\cdot}^{(m)} \sim NB(\sum_k b_{mk}\phi_k, 1/2)$. (The notation $n_k^{(m)}$ denotes the number of words assigned to the kth topic of the mth document, and we use a dot notation to represent summation; that is $n_{\cdot}^{(m)} = \sum_k n_k^{(m)}$.) We note that sampling the total number of words from this distribution and then assigning words according to the multinomial distribution is equivalent to sampling the number of words for each topic as $n_k^{(m)} \sim NB(b_{mk}\phi_k, 1/2)$.

The generative model for M documents is as follows.

1. For $k = 1, 2, \ldots,$
   (a) sample the stick length π_k according to Equation (7.1),
   (b) sample the relative mass parameter $\phi_k \sim Gamma(\gamma, 1)$,
   (c) draw the topic distribution over words, $\boldsymbol{\beta}_k \sim Dirichlet(\eta)$.
2. For $m = 1, \ldots, M$,
   (a) sample a binary vector $\mathbf{b}_m$ according to Equation (7.1),
   (b) draw the total number of words, $n_{\cdot}^{(m)} \sim NB(\sum_k b_{mk}\phi_k, 1/2)$,
   (c) sample the distribution over topics, $\boldsymbol{\theta}_m \sim Dirichlet(\mathbf{b}_m \cdot \boldsymbol{\phi})$,

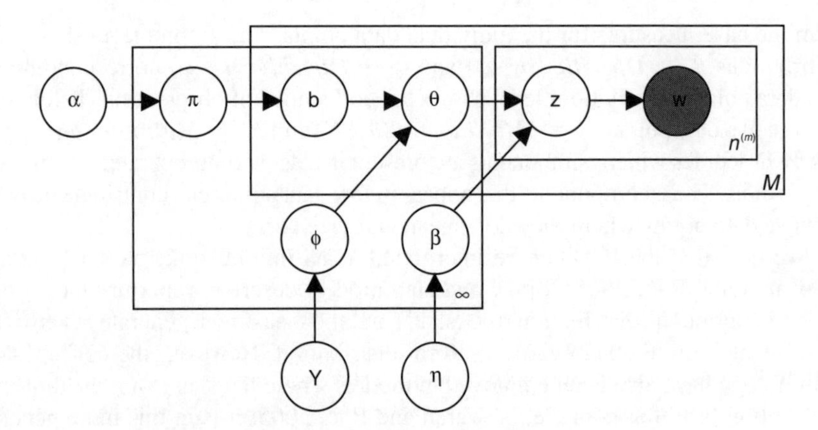

Figure 7.4 Graphical model for the focused topic model.

(d) for each word w_{mi}, $i = 1, \ldots, n_{\cdot}^{(m)}$,
   i. draw the topic index $z_{mi} \sim Discrete(\boldsymbol{\theta}_m)$,
   ii. draw the word $w_{mi} \sim Discrete(\boldsymbol{\beta}_{z_{m_i}})$.

Figure 7.4 shows the graphical model corresponding to this generative process. We call this the *focused topic model* (FTM); the infinite binary matrix **B** serves to focus the distribution over topics on to a finite subset. The number of topics within a single document is almost surely finite, although the total number of topics is unbounded. The topic distribution for the mth document, $\boldsymbol{\theta}_m$, is drawn from a Dirichlet distribution over the topics selected by $\mathbf{b}_m$. This Dirichlet distribution models uncertainty about topic proportions while maintaining the restriction to a sparse set of topics.

The ICD models the distribution over the global topic proportion parameters $\boldsymbol{\phi}$ separately from the distribution over the binary matrix **B**. This captures the idea that a topic may appear infrequently in a corpus, but makes up a high proportion of those documents in which it occurs. Conversely, a topic may appear frequently in a corpus, but only with low proportion.

7.4.4 Inference

We carry out inference in our model using a collapsed Gibbs sampler. The algorithm cyclically samples the value for a single variable from its conditional distribution given the remaining variables. To improve mixing time, we use a collapsed Gibbs sampler, integrating out the topic-specific word distributions $\boldsymbol{\beta}$, the topic mixture distributions $\boldsymbol{\theta}$ and the sparsity pattern **B**.

Integrating out the infinite-dimensional sparsity pattern **B** exactly is intractable, due to the infinite combinatorial sum involved. An approximation to this integral, along with further details of the Gibbs sampling procedure, are described in Williamson *et al.* (2010b).

7.5 Related models

We have discussed the relationship between the ICD and the HDP in Sections 7.3 and 7.4.2. In this section, we will explore the relationship between the ICD and a number of other models found in the literature.

The infinite gamma–Poisson process (Titsias, 2008) defines a distribution over infinite matrices of nonnegative integers. These matrices can be used as the basis for a topic model, with the distribution over components for the mth data point given by the Dirichlet distribution parameterised by the nonnegative elements of the mth row of the matrix. This is superficially similar to the ICD, although the ICD allows general nonnegative values in the underlying matrix. However, closer examination reveals that the infinite gamma–Poisson process does not decouple across-data prevalence and within-data proportions of components. The elements of the kth column are i.i.d. samples from a Poisson distribution with common parameter; if this parameter is small most elements will be zero and the remainder will be small, and if it is large most elements will be large and nonzero. In the ICD the number of zero entries is controlled by a separate process, the IBP, from the values of the nonzero entries, which are controlled by the gamma random variables.

The sparse topic model (SparseTM) (see Wang and Blei, 2009) uses a finite spike and slab model to ensure that each topic is represented by a sparse distribution over words. The spikes are generated by Bernoulli random variables with a single topic-wide parameter. The topic distribution is then drawn from a symmetric Dirichlet distribution defined over these spikes. The ICD also uses a spike-and-slab approach, but allows an unbounded number of 'spikes' (due to the IBP) and a more globally informative 'slab' (due to the shared gamma random variables). While the FTM enforces sparsity in the document-specific distributions over topics, the SparseTM enforces sparsity in the topic-specific distribution over words. It would, of course, be possible to use the ICD to generate unbounded sparse distributions over words in a manner similar to the SparseTM.

The beta process autoregressive hidden Markov model of Fox *et al.* (2009), which is used to model multiple dynamic systems with shared states, can be seen as a special case of the ICD. In this model, the 'slab' provided by the gamma random variables ϕ_k is replaced by a constant value γ plus an additional mass at the location corresponding to self-transition. This model does not therefore allow sharing of information about the within-data probability of a state between data points, which is modelled in the ICD via the gamma-distributed ϕ_k. An alternative inference scheme to that described in Williamson *et al.* (2010b) is used, where the IBP matrix is sampled instead of being integrated out.

A number of other models have been proposed to address the rigid topic correlation structure assumed in the LDA and HDP topic models, including the correlated topic model (CTM) (see Blei and Lafferty, 2006) and the pachinko allocation model (PAM) (see Li and McCallum, 2006). Our aim is different. The FTM reduces undesirable correlations between the prevalence of a topic across the corpus and its proportion within any particular document, rather than adding new correlations between topics. In Section 7.7, we discuss the possibility of extending the FTM to incorporate such correlations.

7.6 Empirical studies

We compared the performance of the FTM to the HDP with the following three datasets.

- *PNAS*: This is a collection of 1766 abstracts from the Proceedings of the National Academy of Sciences (PNAS) from between 1991 and 2001. The vocabulary contains 2452 words.
- *20 Newsgroups*: This is a collection of 1000 randomly selected articles from the 20 newsgroups dataset, available at http://people.csail.mit.edu/jrennie/20Newsgroups/. The vocabulary contains 1407 words.
- *Reuters-21578*: This is a collection of 2000 randomly selected documents from the Reuters-21578 dataset, available at http://kdd.ics.uci.edu/databases/reuters21578/. The vocabulary contains 1472 words.

For each dataset, the vocabulary excluded stop-words (commonly occurring words) and words occurring in fewer than five documents.

In both the FTM and HDP topic models, we fixed the topic distribution hyperparameter η to 0.1. Following Teh *et al.* (2006), we used priors of $\zeta \sim Gamma(5, 0.1)$ and $\tau \sim Gamma(0.1, 0.1)$ in the HDP. In the FTM, we used prior $\gamma \sim Gamma(5, 0.1)$ and fixed $\alpha = 5$.

Test-set perplexity is a measure of how well the models generalise to new data, with lower values indicating better generalisation. The perplexity of a set $\mathbf{D}$ of M documents is given by

$$\text{perp}(\mathbf{D}) = \exp\left(- \frac{\sum_{m=1}^{M} \log p(\mathbf{w}_m)}{\sum_{m=1}^{M} N_m} \right).$$

The top row of Figure 7.5 shows test-set perplexities obtained on each dataset for the two models, using fivefold cross-validation. To obtain these measurements, we ran both Gibbs samplers for 1000 iterations, discarding the first 500. In each case, the FTM achieves better (lower) perplexity on the held-out data.

The aim of the FTM was to decorrelate the probability of a topic being active within a document and its proportion within the documents attributed to it. To consider whether this is observed in the posterior, we next compared two statistics for each topic found. The *topic presence frequency* for a given topic is the fraction of the documents within the corpus that contain at least one incidence of that topic. The *topic proportion* for a topic is the fraction of the words within the corpus attributed to that topic.

The correlation between these two statistics is shown in the bottom row of Figure 7.5. In each case we see lower correlation for the FTM. In fact, the dataset

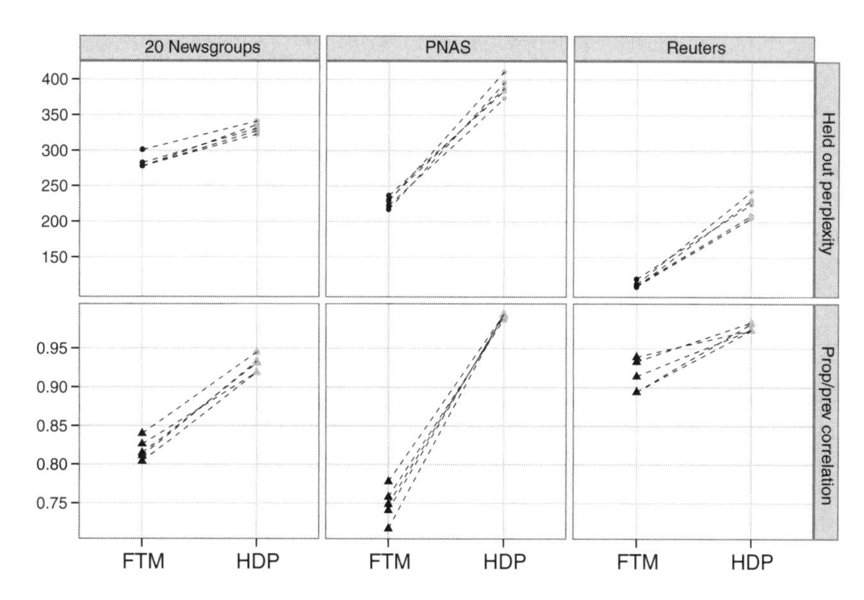

Figure 7.5 Experimental comparison between FTM and HDP on three datasets. Each point represents the result on one fold, and is computed with the other folds as training data. Dashed lines connect the results from the same fold. Test-set perplexities (top). Lower numbers indicate better performance. Correlation between topic presence frequency and topic proportion (bottom). The FTM reduces the correlation between them.

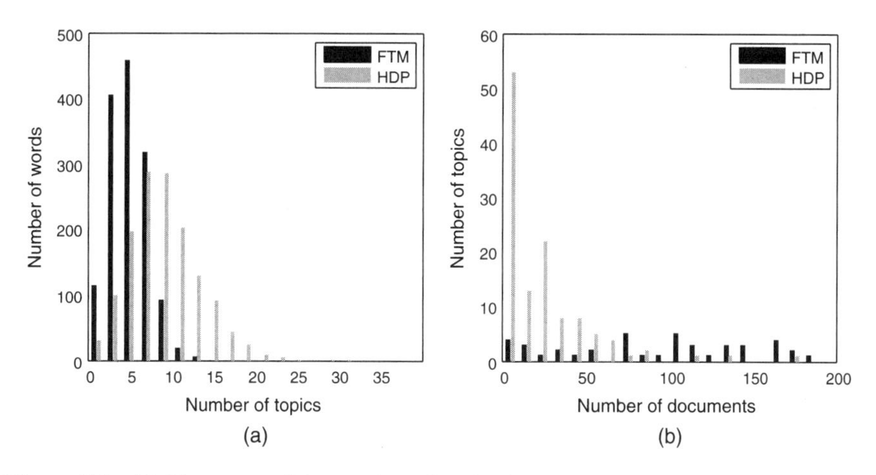

Figure 7.6 (a) Histogram of the number of topics in which a word appears for the FTM (dark, left) and the HDP (pale, right) models on *20 Newsgroups* data. In the FTM, words are generally associated with fewer topics. (b) Histogram of the number of documents in which a topic appears for the FTM (dark, left) and HDP (pale, right) models on *20 Newsgroups* data. In both models, topics appearing in more than 200 documents have been excluded to focus on the low-frequency topics. The HDP has many more topics, which only appear in a very few documents.

which exhibits the greatest improvement in held-out perplexity under the FTM, namely PNAS, also exhibits the greatest decrease in correlation between frequency and proportion.

Figure 7.6 shows histograms of the number of documents in which each topic appears and the number of topics in which each word appears, under the FTM and HDP representations of the 20 newsgroups dataset. The other datasets exhibited similar properties. These histograms show that most of the topics found by the HDP contribute to few documents, and words tend to be spread across multiple topics. The FTM posterior favours a more compact representation, in that it contains fewer topics overall, but the topics are reused more than in the HDP and words are associated with fewer topics.

Finally, we note that the computational complexity of both models grows with the number of topics represented. When the number of topics is equal, a single iteration of the FTM algorithm is more costly than a single iteration of the HDP. However, the more compact representation of the FTM yields a similar runtime. In the data we analysed, the FTM analysis was as fast as the HDP analysis.

7.7 Discussion

In this chapter, we have identified an often-undesirable assumption of the hierarchical Dirichlet process, namely the tendency to correlate the across-data prevalence and within-data proportion of components. We relax this assumption with the *IBP compound Dirichlet process* (ICD), a Bayesian nonparametric prior over discrete, dependent distributions. The ICD is a useful prior for mixed membership modelling where the assumptions of the HDP are inappropriate.

We have concentrated on correlation in HDPs. This correlation does not occur in finite mixed membership models with mixture components drawn according to *Dirichlet*($\boldsymbol{\alpha}$), where $\boldsymbol{\alpha}$ is fixed, since the draws from *Dirichlet*($\boldsymbol{\alpha}$) are i.i.d. However, restriction to a fixed Dirichlet distribution limits the flexibility of mixed membership models and, if $\boldsymbol{\alpha}$ is unknown, there will be a similar correlation bias as with the HDP. Although easily derived, there is currently no parametric counterpart of the ICD.

While the application described herein, the focused topic model, is a generative model for text corpora, we believe the ICD is applicable in a wide variety of situations where the HDP, or finite mixed membership models, are currently used. For example, in a state transition sequence, a certain state may be inaccessible from most other states, but occur with high probability following a small subset of states. Such a relationship will be poorly captured using the HDP-based hidden Markov model (Teh *et al.*, 2006), which will tend to correlate the global occurrence of a state with the probability of moving to it from another state. Exploring which HDP applications benefit most from the use of the ICD is an avenue for further research.

The ICD can be seen as a mixture of Dirichlet distributions, with the IBP providing the mixing measure. More generally, the IBP offers a method for generating mixtures of distributions parameterised by finite vectors of variable length, suggesting a method for nonparametric extensions of a number of existing parametric mixed membership models. Alternative distributions could result in nonparametric mixed membership models with a richer correlation structure. In addition, extensions to the Indian buffet process, such as the correlated IBP (Doshi and Ghahramani, 2009) and the dependent IBP (Williamson *et al.*, 2010a), could be incorporated into the ICD to allow additional structure.

References

Airoldi, E. M., Blei, D. M., Fienberg, E. F. and Xing, E. P. (2008) Mixed membership stochastic blockmodels. *Journal of Machine Learning Research*, **9**, 1981–2014.

Antoniak, C. E. (1974) Mixtures of Dirichlet processes with applications to Bayesian nonparametric problems. *Annals of Statistics*, **2**, 1152–1174.

Blei, D. M. and Lafferty, J. (2006) Correlated topic models. In *Advances in Neural Information Processing Systems*, Vol. 18 (eds Y. Weiss, B. Schölkopf and J. Platt), pp. 147–154. MIT Press.

Blei, D. M., Jordan, M. I. and Ng, A. Y. (2003) Latent Dirichlet allocation. *Journal of Machine Learning Research*, **3**, 993–1022.

Doshi, F. and Ghahramani, Z. (2009) Correlated non-parametric latent feature models. In *Conference on Uncertainty in Artificial Intelligence*, Vol. 25, pp. 143–150. AUAI Press.

Erosheva, E., Fienberg, S. and Lafferty, J. (2004) Mixed-membership models of scientific publications. In *Proceedings of the National Academy of Sciences*, Vol. 101(Suppl. 1), pp. 5220–5227.

Fei-Fei, L. and Perona, P. (2005) A Bayesian hierarchical model for learning natural scene categories. In *IEEE Conference on Computer Vision and Pattern Recognition*, pp. 524–531.

Ferguson, T. S. (1973) A Bayesian analysis of some nonparametric problems. *Annals of Statistics*, **1**, 209–230.

Fox, E. B., Sudderth, E. B., Jordan, M. I. and Willsky, A. S. (2009) Sharing features among dynamical systems with beta processes. In *Advances in Neural Information Processing Systems*, Vol. 22 (eds Y. Bengio, D. Schuurmans, J. Lafferty, C. K. I. Williams and A. Culotta), pp. 549–557.

Ghosal, S. (2010) Dirichlet processes, related priors and posterior asymptotics. In *Bayesian Nonparametrics* (eds N. L. Hjort, C. Holmes, P. Müller and S. G. Walker), pp. 35–79. Cambridge University Press.

Griffiths, T. L., Canini, K. R., Sanborn, A. N. and Navarro, D. J. (2007) Unifying rational models of categorization via the hierarchical Dirichlet process. In *Annual Conference of the Cognitive Science Society*, Vol. 29 (eds D. S. McNamara and J. G. Trafton), pp. 323–328.

Griffiths, T. L. and Ghahramani, Z. (2006) Infinite latent feature models and the Indian buffet process. In *Advances in Neural Information Processing Systems*, Vol. 18 (eds Y. Weiss, B. Schölkopf and J. Platt), pp. 475–482. MIT Press.

Ishwaran, H. and Rao, J. S. (2005) Spike and slab variable selection: frequentist and Bayesian strategies. *Annals of Statistics*, **33**, 730–773.

Li, W. and McCallum, A. (2006) Pachinko allocation: DAG-structured mixture models of topic correlations. In *International Conference on Machine Learning*, Vol. 23 (eds W. W. Cohen and A. Moore), pp. 577–584.

Liang, P., Petrov, S., Jordan, M. I. and Klein, D. (2007) The infinite PCFG using hierarchical Dirichlet processes. In *Proceedings of the Joint Conference on Empirical Methods in*

Natural Language Processing and Computational Natural Language Learning, pp. 688–697.

Lo, A. Y. (1984) On a class of Bayesian nonparametric estimates: I. Density estimates. *Annals of Statistics*, **12**, 351–357.

Teh, Y. W., Jordan, M. I., Beal, M. J. and Blei, D. M. (2006) Hierarchical Dirichlet processes. *Journal of the American Statistical Association*, **101**, 1566–1581.

Teh, Y. W., Görür, D. and Ghahramani, Z. (2007) Stick-breaking construction for the Indian buffet process. In *Proceedings of Artificial Intelligence and Statistics 2007*. Available online.

Titsias, M. K. (2008) The infinite gamma-Poisson feature model. In *Advances in Neural Information Processing Systems*, Vol. 20, (eds J. C., Platt, D. Koller, Y. Singer, and S. Roweis), pp. 1513–1520.

Wang, C. and Blei, D. M. (2009) Decoupling sparsity and smoothness in the discrete hierarchical Dirichlet process. In *Advances in Neural Information Processing Systems*, Vol. 22 (eds Y. Bengio, D. Schuurmans, J. Lafferty, C. K. I. Williams and A. Culotta), pp. 1982–1989.

Williamson, S., Orbanz, P. and Ghahramani, Z. (2010a) Dependent Indian buffet processes. In *Proceedings of Artificial Intelligence and Statistics 2010, Volume 9 of JMLR Workshop and Conference Proceedings* (eds Y. W. Teh and D. M. Titterington), pp. 924–931.

Williamson, S., Wang, C., Heller, K. A. and Blei, D. M. (2010b) The IBP compound Dirichlet process and its application to focused topic modelling. In *International Conference on Machine Learning*, Vol. 27 (eds J. Fürnkranz and T. Joachims), pp. 1151–1158. ACM.

Xing, E. P., Sohn, K., Jordan, M. I. and Teh, Y. W. (2006) Bayesian multi-population haplotype inference via a hierarchical Dirichlet process mixture. In *International Conference on Machine Learning*, Vol. 23 (eds W. W. Cohen and A. Moore), pp. 1049–1056.

8

Discovering nonbinary hierarchical structures with Bayesian rose trees

Charles Blundell, Yee Whye Teh and Katherine A. Heller

8.1 Introduction

Rich hierarchical structures are common across many disciplines, making the discovery of hierarchies a fundamental exploratory data analysis and unsupervised learning problem. Applications with natural hierarchical structure include topic hierarchies in text (Blei *et al.*, 2010), phylogenies in evolutionary biology (Felsenstein, 2003), hierarchical community structures in social networks (Girvan and Newman, 2002) and psychological taxonomies (Rosch *et al.*, 1976).

A large variety of models and algorithms for discovering hierarchical structures have been proposed. These range from the traditional linkage algorithms based on distance metrics between data items (Duda and Hart, 1973), to maximum parsimony and maximum-likelihood methods in phylogenetics (Felsenstein, 2003), to fully Bayesian approaches that compute posterior distributions over hierarchical structures (e.g. Neal, 2003). We will review some of these in Section 8.2.

A common feature of many of these methods is that their hypothesis spaces are restricted to binary trees, where each internal node in the hierarchical structure has exactly two children. This restriction is reasonable under certain circumstances and is a natural output of the popular agglomerative approaches to discovering hierarchies, where each step involves the merger of two clusters of data items into

Mixtures: Estimation and Applications, First Edition. Edited by Kerrie L. Mengersen, Christian P. Robert and D. Michael Titterington.

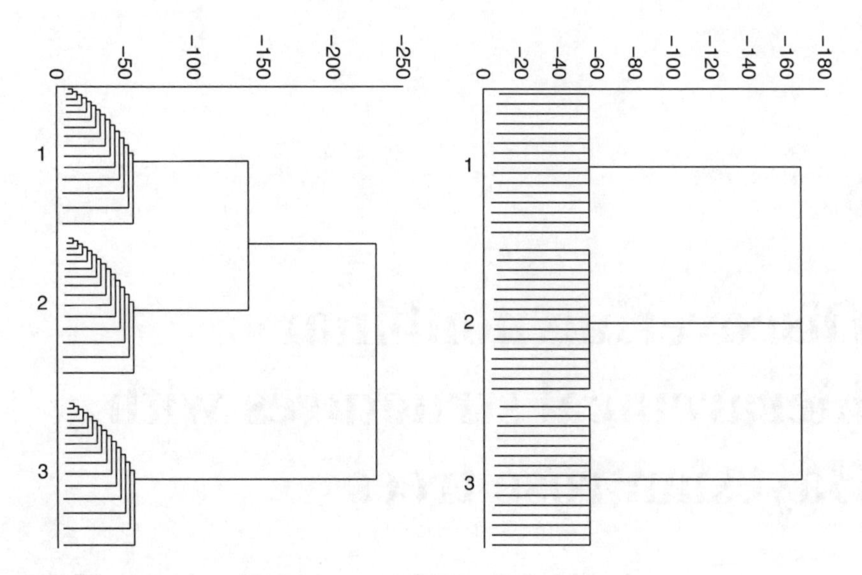

Figure 8.1 Bayesian hierarchical clustering (left) and Bayesian rose trees (right) on the same synthetic dataset. The three groups of 15 similar data items all cluster into groups under both models.

one. However, we believe that there are good reasons why restricting to binary trees is often undesirable. Firstly, we simply do not believe that many hierarchies in real-world applications are binary trees. Secondly, limiting the hypothesis space to binary trees often forces spurious structure to be 'hallucinated' even if this structure is not supported by data, making the practitioner's task of interpreting the hierarchy more difficult. Finally, this spurious structure is also undesirable from an Occam's razor point of view: the methods are not returning the simplest hierarchical structure supported by the data, because the simpler structures that explain the data can involve nonbinary trees and these are excluded from the hypothesis space.

Figure 8.1(left) shows an example whereby spurious structure is imposed on data by a model that assumes binary trees. In this case the model is Bayesian hierarchical clustering (BHC) (Heller and Ghahramani, 2005), a probabilistic model for binary hierarchical clustering. The dataset consists of three clusters of data items, each of which has no further internal substructure. Nevertheless, BHC produced cascades of internal nodes to represent each of the clusters. This is an unnecessarily complex structure for the simple clusters in the dataset, and is a tell-tale sign among probabilistic binary hierarchical clustering algorithms that the tree is not representing large clusters in the data properly. Ideally the tree structure should be simplified by collapsing each cascade into a single node with many children, as in Figure 8.1(right), expressing the lack of substructure among the data items.

In this chapter we describe a probabilistic model for hierarchical structure discovery that operates in a broadened hypothesis space. Here each internal node can have an arbitrary number of children, allowing the model to use simpler, nonbinary trees to describe data if they do so better than binary trees. To help us choose among different trees, we will take a Bayesian model selection approach and choose the tree with the highest marginal data likelihood. Part of the contribution of this paper is the design of a likelihood that conforms to our intuition, such that structures with higher likelihoods also tend to be the ones that are subjectively simpler. We refer to trees in this broadened hypothesis space as *rose trees*, as they are known in the functional programming literature (Bird, 1998; Meertens, 1988) and our model as *Bayesian rose tree mixture models*. The nonbinary tree given in Figure 8.1(right) is in fact the structure discovered by our model.

We take a Bayesian approach to discovering hierarchy structure. At each step in constructing the tree, we perform a Bayesian hypothesis test for each possible merge. The best merge is then greedily selected. This results in the discovery of a single tree structure. analogously to Heller and Ghahramani (2005), as opposed to a fully Bayesian approach, where a prior over trees is defined and the posterior approximated using Monte Carlo sampling (e.g. Felsenstein, 2003; Neal, 2003). A fully Bayesian approach, while in many ways appealing, is computationally very expensive and complex to implement due to the very large (superexponential) number of trees and the complex Metropolis–Hastings moves that are often necessary for Markov chain Monte Carlo methods to mix properly over the posterior. A deterministic, single solution may also make interpreting the results easier for those modellers who are not very familiar with Bayesian methodology.

Therefore in this chapter we opt for a greedy approach, constructing a tree in an agglomerative bottom-up fashion. This gives an efficient algorithm that we find works well empirically.

The remainder of this chapter is organised as follows. In Section 8.2 we briefly review the existing literature on probabilistic hierarchical structure discovery and place Bayesian rose trees within this context. In Section 8.3 we describe our model in detail. In Section 8.5 we describe our greedy agglomerative construction algorithm. In Section 8.6 we discuss relationships with variants and other plausible extensions to BHC. Finally, in Section 8.7 we report experimental results using Bayesian rose trees, and conclude in Section 8.8.

8.2 Prior work

There is a diverse range of methods for hierarchical structure discovery, and unfortunately it is not possible to review every contribution here. Most methods can be construed as methods for hierarchical clustering, where each subtree corresponds to a cluster of data items. Classical introductions to clustering can be found in Hartigan (1975), McLachlan and Basford (1988) and Kaufman and Rousseeuw (1990), while Jain *et al.* (1999) is a more recent survey and Murtagh (1983) is a survey of classic methods for hierarchical clustering.

The most popular methods for hierarchical clustering are probably the agglomerative linkage methods (Duda and Hart, 1973). These start with each data item in its own cluster and iteratively merge the closest pair of clusters together, as determined by some distance metric, until all data belong to a single cluster. A record is kept of the order of merges and used to form a hierarchy, where data items reside on the leaves and branch lengths correspond to distances between clusters. Different methods are determined by different distance metrics among data items and how these distances are combined to define distances between clusters. Popular linkage methods include single, complete and average linkage, where the distance between two clusters is defined to be the minimum, maximum and average intercluster data item distances respectively. While straightforward and computationally efficient, linkage methods are not model-based, making comparisons between the discovered hierarchies based on different distances difficult due to a lack of an objective criterion.

Another important area for hierarchical structure discovery is phylogenetics, where the problem is to infer the phylogenetic tree relating multiple species and where a rich variety of approaches have been explored (Felsenstein, 2003). These include non-model-based methods, e.g. linkage algorithms based on distances among species (Fitch and Margoliash, 1967; Saitou and Nei, 1987; Studier and Keppler, 1988) and parsimony-based methods (Camin and Sokal, 1965), as well as model-based maximum-likelihood methods (Felsenstein, 1973, 1981). Consistency has been shown of maximum-likelihood estimators (Rogers, 1997; Yang, 1994). Another approach taken in phylogenetics is the Bayesian one, where a prior over phylogenetic trees is defined and the posterior distribution over trees estimated (Huelsenbeck and Ronquist, 2001; Yang and Rannala, 1997). This ensures that uncertainty in inferred hierarchical structures is accounted for, but is significantly more complex and computationally expensive.

Similarly to model-based phylogenetics, recent machine learning approaches to model-based hierarchical structure discovery have a dichotomy between maximum-likelihood estimation (Friedman, 2003; Heller and Ghahramani, 2005; Segal *et al.*, 2002; Vaithyanathan and Dom, 2000) and Bayesian posterior inference (Kemp *et al.*, 2004; Neal, 2003; Roy *et al.*, 2007; Teh *et al.*, 2008; Williams, 2000), reflecting the trade-off between simplicity and computational efficiency, on the one hand, and knowledge of structural uncertainty, on the other. The approach taken in this paper is a direct extension of those in Friedman (2003) and Heller and Ghahramani (2005); we will discuss these in more detail in Section 8.6.

There are few hierarchical clustering methods that directly produce nonbinary hierarchies. Williams (2000) fixes the maximum number of layers and nodes per layer, and defines a prior over trees whereby each node picks a node in the layer above independently. Blei *et al.* (2010) use a nested Chinese restaurant process to define a prior over layered trees. Both are Bayesian methods, which are quite computationally complex and use Monte Carlo sampling for inference. This is to be expected since with nonbinary hierarchies the number of internal nodes inferred can vary, and methods that cannot account for varying numbers of parameters can easily overfit. In methods that infer branch lengths as well as binary tree structures, nonbinary hierarchies can be obtained by visual inspection and by heuristics for collapsing short branches.

8.3 Rose trees, partitions and mixtures

In this section we describe our probabilistic model, as well as the terminology used in the subsequent sections. We shall refer to the hierarchical clustering structure describing a data set as a rose tree and to the probabilistic model based on a rose tree as a Bayesian rose tree mixture model, or BRT for short. Figure 8.2 gives two examples of rose trees over data items labelled a to e, one with all binary internal nodes and one with a ternary node. We will use these as running examples throughout this section. Leaves correspond to data items and every node of a tree corresponds to a cluster of its leaves, with nodes higher up the tree corresponding to larger clusters.

Let $\mathcal{D}$ be a set of data items. Our probabilistic model for these data items $\mathcal{D}$ under a rose tree T, $p(\mathcal{D}|T)$, is a mixture model where each mixture component is a partition of the dataset, which is in turn a disjoint set of clusters of data items. The role of the rose tree T is as a model index dictating which partitions of the dataset are included in the mixture model; in particular, the partitions that are included are those that are 'consistent' with the rose tree T. In the following, we will elaborate on the key concepts of clusters, partitions and rose trees, and how each of these is modelled in our probabilistic model.

Definition 1: A *rose tree* T either consists of a single leaf $x \in \mathcal{D}$ or consists of a root node along with n_T children, $T_1, \ldots, T_{n_T}$ say, each of which is a rose tree whose leaves are disjoint. We write children(T) for the set of *children* of T and pa(T_i) for the *parent* of T_i. A *node* or *subtree*, identified by its root, T', of a rose tree T is either T or one of its descendants under the child relation. The *ancestors* of T', ancestor(T'), consist of the nodes of T that have T' as a descendant. The leaves leaves(T') of a node T' are the set of data items that are descendants of T'.

For example, the tree in Figure 8.2(right) consists of two children, each being itself a tree, one with leaves a, b and c, and one with leaves d and e. Each node of a rose tree can be taken to mean that its leaves form a cluster of data items that are more similar among themselves than to other data items. To make this precise, we

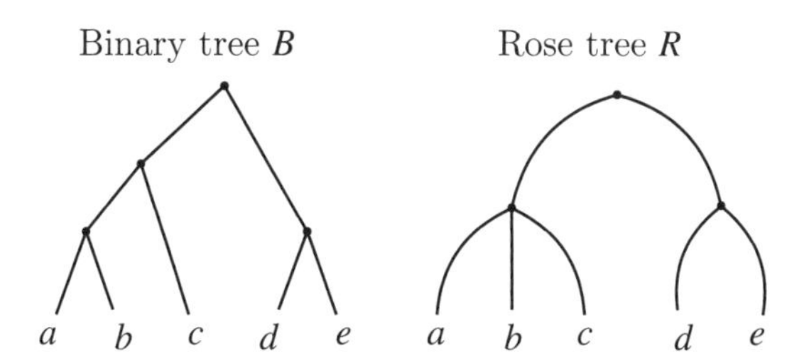

Figure 8.2 Examples of rose trees with five leaves labelled a to e. Left: a rose tree with only binary internal nodes and a cascading subtree (on leaves a, b and c). Right: a rose tree with a ternary node resulting from collapsing the cascading subtree.

will model each cluster of data items using a single shared parameter with different clusters using different parameters.

Definition 2: A *cluster* is a set of data items $D \subseteq \mathcal{D}$. Let θ parameterise a probability distribution for a data item $x \in D$, $f(x|\theta)$, with a corresponding prior $f(\theta|\eta)$ and hyperparameter η. If we marginalise out θ, the probability of a cluster of data items D is

$$f(D) = \int f(\theta|\eta) \prod_{x \in D} f(x|\theta) d\theta. \tag{8.1}$$

In Section 8.4.1 we describe two models for clusters used in our experiments (Section 7.6): a beta-Bernoulli model and a Gaussian process model.

Definition 3: A partition ϕ is a disjoint set of clusters whose union is the whole dataset $\mathcal{D}$. We shall write partitions using the '|' symbol; for example $ab|c$ denotes a partition of the set $\{a, b, c\}$ into clusters $\{a, b\}$ and $\{c\}$. We model the probability of $\mathcal{D}$ under a partition as the product of the probabilities of its constituent clusters:

$$g(\phi = \{D_1| \cdots |D_J\}) = \prod_{j=1}^{J} f(D_j), \tag{8.2}$$

where $f(D_j)$ is the probability of the jth cluster of the partition.

Since each cluster in a partition is modelled independently using one parameter, the likelihood of a partition will be high if data items have high intracluster similarities and low intercluster similarities.

Having shown how partitions are constructed from clusters, we now turn to showing how a Bayesian rose tree is constructed as a mixture over partitions. Each node of the rose tree is meant to represent a group of data items, at the leaves, which share some element of similarity. Therefore, it seems reasonable to assume that each cluster of each partition in the mixture corresponds to one subtree of the rose tree. Partitions consisting of such clusters are called tree consistent partitions, and will constitute the components of the mixture model.

Definition 4: A partition is consistent with a rose tree T if each cluster in the partition corresponds to the leaves of some subtree in T. Denote the set of all partitions consistent with T by $\mathbf{\Phi}(T)$.

Our definition of $\mathbf{\Phi}(T)$ is a straightforward generalisation to rose trees of the definition of tree-consistent partitions found in Heller and Ghahramani (2005). It is easy to see that $\mathbf{\Phi}(T)$ can be constructed explicitly by recursion as follows:

$$\mathbf{\Phi}(T) = \{\text{leaves}(T)\} \cup \left\{\phi_1| \cdots |\phi_{n_T} : \phi_i \in \mathbf{\Phi}(T_i), T_i \in \text{children}(T)\right\}, \tag{8.3}$$

where children(T) are the children of T, n_T is the number of children of T and {leaves(T)} represents the partition where all data items at the leaves of T are clustered together. Roughly speaking, each partition starts at the root of the tree, and either keeps the leaves in one cluster or partitions the leaves into the subtrees, the process repeating on each subtree. The end result is that each $\phi \in \mathbf{\Phi}(T)$ consists of disjoint clusters, each of which consists of the leaves of some subtree in T. For example, the partitions consistent with the rose trees in Figure 8.2 are given in the middle column of Figure 8.3.

All rose trees include the complete partition {leaves(T)} and (by recursion) the completely discriminating partition where each data item in $\mathcal{D}$ is in its own cluster. Different trees give rise to different sets of partitions between these two extremities. The binary tree with the fewest tree consistent partitions between these extremities is a cascading binary tree, where at each internal node one leaf is separated from the other leaves. On the other hand, the simplest rose tree has just two consistent partitions: the complete partition consisting of all leaves in one cluster and the completely discriminating partition. In general, a rose tree will have at most the same number of partitions as a binary tree. The rose tree allows us to represent

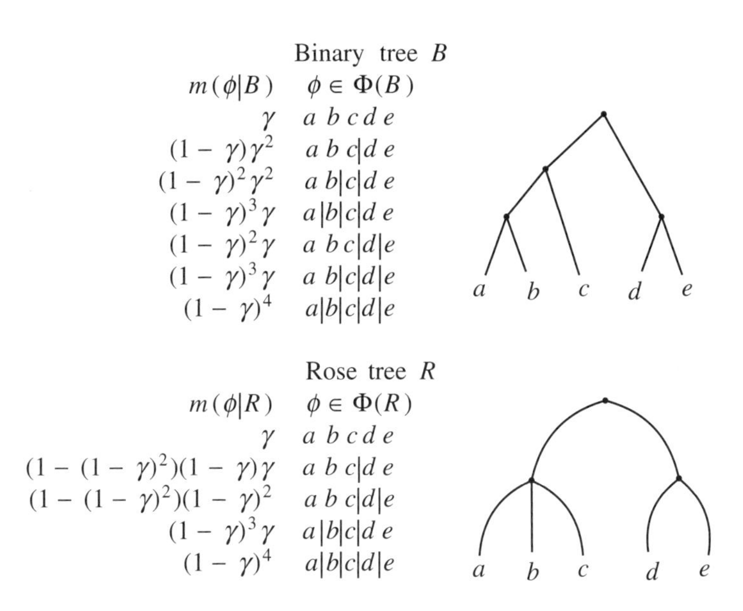

Binary tree B

$m(\phi \mid B)$	$\phi \in \Phi(B)$
γ	$a\ b\ c\ d\ e$
$(1-\gamma)\gamma^2$	$a\ b\ c \mid d\ e$
$(1-\gamma)^2\gamma^2$	$a\ b \mid c \mid d\ e$
$(1-\gamma)^3\gamma$	$a \mid b \mid c \mid d\ e$
$(1-\gamma)^2\gamma$	$a\ b\ c \mid d \mid e$
$(1-\gamma)^3\gamma$	$a\ b \mid c \mid d \mid e$
$(1-\gamma)^4$	$a \mid b \mid c \mid d \mid e$

Rose tree R

$m(\phi \mid R)$	$\phi \in \Phi(R)$
γ	$a\ b\ c\ d\ e$
$(1-(1-\gamma)^2)(1-\gamma)\gamma$	$a\ b\ c \mid d\ e$
$(1-(1-\gamma)^2)(1-\gamma)^2$	$a\ b\ c \mid d \mid e$
$(1-\gamma)^3\gamma$	$a \mid b \mid c \mid d\ e$
$(1-\gamma)^4$	$a \mid b \mid c \mid d \mid e$

Figure 8.3 Examples of (top) a binary tree B with a cascading subtree (on leaves a, b and c) and (bottom) a rose tree R with the cascade collapsed into one node. All tree-consistent partitions for each tree, as well as the associated mixture proportions, are listed to the left of the corresponding tree. Note that $m(\{a\,b\,c|d\,e\}|R) = m(\{a\,b\,c|d\,e\}|B) + m(\{a\,b|c|d\,e\}|B)$ and $m(\{a\,b\,c|d|e\}|R) = m(\{a\,b\,c|d|e\}|B) + m(\{a\,b|c|d|e\}|B)$; the mixing proportion assigned to each of the partitions in R is the sum of those partitions in B that are refinements of the partition in R.

simple hierarchical structure without having to introduce spurious partitions, as in the cascading binary tree.

We are now ready to define our Bayesian rose tree mixture model.

Definition 5: Given a rose tree T, a Bayesian rose tree is a mixture over partitions in $\mathbf{\Phi}(T)$ of the cluster of data items at its leaves $\mathcal{D} = \text{leaves}(T)$:

$$p(\mathcal{D}|T) = \sum_{\phi \in \mathbf{\Phi}(T)} m(\phi|T) g(\phi) \tag{8.4}$$

where $m(\phi|T)$ is the mixing proportion of partition ϕ and $g(\phi)$ is the data likelihood term for partition ϕ given in (8.2).

In general, the number of partitions consistent with T can be exponentially large in the number of leaves, making (8.4) computationally intractable for large datasets. Instead, we propose a specific factorised form for the mixture proportions, m, that allows for (8.4) to be computed efficiently by recursion as well as for the efficient agglomerative tree construction algorithm in Section 8.5. For each subtree T' of T, let $\pi_{T'}$ be a parameter between 0 and 1. We shall discuss the choice of these parameters at length in Section 8.4. We will compute (8.4) recursively as follows:

$$p(\text{leaves}(T)|T) = \pi_T f(\mathcal{D}) + (1 - \pi_T) \prod_{T_i \in \text{children}(T)} p(\text{leaves}(T_i)|T_i), \tag{8.5}$$

where $f(\mathcal{D})$ is the probability of the cluster $\mathcal{D}$. Equation (8.5) corresponds to the following generative process. Beginning at the root of the tree, with probability π_T, generate θ according to the prior $f(\theta|\eta)$ and all data items according to $f(\cdot|\theta)$. Otherwise, recurse down the tree, with each subtree independently generating the data items at its leaves according to the same process. If a leaf is reached the recursion stops and the data item is generated. Note that it follows from this narrative that we can identify each cluster of the data with a node of the tree.

Comparing (8.5) and (8.4), we find that

$$m(\phi|T) = \prod_{S \in \text{ancestor}_T(\phi)} (1 - \pi_S) \prod_{S \in \text{subtree}_T(\phi)} \pi_S, \tag{8.6}$$

where $\text{subtree}_T(\phi)$ are the subtrees in T corresponding to the clusters in the partition ϕ and $\text{ancestor}_T(\phi)$ are the ancestors of subtrees in $\text{subtree}_T(\phi)$. Figure 8.3 gives these mixing proportions for the two example rose trees.

In summary, the marginal probability of a data set $\mathcal{D}$ under a rose tree T is a mixture over the partitions consistent with T, with the probability of $\mathcal{D}$ under a partition $\phi \in \mathbf{\Phi}(T)$ being a product of the probabilities of clusters in ϕ. We call our mixture a *Bayesian rose tree* (BRT) mixture model. In Section 8.4 we motivate using a particular choice of the mixing proportions π_T and in Section 8.6 we contrast our Bayesian rose tree mixture over partitions with a number of related models.

8.4 Avoiding needless cascades

In this section we propose a particular choice for the mixing proportions π_T given by

$$\pi_T = 1 - (1 - \gamma)^{n_T - 1}, \tag{8.7}$$

where $0 \leq \gamma \leq 1$ is a hyperparameter of the model controlling the relative proportion of coarser partitions of the data as opposed to finer ones. When restricted to just binary trees, $\pi_T = \gamma$ and the BRT reduces to one of the models in Heller and Ghahramani (2005). Subtrees T with more children are assigned a larger π_T and so the likelihood of the cluster at the root of the subtree is more highly weighted than those of smaller clusters further down the subtree.

As we will see, this choice of π_T is intimately related to our maxim that the maximum-likelihood tree should be simple if the data are unstructured. We will start by considering the simple situation given by the running examples in Figure 8.3 before looking at the more general case. The two rose trees in Figure 8.3 differ from each other only in that B places the data items a, b, c into a cascading structure with two binary nodes while R uses a single ternary node. The figure also shows the set of partitions and their mixing proportions given by (8.6) and (8.7).

Suppose that a, b and c are similar to each other but are otherwise indistinguishable, yet are distinguishable from d and e. We would like a model that prefers the rose tree R over the binary tree B, since its structure matches the similarity relationships among the data items. To do this, note that, because the data items a, b, c belong together in one cluster, we can expect the inequality $f(\{a, b, c\}) > f(\{a, b\})f(\{c\})$ to hold. This implies the following inequalities among the partition likelihoods:

$$\begin{aligned} g(\{a\,b\,c|d\,e\}) &> g(\{a\,b|c|d\,e\}), \\ g(\{a\,b\,c|d|e\}) &> g(\{a\,b|c|d|e\}). \end{aligned} \tag{8.8}$$

Expanding the likelihoods for R and B as a mixture of the likelihoods under each partition, using the inequalities (8.8) and using the equalities in the mixing proportions noted in the caption of Figure 8.3, we find that

$$p(\{a, b, c, d, e\}|R) > p(\{a, b, c, d, e\}|B). \tag{8.9}$$

In other words, the collapsed rose tree R is assigned a higher likelihood than the binary tree B. Therefore if we were to select the rose tree with higher likelihood we would have chosen the one that better describes the data, i.e. R.

In the general case, if we have a cluster of n indistinguishable data items, we can guarantee preferring a rose tree R consisting of a single internal node with n children over a tree B with multiple internal nodes, such as a cascading binary tree, if the mixing proportion of the complete partition in R is the sum over the mixing proportions of all partitions consistent with B except the most discriminating partition. Fortunately, under (8.7) this sum turns out to be the same regardless of the structure of B, and equals

$$\pi_R = 1 - (1 - \gamma)^{n-1}. \tag{8.10}$$

Equation (8.10) can be easily proven by induction on the number of internal nodes of B. Using this, we can now show that the collapsed rose tree R will be more likely than a rose tree which introduces spurious structure (e.g. cascades).

Proposition 1: Let B be a rose tree with $n_B > 1$ children $T_1, \ldots, T_{n_B}$, with T_1 being an internal node. Let R be a rose tree obtained by collapsing the B and T_1 nodes into one, i.e. R has children$(R) = \text{children}(T_1) \cup T_2 \cup \cdots \cup T_{n_B}$. Suppose the data items of T are indistinguishable, i.e. the likelihoods of the noncomplete partitions are smaller than for the complete partition:

$$g(\{\text{leaves}(T_1)|\phi_2| \cdots |\phi_{n_B}\}) < g(\{\text{leaves}(R)\}) \tag{8.11}$$

for every $\phi_i \in \boldsymbol{\Phi}(T_i)$, $i = 2, \ldots, n_B$. Then the likelihood of B is lower than for R:

$$p(\text{leaves}(B)|B) < p(\text{leaves}(R)|R). \tag{8.12}$$

Proof. By construction, we have $\boldsymbol{\Phi}(R) \subset \boldsymbol{\Phi}(B)$. Let $\psi = \{\text{leaves}(R)\}$ be the complete partition and $\delta(B, R) = \boldsymbol{\Phi}(B) \backslash \boldsymbol{\Phi}(R)$ be those partitions of B not in R. It is straightforward to see that

$$\delta(B, R) = \{\text{leaves}(T_1)|\phi_2| \cdots |\phi_{n_B} : \phi_i \in \boldsymbol{\Phi}(T_i)\}. \tag{8.13}$$

The mixture likelihood of B can now be decomposed as

$$p(\text{leaves}(B)|B) = \pi_B g(\psi) + \sum_{\phi \in \delta(B,R)} m(\phi|B) g(\phi) + \sum_{\phi \in \boldsymbol{\Phi}(R) \backslash \{\psi\}} m(\phi|B) g(\phi) \tag{8.14}$$

From the premise (8.11) we find that

$$p(\text{leaves}(B)|B) < g(\{\psi\}) \left(\pi_B + \sum_{\phi \in \delta(B,R)} m(\phi|B) \right) + \sum_{\phi \in \boldsymbol{\Phi}(R) \backslash \{\psi\}} m(\phi|B) g(\phi). \tag{8.15}$$

We now turn to evaluating the summation over $\delta(B, R)$ in (8.15). From (8.13) and (8.5) we see that the mixture proportion assigned to each $\phi \in \delta(B, R)$ is

$$m(\phi|B) = (1 - \pi_B) \pi_{T_1} \prod_{i=2}^{n_B} m(\phi_i|T_i). \tag{8.16}$$

Since (8.16) decomposes into a product over the partition of each subtree T_i, $i = 2, \ldots, n_B$, and $\sum_{\phi_i \in \boldsymbol{\Phi}(T_i)} m(\phi_i|T_i) = 1$, we see that

$\sum_{\phi \in \delta(B,R)} m(\phi|B) = (1 - \pi_B)\pi_{T_1}$. Now from (8.7) we see that the term in the parentheses of (8.15) is

$$\pi_B + (1 - \pi_B)\pi_{T_1} = \pi_R = m(\psi|R). \tag{8.17}$$

On the other hand $m(\phi|B) = m(\phi|R)$ for each $\phi \in \mathbf{\Phi}(R)\backslash\{\psi\}$. We have now established the right-hand side of (8.15) as $p(\text{leaves}(R)|R)$.

The Proposition applies when the root of the tree along with one of its children are collapsed into one node. This can be trivially generalised to collapsing any subtree.

Corollary 1: Let S be a rose tree with a subtree B and let T be constructed by collapsing all internal nodes of B into one node. If the data items under B are indistinguishable, i.e. noncomplete partitions in B have lower likelihoods than the complete one, then

$$p(\text{leaves}(S)|S) < p(\text{leaves}(T)|T). \tag{8.18}$$

8.4.1 Cluster models

Each cluster D of data items has an associated marginal likelihood $f(D)$ defined according to (8.1). In this chapter we consider two families of parameterised cluster models: for d-dimensional binary data, we use a product of beta-Bernoulli distributions and, for curves in $\mathbb{R}^2$, we use Gaussian processes. Other cluster models may be considered, e.g. other exponential families with conjugate priors.

Binary data clusters

For d-dimensional binary data, we model the ith dimension independently using a Bernoulli distribution with parameter θ_i and use a beta prior for θ_i with hyperparameters (α_i, β_i). Integrating out the parameters, the cluster likelihood $f(D)$ is then the probability mass formed by a product of independent beta-Bernoulli distributions in each dimension:

$$\begin{aligned} f(D) &= \prod_{i=1}^{d} \int f(D_i|\theta_i) f(\theta_i|\alpha_i, \beta_i) \mathrm{d}\theta_i \\ &= \prod_{i=1}^{d} \frac{\text{Beta}(\alpha_i + n_i, \beta_i + |D| - n_i)}{\text{Beta}(\alpha_i, \beta_i)}, \end{aligned} \tag{8.19}$$

where D_i consists of the ith entry of each data item in D, n_i is the number of ones in D_i and $|D|$ is the total number of data items in cluster D. The hyperparameters of the entire cluster model are thus $\eta = \{(\alpha_i, \beta_i)\}_{i=1}^{d}$.

Gaussian process expert clusters

Here we consider data items consisting of input–output pairs, $D = \{(x_i, y_i)\}_{i=1}^n$, and are interested in modelling the conditional distribution over outputs given inputs. Rasmussen and Ghahramani (2002) proposed a Dirichlet process (DP) mixture of Gaussian process (GP) experts where a dataset is partitioned, via the DP mixture, into clusters, each of which is modelled by a GP. Such a model can be used for nonparametric density regression, where a full conditional density over an output space is estimated for each value of input. This allows generalisation of GPs allowing for multimodality and nonstationarity. The original model in Rasmussen and Ghahramani (2002) had mixing proportions that do not depend on input values; this was altered in the paper in an ad hoc manner using radial basis function kernels. Later Meeds and Osindero (2006) extended the model by using a full joint distribution over both inputs and outputs, allowing for properly defined input-dependent mixing proportions.

With both approaches MCMC sampling was required for inference, and this might be slow to converge. Here we consider using Bayesian rose trees instead. The joint distribution of each cluster is modelled using a Gaussian over the inputs and a GP over the outputs given the inputs:

$$f(D) = f(\{x_i\})f(\{y_i\}|\{x_i\}), \tag{8.20}$$

where

$$\begin{aligned} f(\{x_i\}) &= \int\int \left[\prod_{i=1}^n \mathcal{N}(x_i|\mu, \boldsymbol{R}^{-1})\right] \mathcal{N}(\mu|m, (r\boldsymbol{R})^{-1})\mathcal{W}(\boldsymbol{R}|\boldsymbol{S}, \nu)\mathrm{d}\mu\mathrm{d}R \\ f(\{y_i\}|\{x_i\}) &= \mathcal{N}(\{y_i\}|\boldsymbol{0}, \boldsymbol{K}), \end{aligned} \tag{8.21}$$

in which $\mathcal{N}(x|\boldsymbol{\mu}, \boldsymbol{\Sigma})$ is the normal density with mean $\boldsymbol{\mu}$ and covariance matrix $\boldsymbol{\Sigma}$, $\mathcal{W}(\boldsymbol{R}|\boldsymbol{S}, \nu)$ is a Wishart density with degrees of freedom ν and scale matrix $\boldsymbol{S}$ and the matrix $\boldsymbol{K}$ is a Gram matrix formed by the covariance function of the GP (we used the squared exponential). The normal inverse Wishart prior over the parameters μ and $\boldsymbol{R}$ is conjugate to the normal likelihood, so $f(D)$ can be computed analytically. It follows that for a Gaussian process expert cluster the hyperparameters η are $(r, \nu, \boldsymbol{S})$, where r is the scaling parameter of the normal inverse Wishart prior.

8.5 Greedy construction of Bayesian rose tree mixtures

We take a model selection approach to finding a rose tree structure given data. Ideally we wish to find a rose tree T^* maximising the marginal probability of the data $\mathcal{D}$:

$$T^* = \mathrm{argmax}_T\, p(\mathcal{D}|T). \tag{8.22}$$

This is intractable since there is a superexponential number of rose trees.

input: data $\mathcal{D} = \{x_1 \ldots x_n\}$,
cluster model $p(x|\theta)$,
cluster parameter prior $p(\theta|\eta)$,
cluster hyperparameters η
initialise: $T_i = \{x_i\}$ for $i = 1 \ldots n$
for $c = n$ to 2 **do**
Find the pair of trees T_i and T_j, and merge operation m with the highest likelihood ratio:

$$L(T_m) = \frac{p(\text{leaves}(T_m)|T_m)}{p(\text{leaves}(T_i)|T_i)p(\text{leaves}(T_j)|T_j)}$$

Merge T_i and T_j into T_m using operation m
$T_{n+c-1} \leftarrow T_m$
Delete T_i and T_j
end for
output: Bayesian rose tree T_{n+1}, a mixture over partitions of $\mathcal{D}$

Figure 8.4 Agglomerative construction algorithm for Bayesian rose trees.

Inspired by the success of other agglomerative clustering algorithms, we instead consider constructing rose trees in a greedy agglomerative fashion as follows. Initially every data item is assigned to its own rose tree: $T_i = \{x_i\}$ for all data items x_i. At each step of our algorithm we pick two rose trees T_i and T_j and merge them into one tree T_m using one of a few merge operations. This procedure is repeated until just one tree remains (for n data items this will occur after $n - 1$ merges), and is illustrated in Figure 8.4.

Each step of the algorithm consists of picking a pair of trees as well as a merge operation. We use a maximum likelihood ratio criterion, picking the combination that maximises

$$L(T_m) = \frac{p(\text{leaves}(T_m)|T_m)}{p(\text{leaves}(T_i)|T_i)p(\text{leaves}(T_j)|T_j)}. \tag{8.23}$$

We use the likelihood ratio rather than simply the likelihood $p(\text{leaves}(T_m)|T_m)$ because the denominator makes $L(T_m)$ comparable across choices of trees T_i and T_j of differing sizes (Friedman, 2003; Heller and Ghahramani, 2005).

We considered a number of merge operations to allow for nodes with more than two children to be constructed: a *join*, an *absorb* and a *collapse* operation (see Figure 8.5). In all operations the merged rose tree T_m has $\text{leaves}(T_m) = \text{leaves}(T_i) \cup \text{leaves}(T_j)$, the difference being the resulting structure at the root of the merged tree. For a *join*, a new node T_m is created with children T_i and T_j. A join is chosen if the children of T_i and T_j are related, but are sufficiently distinguishable to keep both subtrees separated. For an *absorb* the children of the resulting tree T_m are $\text{children}(T_i) \cup \{T_j\}$; that is tree T_j is absorbed as a child

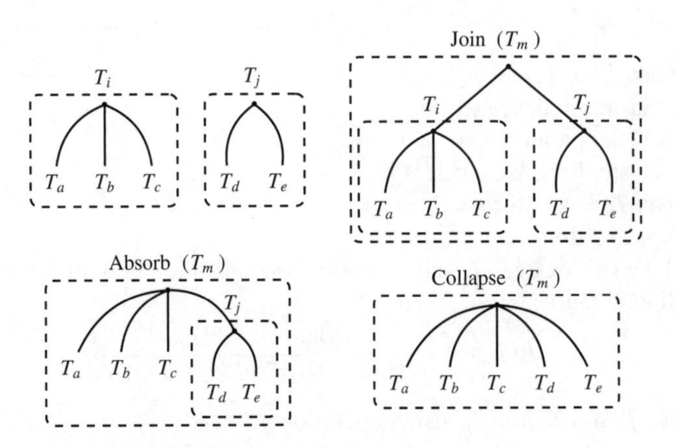

Figure 8.5 Merges considered during greedy search.

of T_i. This operation is chosen if the children are related, but there exists finger-distinguishing structure already captured by T_j. This operation is not symmetric so we also consider the converse, where the children are $\{T_i\} \cup \text{children}(T_j)$. Finally, a *collapse* merges the roots of both trees, so that the resulting children of T_m are $\text{children}(T_i) \cup \text{children}(T_j)$. This is performed when the children of T_i and T_j are indistinguishable and so may be combined and treated similarly.

Binary hierarchical clustering algorithms such as in Heller and Ghahramani (2005) only need to consider the join operation. To be able to construct every possible rose tree the absorb operation is necessary as well. Intuitively, a join merge makes the tree taller while an absorb merge makes the tree wider. The collapse operation is not technically necessary, but we found that including it allowed us to find better rose trees.

In general the computational complexity of the greedy agglomerative clustering algorithm of Section 8.5 is in $\Omega(n^2 \log nL)$ (where L is a contribution to the complexity due to the particular cluster likelihood). First, for every pair of data items we must calculate the likelihood of a merged tree; there are $O(n^2)$ such pairs. Second, these pairs must be sorted, requiring $O(n^2 \log n)$ computational complexity. The data structure we use is simply a binary heap.

If the cluster marginal likelihood is a d-dimensional product of beta-Bernoulli distributions (i.e. for d-dimensional binary valued data) then $L = O(d)$. Instead of keeping track of every data item, it is sufficient to keep track of the sufficient statistics (counts of zeros and ones) of each cluster. The same argument applies to any conjugate exponential family cluster model. For a BRT mixture of Gaussian process experts, $L = O(n^3)$, which comes from performing Cholesky decompositions of Gram matrices.

There are several opportunities for approximations to improve the computational complexity of the greedy agglomerative algorithm. Heller (2008) explored using randomisation to sample random subsets of the dataset in order to make the computational complexity scale more favourably, and Xu *et al.* (2009) use a flat clustering to constrain one level of the hierarchy, thereby reducing the complexity of discovering the remaining structure. Low-rank approximations (such as in Williams and Seeger, 2001) could also be used to reduce the computational complexity of the Gaussian process expert variant.

8.5.1 Prediction

Two kinds of prediction are possible with BRT: predicting partially observed data items and predicting unobserved data items themselves.

The predictive probability of a Bayesian rose tree for partially observed data is

$$p(\mathcal{D}^m|\mathcal{D}, T) = \frac{p(\mathcal{D}^m, \mathcal{D}|T)}{p(\mathcal{D}|T)}, \tag{8.24}$$

where $\mathcal{D}^m$ are the unobserved parts of data items in $\mathcal{D}$. The denominator of (8.24) is the quantity optimised to find T, and calculating it is tractable if marginalising components of the cluster likelihood is tractable.

As the rose tree T only accounts for observed data items, predicting unobserved data requires additional assumptions about the location of unobserved data within the rose tree. The assumption we make is the same as in Heller (2008): the probability of an unobserved data item being in a particular cluster is proportional to the number of observed data items in that cluster and also the number of observed data items in any cluster above it in the tree. Intuitively this assumption means that an unobserved data item is more likely to come from a larger cluster than from a smaller cluster and it is more likely to come from a cluster higher up the tree than from one further down the tree. The predictive distribution of an unobserved data item is then a mixture over clusters in T:

$$p(x|\mathcal{D}, T) = \sum_{S \in \text{subtree}(T)} w_S f(x|\text{leaves}(S)), \tag{8.25}$$

where

$$\begin{aligned} w_S &= r_S \left(\prod_{A \in \text{ancestor}(S)} (1 - r_A) \frac{n_{A \to S}}{|\text{leaves}(S)|} \right), \\ r_S &= \frac{\pi_S f(\text{leaves}(S))}{p(\mathcal{D}|S)}, \end{aligned} \tag{8.26}$$

in which subtree(T) are the subtrees of T corresponding to each cluster in T, ancestor(S) are the ancestors of the subtree S and $n_{A \to S}$ is the number of data items in the subtree of A containing all the leaves of S. Here $f(x|\text{leaves}(S))$ is the predictive cluster distribution of the corresponding cluster model. Since in (8.25) x belongs to every cluster of T with some probability, (8.25) does not describe a Bayesian rose tree mixture model.

8.5.2 Hyperparameter optimisation

We optimise the hyperparameters of the cluster marginal likelihood, η, and the mixture proportion parameter, γ, by gradient ascent on the log marginal likelihood $\log p(\mathcal{D}|T)$. From (8.5), the gradient of the marginal log-likelihood $\log p(\mathcal{D}|T)$ with respect to the cluster hyperparameters can be efficiently computed recursively:

$$\frac{\partial \log p(\mathcal{D}|T)}{\partial \eta} = r_T \frac{\partial \log f(\text{leaves}(T))}{\partial \eta} + (1 - r_T) \sum_{T_i \in \text{children}(T)} \frac{\partial \log p(\text{leaves}(T_i)|T_i)}{\partial \eta}, \tag{8.27}$$

where r_T is given by (8.26).

Similarly, the gradient for the mixture proportion parameter γ is

$$\frac{\partial \log p(\mathcal{D}|T)}{\partial \gamma} = r_T \frac{\partial \log \pi_T}{\partial \gamma} + (1 - r_T) \left(\frac{\partial \log(1 - \pi_T)}{\partial \gamma} + \sum_{T_i \in \text{children}(T)} \frac{\partial \log p(\text{leaves}(T_i)|T_i)}{\partial \gamma} \right). \tag{8.28}$$

After optimising the hyperparameters for a particular tree, the marginal log-likelihood can be optimised further using these hyperparameters in a coordinate ascent procedure: greedily find a better tree given the current hyperparameters (Figure 8.4), then find the best hyperparameters for that tree (via (8.27) and (8.28)) and repeat until convergence, alternating between optimising the hyperparameters and the tree. This optimisation procedure is not guaranteed to find a global optimum of the marginal likelihood, as the marginal likelihood is typically not convex in its cluster hyperparameters. However, the optimisation procedure will eventually converge to a local optimum for the hyperparameters and the tree, if the cluster likelihood is bounded, as both steps optimise the same objective function.

We found that, particularly where binary data are missing, optimising the beta-Bernoulli hyperparameters is sensitive to initial conditions. Figure 8.6 shows the effect of changing the value of just two of the hyperparameters of the beta-Bernoulli cluster models on the optimised log marginal likelihood $\log p(\mathcal{D}|T)$ of BRT. All other hyperparameters were held fixed. Consequently, we used 10 restarts at random points around a MAP estimate of the cluster hyperparameters, and, for missing data, we averaged the hyperparameters of the beta-Bernoulli model over a small region around the optimum found.

8.6 Bayesian hierarchical clustering, Dirichlet process models and product partition models

In this section we describe a number of models related to Bayesian rose trees: finite mixture models, product partition models (Barry and Hartigan, 1992), PCluster

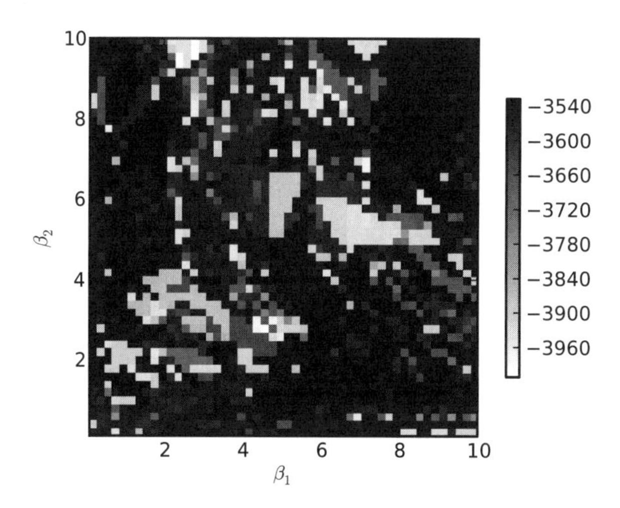

Figure 8.6 Sensitivity of optimising the marginal likelihood to initial conditions, with β_1 and β_2 of the beta-Bernoulli cluster model initialised as above and all other hyperparameters held fixed. Background level of darkness indicates the marginal log-likelihood at convergence according to the scale at the right of the plot.

(Friedman, 2003), Bayesian hierarchical clustering (BHC) (Heller and Ghahramani, 2005) and Dirichlet process mixture models.

8.6.1 Mixture models and product partition models

A Bayesian rose tree is a mixture model over partitions of the data. This is an unusual way to model how data items cluster and it may be beneficial to consider how other clustering models relate to mixtures over partitions.

We start by considering a typical Bayesian mixture model consisting of K components. Such a model associates each data item, x_i say, with a latent indicator variable, $z_i \in \{1, \ldots, K\}$ say, for the cluster to which it belongs. We can express this as a mixture over partitions as follows:

$$p(\mathcal{D}) = \prod_{i=1}^{n} \sum_{z_i=1}^{K} p(z_i) p(x_i|z_i) = \sum_{\phi} m(\phi) \prod_{D_j \in \phi} f(D_j) = \sum_{\phi} m(\phi) g(\phi), \quad (8.29)$$

where the component parameters have been marginalised out and ϕ ranges over all possible partitions of the data with up to K clusters. This is a consequence of interchanging the product and summation and using the commonality among the values of the z_is to form partitions. The likelihood $g(\phi)$ is the probability of the data under the partition ϕ given in (8.2), while $m(\phi)$ is the mixing proportion over partitions and is obtained by summing over all assignments $\{z_i\}$ giving rise to partition ϕ.

If the mixing proportion factorises into separate terms for each cluster, $m(\phi) = \prod_{D \in \phi} m(D)$ say, the last term of (8.29) is the marginal probability of a product partition model (Barry and Hartigan, 1992). An example of a product partition model with an unbounded number of mixture components is the Dirichlet process (DP) mixture model, which has $m(\phi)$ corresponding to the probability of a particular seating arrangement under the Chinese restaurant process:

$$m(\phi) = \frac{\prod_{D \in \phi} \alpha \Gamma(|D|)}{\alpha(\alpha + 1) \cdots (\alpha + n - 1)}. \tag{8.30}$$

In the context of regression, our Bayesian rose tree mixture of Gaussian process experts also bears some resemblance to an extension of product partition models described by Müller and Quintana (2010). There, the mixing proportions $m(\phi)$ are allowed to depend on covariates in a form functionally similar to one we would obtain if we condition on the inputs in (8.21).

Compared to the finite mixture model and product partition model, the Bayesian rose tree mixture model allows a larger range over the number of components in its partitions, ranging from those with a single cluster to those with as many clusters as there are data items. On the other hand, all the partitions in a Bayesian rose tree have to be consistent with the tree structure. A Bayesian rose tree can be interpreted as follows: the data are partitioned in some unknown way, but all the reasonable ways in which the data could be partitioned are consistent with some rose tree.

8.6.2 PCluster and Bayesian hierarchical clustering

Our work on Bayesian rose trees is directly motivated by issues arising from previous work on PCluster and Bayesian hierarchical clustering (BHC). The first model, PCluster (Friedman, 2003), is a direct probabilistic analogue of linkage methods where a likelihood ratio similar to (8.23) is used to measure the distance between two clusters. Each iteration of the agglomeration procedure thus produces a partition of the data with a likelihood similar to (8.2). However, the resulting tree structure does not itself correspond to a probabilistic model of the data; rather it is simply a trace of the series of partitions discovered by the procedure.

Addressing the lack of a probabilistic model in PCluster, Heller and Ghahramani (2005) proposed a probabilistic model called BHC, indexed by binary trees called BHC. The BHC model is a mixture over partitions consistent with the binary tree, and the BRT approach described in the present chapter is a generalisation and reinterpretation of BHC. There are three distinct differences between BRT and BHC. First, the likelihood and agglomerative construction of BHC only accounts for binary trees. Second, Heller and Ghahramani (2005) considered two alternative parameterisations of the mixing proportions π_T. In the first parameterisation, which we shall call BHC-γ, π_T equals some constant γ, and is in agreement with BRT on binary trees. The second parameterisation of BHC, which we shall call BHC-DP, uses a π_T that leads to $m(\phi)$ being precisely the mixing proportion of ϕ under the DP mixture (8.30). Since the marginal probability of the data under the DP mixture is a sum over all partitions, while that for BHC is only over those consistent with the tree, (8.4) gives a lower bound on the DP mixture marginal probability. Note that a similar setting of π_T in BRT allows it also to produce lower bounds on the

DP mixture marginal probability. However, since the set of partitions consistent with a BRT is always a subset of those consistent with some binary tree where we replace each nonbinary internal node on the rose tree with a cascade of binary nodes, the BRT lower bound will always be no higher than that for the BHC. This argument obviates the use of BRT as approximate inference for DP mixtures. In fact our reason for using rose trees is precisely that the set of partitions is smaller; all else being equal, we should prefer simpler models by Occam's razor. This view of hierarchical clustering is very different from the one expounded by Heller and Ghahramani (2005), and is the third distinction between BHC and BRT. In the next section we will compare the parameterisation of BRTs described in Section 8.3 against BHC as well as other parameterisations of BRT inspired by BHC.

8.7 Results

We compared BRT with several alternative probabilistic hierarchical clustering models: two binary hierarchical clustering algorithms, BHC-γ and BHC-DP, and two other rose tree hierarchical clustering algorithms. We shall call the model where BRT has $\pi_T = \gamma$, BRT-γ. The model BRT-γ differs from BHC-γ only in the number of possible children. Furthermore, we shall call the model where BRT has π_T configured in a similar fashion to BHC-DP, BRT-DP. In this way, all models with prefix 'BRT' shall have rose trees and all models with prefix 'BHC' shall have binary trees, while the suffix denotes how π_T is parameterised. The cluster likelihood models used are described in Section 8.4.1.

For BHC-DP and BRT-DP we report its marginal likelihood $p(\mathcal{D}|T)$, not the lower bound on the DP mixture, which is $p(\mathcal{D}|T)$ multiplied by a factor that is less than one.

8.7.1 Optimality of tree structure

The agglomerative scheme described in Section 8.5 is a greedy algorithm that is not guaranteed to find the optimal tree. Here we compare the trees found by BRT, BHC-γ and BHC-DP against the optimal (maximum-likelihood) rose tree T^* found by exhaustive search. We generated datasets of sizes ranging from 2 to 8, each consisting of binary vectors of dimension 64, from a BRT mixture with randomly chosen rose tree structures. On each of the n datasets we compare the performances in terms of the average $\log_2$ probability of the data assigned by the three greedily found trees T relative to the maximum-likelihood Bayesian rose tree T^*,

$$\Delta_l = \frac{1}{ln} \sum_{i=1}^{n} \log_2 p(\mathcal{D}_i|T_i^*) - \log_2 p(\mathcal{D}_i|T_i), \tag{8.31}$$

where l is the number of data vectors in the dataset. Here Δ_l measures the average number of bits required to code for a data vector under T, in excess of the same under T^*. The results, averaged over 100 datasets per dataset size are shown in Figure 8.7.

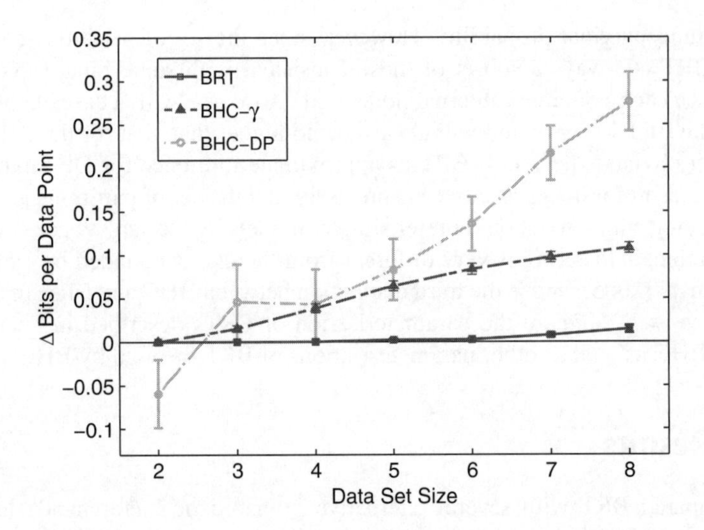

Figure 8.7 Per data item $\log_2$ probability of trees found greedily by BRT, BHC-γ and BHC-DP, relative to the optimal Bayesian rose tree. Error bars are one standard error. It is better to be lower in the graph.

We see that BRT finds significantly better trees than either BHC algorithm. We also found that BRT frequently finds the optimal tree, for example when $l = 8$ BRT found the optimum 70 % of the time. Note that when $l = 2$ BHC-DP produced higher log probability than the optimal BRT T^*, although it performed significantly worse than BHC-γ and BRT for larger l. This is because the BHC-DP and BRT models are not nested so BHC-DP need not perform worse than T^*.

8.7.2 Hierarchy likelihoods

We compared the marginal likelihoods of trees found by BHC-γ, BHC-DP, BRT-γ, BRT-DP and BRT on five binary-valued datasets. Some are the same datasets as those used in Heller and Ghahramani (2005). The characteristics of the datasets are summarised in Table 8.1, where `toy` is a synthetic dataset constructed so that ones only appear in three disjoint parts of the binary vector, with each class having ones in a different part (the hierarchies in Figure 8.1 were found by BHC-γ

Table 8.1 Characteristics of datasets.

Dataset	Attributes	Classes	Binarisation
`toy`	12	3	Handcrafted
`spambase`	57	2	Zero or nonzero
`newgroups`	485	4	Word presence/absence
`digits`	64	10	Threshold at 32
`digits024`	64	3	Threshold at 32

Table 8.2 Log marginal likelihoods and standard errors.

Dataset	BHC-DP	BRT-DP	BHC-γ	BRT-γ	BRT
toy	-215 ± 0.0	-215 ± 0.0	-168 ± 0.1	-167 ± 0.2	$\mathbf{-166} \pm 0.1$
spambase	-2258 ± 7.3	-2259 ± 7.2	-1980 ± 7.0	-2006 ± 8.0	$\mathbf{-1973} \pm 7.6$
digits024	-4010 ± 6.8	-4015 ± 6.8	-3711 ± 6.9	-3726 ± 6.9	$\mathbf{-3702} \pm 7.0$
digits	-4223 ± 6.9	-4216 ± 6.9	-3891 ± 6.7	-3916 ± 6.6	$\mathbf{-3888} \pm 6.8$
newsgroups	-10912 ± 61	-10937 ± 55	$\mathbf{-10606} \pm 63$	-10807 ± 59	-10645 ± 60

and BRT on this dataset); `spambase` is the UCI repository dataset of the same name (Frank and Asuncion, 2010); `newsgroups` is the CMU 20newsgroups dataset restricted to the news groups `rec.autos`, `rec.sport.baseball`, `rec.sport.hockey` and `sci.space`, and preprocessed using Rainbow (McCallum, 1996); `digits` is a subset of the CEDAR Buffalo digits dataset and `digits024` is the same dataset with only those samples corresponding to the digits 0, 2 and 4.

Each dataset consists of 120 data vectors split equally among the classes, except for `toy`, which has only 48 data vectors. When the original datasets are larger the 120 data vectors are subsampled from the original.

Table 8.2 shows the average log likelihoods of the trees found on these datasets. BRT typically finds a more likely tree; the difference between the DP-approximating models and the others is significant, while the difference between the $\pi_T = \gamma$ models and BRT is often significant.

Table 8.3 shows the logarithm (base 10) of the number of partitions represented by the maximum-likelihood trees. Model BRT typically finds a tree with far fewer partitions than the other models, corresponding to a simpler model of the data. The other rose tree-based models (BRT-DP and BRT-γ) have the same or only slightly fewer partitions than their corresponding binary tree equivalents (BHC-DP and BHC-γ, respectively). This reflects our design of the mixing proportion γ_T of BRT: partitions should only be added to the mixture where doing so produces a more likely model. The resulting BRT model is easier to interpret by the practitioner than the alternatives.

Table 8.3 Log_{10} of the number of partitions used by the maximum-likelihood tree, with standard errors.

Dataset	BHC-DP	BRT-DP	BHC-γ	BRT-γ	BRT
toy	4 ± 0.0	4 ± 0.0	4 ± 0.0	4 ± 0.0	$\mathbf{1} \pm 0.0$
spambase	14 ± 0.1	14 ± 0.1	14 ± 0.1	14 ± 0.1	$\mathbf{7} \pm 0.1$
digits024	15 ± 0.1	15 ± 0.1	15 ± 0.1	13 ± 0.1	$\mathbf{6} \pm 0.1$
digits	17 ± 0.1	17 ± 0.1	17 ± 0.1	16 ± 0.1	$\mathbf{8} \pm 0.1$
newsgroups	14 ± 0.1	14 ± 0.1	14 ± 0.1	13 ± 0.1	$\mathbf{7} \pm 0.1$

Table 8.4 Log predictive probabilities on 10 % missing data and standard errors.

Dataset	BHC-DP	BRT-DP	BHC-γ	BRT-γ	BRT
`toy`	-14.7 ± 0.7	-14.7 ± 0.7	-14.4 ± 0.6	-14.6 ± 0.6	$\mathbf{-14.3} \pm 0.6$
`spambase`	-190 ± 1.7	$\mathbf{-187} \pm 1.9$	-192 ± 1.7	-192 ± 1.7	-190 ± 1.6
`digits024`	-347 ± 2.1	-345 ± 2.3	-345 ± 2.2	-345 ± 2.2	$\mathbf{-343} \pm 2.1$
`digits`	-372 ± 2.6	-371 ± 2.7	$\mathbf{-369} \pm 2.7$	-371 ± 2.6	-370 ± 2.8
`newsgroups`	-1122 ± 11	-1122 ± 11	$\mathbf{-1114} \pm 11$	$\mathbf{-1114} \pm 11$	$\mathbf{-1114} \pm 11$

8.7.3 Partially observed data

We compared the predictive probabilities of trees found by BHC-γ, BHC-DP, BRT-γ, BRT-DP and BRT on the same five binary-valued datasets as in the previous section, but with partially observed data; 10 % of the data were removed at random. The predictive probabilities, as calculated as in (8.24), are shown in Table 8.4 along with the standard errors. The predictive performance of all five models is similar, with BRT performing slightly better on `toy` and `digits024`, while BRT-DP and BHC-γ perform slightly better on `spambase` and `digits`, respectively.

8.7.4 Psychological hierarchies

The dataset of Figure 8.8 is from Cree and McRae (2003). The dataset consists of a matrix whose rows correspond to objects and whose columns correspond to attributes. The elements of the matrix are binary and indicate whether or not a particular object has a particular attribute. There are 60 objects and 100 attributes (such as used for transportation, has legs, has seeds, is cute). Figure 8.8 shows the trees found by BRT and BHC-γ by clustering the 60 objects.

One noticeable oddity of the hierarchical clustering produced by BRT is that lions and tigers inhabit their own cluster that is divorced from the other animals. The features of the dataset also include is it ferocious?, does it roar?, which only lions and tigers have, while they share few attributes in common with other animals in this dataset: this is why they lie on a distinct branch. Removing the attributes is it ferocious?, does it roar? and lives in the wilderness (shared only with deer) causes lions and tigers to be included along with other animals.

This figure shows how BRT not only finds simpler, easier-to-interpret hierarchies than BHC-γ but also more probable explanations of the data. BHC-DP, BRT-DP and BRT-γ also find similarly less probable and more complicated hierarchies than BRT.

8.7.5 Hierarchies of Gaussian process experts

Figure 8.9 shows fits by a single Gaussian process, a BHC-γ mixture of Gaussian process experts and a BRT mixture of Gaussian process experts to multimodal data consisting of two noisy interlaced sine waves. The background of the figure is shaded according to the predictive density of the model: for BHC-γ and BRT this

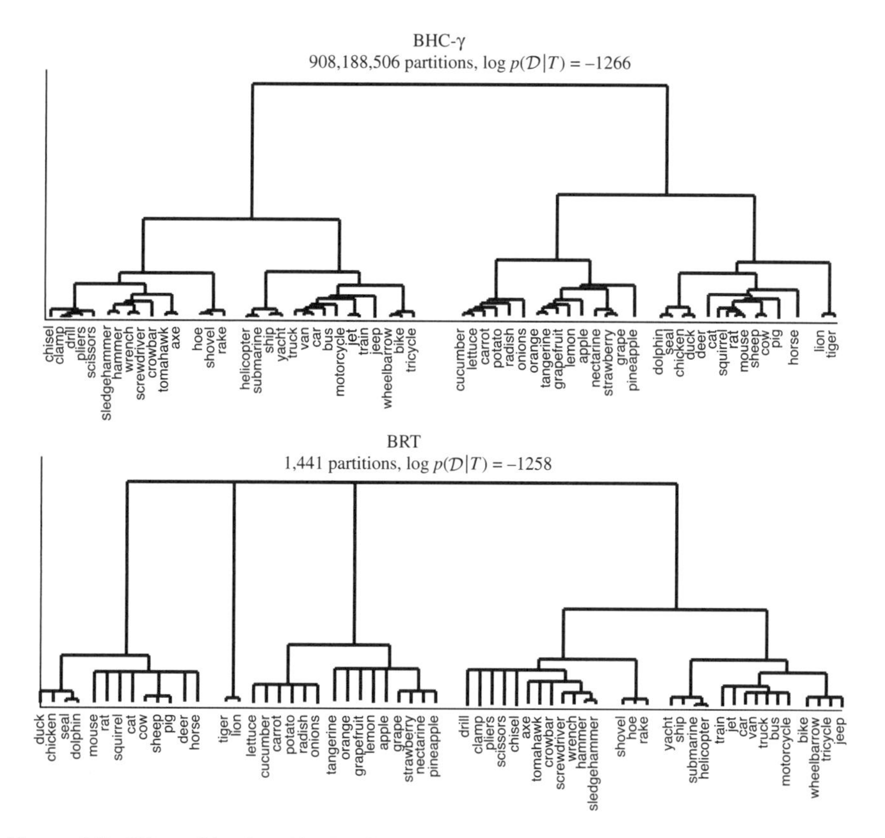

Figure 8.8 Hierarchies found by BHC-γ and BRT on a 60 objects (with 100 binary features) dataset.

is calculated as in (8.25). The solid lines on the Gaussian process and BRT plots correspond to the posterior means of the clusters. The BRT model finds a simpler and more probable explanation of the data than the alternative models: two primary clusters identify with each sine wave and a root cluster ties the data together.

8.8 Discussion

We have described a model and developed an algorithm for performing efficient, nonbinary hierarchical clustering. Our Bayesian rose tree approach is based on model selection: each tree is associated with a mixture of partitions of the dataset, and a greedy agglomerative algorithm finds trees that have high marginal likelihood under the data.

Bayesian rose trees are a departure from the common binary trees for hierarchical clustering. The flexibility implied by a mixture model over partitions with tree consistency is used in Bayesian rose trees to allow mixture models with fewer components, and thus simpler explanations of the data, than those afforded by BHC.

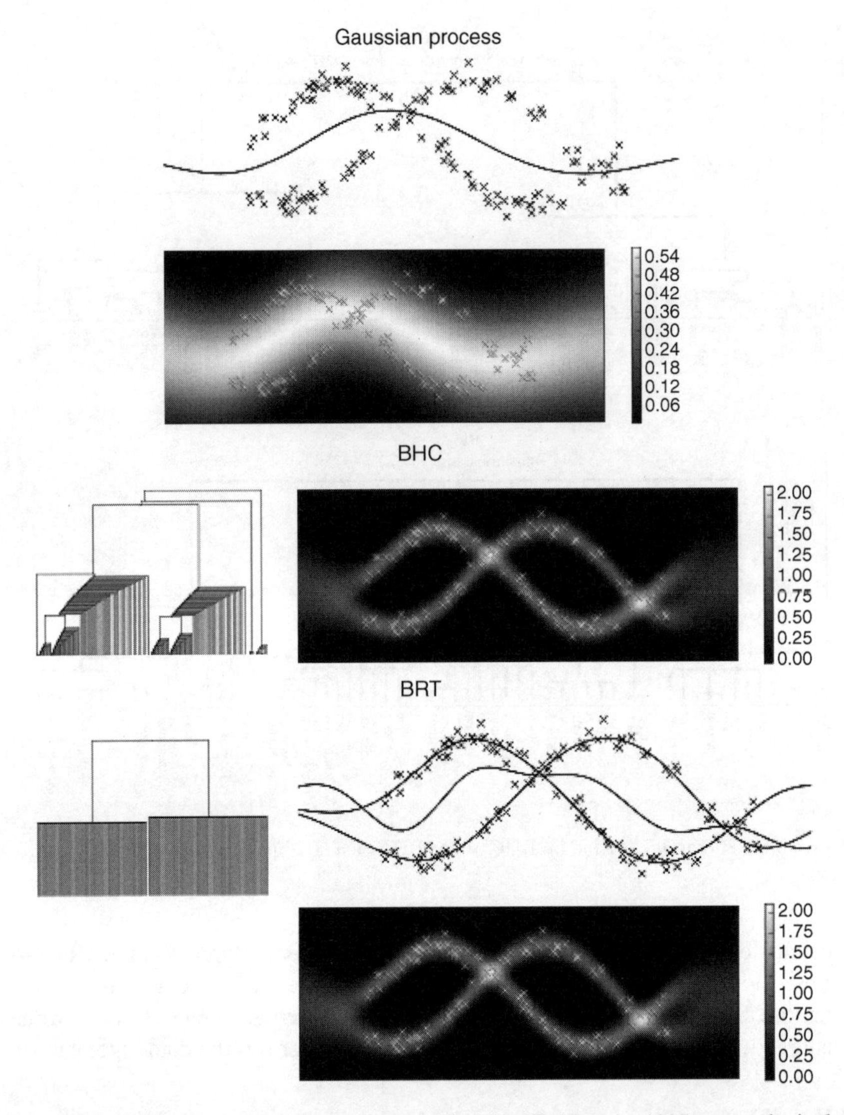

Figure 8.9 A Gaussian process (top) with marginal log likelihood −1037, on synthetic data of two interlaced curves. Observations are crosses, and the line is the posterior mean function of the GP. The background grey scale indicates the predictive density of the GP (scale on the left of the predictive plot). A BHC mixture of GP experts (middle) with marginal log likelihood: −801, consisting of 149 nonsingleton clusters in 7 527 281 partitions. To the left is the tree found by BHC. Finally, a BRT mixture of GP experts (bottom) with marginal log-likelihood 59, consisting of 3 nonsingleton clusters in five partitions. The background grey scale indicates corresponding density $p(\mathcal{D}^*|\mathcal{D})$ (as defined in Heller, 2008) via the scale on the corresponding left of the plot. The posterior mean functions passing through the data points correspond to the two subtrees found by BRT, while the third posterior mean function corresponds to the GP at the root of the tree.

The BRT mixture proportions are designed so that simpler models that explain the data are favoured over more complicated ones: this is in contrast to BHC-DP, where forced binary merges create an extra, spurious structure that is not supported by the data.

We have demonstrated in our experiments that this algorithm finds simple models that explain both synthetic and real-world data. On all datasets considered, our Bayesian rose tree algorithm found a rose tree with higher marginal likelihood under the data than Bayesian hierarchical clustering (BHC), as well as significantly simpler explanations of the data, in terms of the number of partitions. We built BRTs using two likelihood models, a beta-Bernoulli and a Gaussian process expert. In both cases the model yielded reasonable mixtures of partitions of the data.

Our use of BRT for nonparametric conditional density estimation is a proof of concept. BRT offers an attractive means of fitting a mixture of GP experts compared to sampling (Meeds and Osindero, 2006; Rasmussen and Ghahramani, 2002): with sampling one is never sure when the stationary distribution is attained, while the BRT algorithm is guaranteed to terminate after a greedy pass through the dataset, constructing a reasonably good estimate of the conditional density. Note, however, that the run time of the current algorithm is $O(n^5 \log n)$, where n is the number of data items. The additional $O(n^3)$ factor is due to the unoptimised GP computations. An interesting future project would be to make the computations more efficient using recent advanced approximations.

Another direction for BRT is to adapt the model and inference to produce more than one tree. For example, co-clustering both data items and their features simultaneously (a probabilistic version of Hartigan, 1972) might allow a richer interpretation of the data. Alternatively, one might imagine that for some data it is possible to extract multiple hierarchical facets by producing several equally plausible (though different) hierarchies of the same data.

It might also be possible to simplify further the tree produced by BRT: for large amounts of data it can be unwieldy to interpret the hierarchy produced. A common approach is to 'cut' the hierarchy to produce a flat clustering, but this necessarily removes the hierarchical information found. Instead one might adapt an approach similar to that of Aldous *et al.* (2008), where a coarser subhierarchy is extracted in a probabilistic fashion.

Nonparametric Bayesian priors also exist for rose trees, such as Pitman (1999) and Bertoin (2001), and in future these could be used for nonbinary hierarchical clustering that is able to take structure uncertainty into account as well as leverage the nonparametric nature to predict unobserved data items directly.

References

Aldous, D., Krikun, M. and Popovic, L. (2008) Stochastic models for phylogenetic trees on higher-order taxa. *Journal of Mathematical Biology*, **56**, 525–557.

Barry, D. and Hartigan, J. A. (1992) Product partition models for change point problems. *Annals of Statistics*, **20**, 260–279.

Bertoin, J. (2001) Homogeneous fragmentation processes. *Probability Theory and Related Fields*, **121**, 301–318.

Bird, R. (1998) *Introduction to Functional Programming using Haskell*, 2nd edn. Prentice Hall.

Blei, D. M., Griffiths, T. L. and Jordan, M. I. (2010) The nested Chinese restaurant process and Bayesian nonparametric inference of topic hierarchies. *Journal of the Association for Computing Machines*, **57**, 1–30.

Camin, J. H. and Sokal, R. R. (1965) A method for deducing branching sequences in phylogeny. *Evolution*, **19**, 311–326.

Cree, G. S. and McRae, K. (2003) Analyzing the factors underlying the structure and computation of the meaning of chipmunk, cherry, chisel, cheese and cello (and many other such concrete nouns). *Journal of Experimental Psychology: General*, **132**, 163–201.

Duda, R. O. and Hart, P. E. (1973) *Pattern Classification and Scene Analysis*. John Wiley & Sons, Inc., New York.

Felsenstein, J. (1973) Maximum likelihood estimation of evolutionary trees from continuous characters. *American Journal of Human Genetics*, **25**, 471–492.

Felsenstein, J. (1981) Evolutionary trees from DNA sequences: a maximum likelihood approach. *Journal of Molecular Evolution*, **17**, 368–376.

Felsenstein, J. (2003) *Inferring Phylogenies*. Sinauer Associates.

Fitch, W. M. and Margoliash, E. (1967) Construction of phylogenetic trees. *Science*, **155**, 279–284.

Frank, A. and Asuncion, A. (2010) UCI machine learning repository.

Friedman, N. (2003) Pcluster: Probabilistic agglomerative clustering of gene expression profiles. Technical Report 2003-80, Hebrew University.

Girvan, M. and Newman, M. E. J. (2002) Community structure in social and biological networks. In *Proceedings of the National Academy of Sciences of the USA*, Vol. 99, pp. 7821–7826.

Hartigan, J. A. (1972) Direct clustering of a data matrix. *Journal of the American Statistical Association*, **67**, 123–129.

Hartigan, J. A. (1975) *Clustering Algorithms*. John Wiley & Sons, Ltd.

Heller, K. A. (2008) Efficient Bayesian methods for clustering. PhD thesis, Gatsby Computational Neuroscience Unit, University College London.

Heller, K. A. and Ghahramani, Z. (2005) Bayesian hierarchical clustering. In *International Conference on Machine Learning*, Vol. 22.

Huelsenbeck, J. P. and Ronquist, F. (2001) MRBAYES: Bayesian inference of phylogenetic trees. *Bioinformatics*, **17**, 754–755.

Jain, A. K., Murty, M. N. and Flynn, P. J. (1999) Data clustering: a review. *ACM Computer Surveys*, **31**, 264–323.

Kaufman, L. and Rousseeuw, P. J. (1990) *Finding Groups in Data*. John Wiley & Sons, Ltd.

Kemp, C. C., Griffiths, T. L., Stromsten, S. and Tenenbaum, J. B. (2004) Semi-supervised learning with trees. In *Advances in Neural Information Processing Systems*, Vol. 16 (eds S. Thrun, L. Saul and B. Schölkopf). MIT Press.

McCallum, A. K. (1996) Bow: a toolkit for statistical language modelling, text retrieval, classification and clustering. http://www.cs.cmu.edu/ mccallum/bow.

McLachlan, G. J. and Basford, K. E. (1988) *Mixture Models: Inference and Applications to Clustering*. Marcel Dekker.

Meeds, E. and Osindero, S. (2006) An alternative infinite mixture of Gaussian process experts. In *Advances in Neural Information Processing Systems*, Vol. 18 (eds Y. Weiss, B. Schölkopf and J. Platt), pp. 883–890. MIT Press.

Meertens, L. (1988) First steps towards the theory of rose trees. Working Paper 592 ROM-25, IFIP Working Group 2.1.

Müller, P. and Quintana, F. A. (2010) Random partition models with regression on covariates. *Journal of Statistical Inference and Planning*, **140**, 2801–2808.

Murtagh, F. (1983) A survey of recent advances in hierarchical clustering algorithms. *Computer Journal*, **26**, 254–259.

Neal, R. M. (2003) Density modelling and clustering using Dirichlet diffusion trees. In *Bayesian Statistics 7* (eds J. M. Bernardo, M. J. Bayarri, J. O. Berger, A. P. Dawid, D. Heckerman, A. F. M. Smith and M. West), pp. 619–629. Oxford University Press.

Pitman, J. (1999) Coalescents with multiple collisions. *Annals of Probability*, **27**, 1870–1902.

Rasmussen, C. E. and Ghahramani, Z. (2002) Infinite mixtures of Gaussian process experts. In *Advances in Neural Information Processing Systems*, Vol. 14 (eds T. G. Dietterich, S. Becker and Z. Ghahramani), pp. 881–888. MIT Press.

Rogers, J. S. (1997) On the consistency of maximum likelihood estimation of phylogenetic trees from nucleotide sequences. *Systems Biology*, **42**, 354–357.

Rosch, E., Mervis, C., Gray, W., Johnson, D. and Boyes-Braem, P. (1976) Basic objects in natural categories. *Cognitive Psychology*, **8** 382–439.

Roy, D. M., Kemp, C., Mansinghka, V. and Tenenbaum, J. B. (2007) Learning annotated hierarchies from relational data. In *Advances in Neural Information Processing Systems*, Vol. 19 (eds B. Chölkopf, J. Platt and T. Hoffman), pp. 1185–1192. MIT Press.

Saitou, N. and Nei, M. (1987) The neighbor-joining method: a new method for reconstructing phylogenetic trees. *Molecular Biology and Evolution*, **4**, 406–425.

Segal, E., Koller, D. and Ormoneit, D. (2002) Probabilistic abstraction hierarchies. In *Advances in Neural Information Processing Systems*, Vol. 14 (eds T. G. Dietterich, S. Becker and Z. Ghahramani). MIT Press.

Studier, J. A. and Keppler, K. J. (1988) A note on the neighbor-joining algorithm of Saitou and Nei. *Molecular Biology and Evolution*, **14**, 210–211.

Teh, Y. W., Daume III, H. and Roy, D. M. (2008) Bayesian agglomerative clustering with coalescents. In *Advances in Neural Information Processing Systems*, Vol. 20 (eds J. C. Platt, D. Koller, Y. Singer and S. Roweis), pp. 1473–1480. MIT Press.

Vaithyanathan, S. and Dom, B. (2000) Model-based hierarchical clustering. In *Proceedings of the 16th Conference on Uncertainty in Artificial Intelligence*, pp. 599–608. UAI Press.

Williams, C. K. I. (2000) A MCMC approach to hierarchical mixture modelling. In *Advances in Neural Information Processing Systems*, Vol. 12 (eds S. A. Solla, T. K. Leen and K.-R. Muller). MIT Press.

Williams, C. and Seeger, M. (2001) Using the Nyström method to speed up kernel methods. In *Advances in Neural Information Processing*, Vol. 13 (eds T. K. Leen, T. G. Diettrich, and V. Tresp). MIT Press.

Xu, Y., Heller, K. A. and Ghahramani, Z. (2009) Tree-based inference for Dirichlet process mixtures. In *JMLR Workshop and Conference Proceedings Volume 5: AISTATS 2009*, (eds D. van Dyk and M. Welling), pp. 623–630.

Yang, Z. (1994) Statistical properties of the maximum likelihood method of phylogenetic estimation and comparison with distance matrix methods. *Systematic Biology*, **43**, 329–342.

Yang, Z. and Rannala, B. (1997) Bayesian phylogenetic inference using DNA sequences: a Markov chain Monte Carlo method. *Molecular Biology and Evolution*, **14**, 717–724.

9

Mixtures of factor analysers for the analysis of high-dimensional data

Geoffrey J. McLachlan, Jangsun Baek and Suren I. Rathnayake

9.1 Introduction

Finite mixture models are being increasingly used to model the distributions of a wide variety of random phenomena and to cluster datasets; see, for example, McLachlan and Peel (2000a). Let

$$\mathbf{Y} = (Y_1, \ldots, Y_p)^T \tag{9.1}$$

be a p-dimensional vector of feature variables. For continuous features Y_j, the density of $\mathbf{Y}$ can be modelled by a mixture of a sufficiently large enough number g of multivariate normal component distributions,

$$f(\mathbf{y}; \boldsymbol{\Psi}) = \sum_{i=1}^{g} \pi_i \, \phi(\mathbf{y}; \boldsymbol{\mu}_i, \boldsymbol{\Sigma}_i), \tag{9.2}$$

where $\phi(\mathbf{y}; \boldsymbol{\mu}, \boldsymbol{\Sigma})$ denotes the p-variate normal density function with mean $\boldsymbol{\mu}$ and covariance matrix $\boldsymbol{\Sigma}$. Here the vector $\boldsymbol{\Psi}$ of unknown parameters consists of the mixing proportions π_i, the elements of the component means $\boldsymbol{\mu}_i$ and the distinct elements of the component-covariance matrices $\boldsymbol{\Sigma}_i$ $(i = 1, \ldots, g)$.

Mixtures: Estimation and Applications, First Edition. Edited by Kerrie L. Mengersen, Christian P. Robert and D. Michael Titterington.

The parameter vector $\mathbf{\Psi}$ can be estimated by maximum likelihood. For an observed random sample, $\mathbf{y}_1, \ldots, \mathbf{y}_n$, the log-likelihood function for $\mathbf{\Psi}$ is given by

$$\log L(\mathbf{\Psi}) = \sum_{j=1}^{n} \log f(\mathbf{y}_j;\ \mathbf{\Psi}). \tag{9.3}$$

The maximum-likelihood estimate (MLE) of $\mathbf{\Psi}$, $\hat{\mathbf{\Psi}}$ is given by an appropriate root of the likelihood equation,

$$\partial \log L(\mathbf{\Psi})/\partial \mathbf{\Psi} = \mathbf{0}. \tag{9.4}$$

Solutions of (9.4) corresponding to local maximisers of $\log L(\mathbf{\Psi})$ can be obtained via the expectation maximisation (EM) algorithm (Dempster *et al.*, 1977; see also McLachlan and Krishnan, 2008).

Besides providing an estimate of the density function of $\mathbf{Y}$, the normal mixture model (9.2) provides a probabilistic clustering of the observed data $\mathbf{y}_1, \ldots, \mathbf{y}_n$ into g clusters in terms of their estimated posterior probabilities of component membership of the mixture. The posterior probability $\tau_i(\mathbf{y}_j;\ \mathbf{\Psi})$ that the jth feature vector with observed value $\mathbf{y}_j$ belongs to the ith component of the mixture can be expressed by Bayes' theorem as

$$\tau_i(\mathbf{y}_j;\ \mathbf{\Psi}) = \frac{\pi_i \phi(\mathbf{y}_j;\ \boldsymbol{\mu}_i, \mathbf{\Sigma}_i)}{\sum_{h=1}^{g} \pi_h \phi(\mathbf{y}_j;\ \boldsymbol{\mu}_h, \mathbf{\Sigma}_h)} \qquad (i = 1, \ldots, g;\ j = 1, \ldots, n). \tag{9.5}$$

An outright assignment of the data is obtained by assigning each data point $\mathbf{y}_j$ to the component to which it has the highest estimated posterior probability of belonging.

The g-component normal mixture model (9.2) with unrestricted component-covariance matrices is a highly parameterised model with $d = \frac{1}{2}p(p+1)$ parameters for each component-covariance matrix $\mathbf{\Sigma}_i$ $(i = 1, \ldots, g)$. Banfield and Raftery (1993) introduced a parameterisation of the component-covariance matrix $\mathbf{\Sigma}_i$ based on a variant of the standard spectral decomposition of $\mathbf{\Sigma}_i$ $(i = 1, \ldots, g)$. However, if p is large relative to the sample size n, it may not be possible to use this decomposition to infer an appropriate model for the component-covariance matrices. Even if it is possible, the results may not be reliable due to potential problems with near-singular estimates of the component-covariance matrices when p is large relative to n.

A common approach to reducing the the number of dimensions is to perform a principal component analysis (PCA). As is well known, projections of the feature data $\mathbf{y}_j$ on to the first few principal axes are not always useful in portraying the group structure (see McLachlan and Peel, 2000a, Chapter 8). This point was also stressed by Chang (1983), who showed in the case of two groups that the principal component of the feature vector that provides the best separation between groups in terms of Mahalanobis distance is not necessarily the first component.

Another approach for reducing the number of unknown parameters in the forms for the component-covariance matrices is to adopt the mixture of factor analysers model, as considered in McLachlan and Peel (2000a, 2000b). This model was originally proposed by Ghahramani and Hinton (1997) and Hinton *et al.* (1997) for the purposes of visualising high-dimensional data in a lower-dimensional space to explore for group structure; see also Tipping and Bishop (1997) who considered the related model of mixtures of principal component analysers for the same purpose. Further references may be found in McLachlan *et al.* (2003, 2007) and Baek *et al.* (2010).

In this chapter, we focus on the use of mixtures of factor analysers to reduce the number of parameters in the specification of the component-covariance matrices.

9.2 Single-factor analysis model

Factor analysis is commonly used for explaining data, in particular correlations between variables in multivariate observations. It can be used also for dimensionality reduction. In a typical factor analysis model, each observation $\mathbf{Y}_j$ is modelled as

$$\mathbf{Y}_j = \boldsymbol{\mu} + \boldsymbol{B}\boldsymbol{U}_j + \boldsymbol{e}_j \quad (j = 1, \ldots, n), \tag{9.6}$$

where $\boldsymbol{U}_j$ is a q-dimensional $(q < p)$ vector of latent or unobservable variables called factors and $\boldsymbol{B}$ is a $p \times q$ matrix of factor loadings (parameters). The $\boldsymbol{U}_j$ are assumed to be i.i.d. as $N(\mathbf{0}, \boldsymbol{I}_q)$, independently of the errors $\boldsymbol{e}_j$, which are assumed to be i.i.d. as $N(\mathbf{0}, \boldsymbol{D})$, where $\boldsymbol{D}$ is a diagonal matrix,

$$\boldsymbol{D} = \text{diag}\,(\sigma_1^2, \ldots, \sigma_p^2),$$

and where $\boldsymbol{I}_q$ denotes the $q \times q$ identity matrix. Thus, conditional on $\boldsymbol{U}_j = \boldsymbol{u}_j$, the $\mathbf{Y}_j$ are independently distributed as $N(\boldsymbol{\mu} + \boldsymbol{B}\,\boldsymbol{u}_j,\ \boldsymbol{D})$. Unconditionally, the $\mathbf{Y}_j$ are i.i.d. according to a normal distribution with mean $\boldsymbol{\mu}$ and covariance matrix

$$\boldsymbol{\Sigma} = \boldsymbol{B}\boldsymbol{B}^T + \boldsymbol{D}. \tag{9.7}$$

If q is chosen sufficiently smaller than p, the representation (9.7) imposes some constraints on the component-covariance matrix $\boldsymbol{\Sigma}$ and thus reduces the number of free parameters to be estimated. Note that in the case of $q > 1$, there is an infinity of choices for $\boldsymbol{B}$, since (9.7) is still satisfied if $\boldsymbol{B}$ is replaced by $\boldsymbol{B}\boldsymbol{C}$, where $\boldsymbol{C}$ is any orthogonal matrix of order q. One (arbitrary) way of uniquely specifying $\boldsymbol{B}$ is to choose the orthogonal matrix $\boldsymbol{C}$ so that $\boldsymbol{B}^T\boldsymbol{D}^{-1}\boldsymbol{B}$ is diagonal (with its diagonal elements arranged in decreasing order); see Lawley and Maxwell (1971, Chapter 1). Assuming that the eigenvalues of $\boldsymbol{B}\boldsymbol{B}^T$ are positive and distinct, the condition that $\boldsymbol{B}^T\boldsymbol{D}^{-1}\boldsymbol{B}$ is diagonal as above imposes $\frac{1}{2}q(q-1)$ constraints on the parameters. Hence then the number of free parameters is $pq + p - \frac{1}{2}q(q-1)$.

The factor analysis model (9.6) can be fitted by the EM algorithm and its variants, which is discussed in the Appendix for the more general case of mixtures of such models. Note that with the factor analysis model, we avoid having to compute the inverses of iterates of the estimated $p \times p$ covariance matrix $\boldsymbol{\Sigma}$, which may be singular for large p relative to n. This is because the inversion of the current value of the $p \times p$ matrix $(\boldsymbol{BB}^T + \boldsymbol{D})$ on each iteration can be undertaken using the result that

$$(\boldsymbol{BB}^T + \boldsymbol{D})^{-1} = \boldsymbol{D}^{-1} - \boldsymbol{D}^{-1}\boldsymbol{B}(\boldsymbol{I}_q + \boldsymbol{B}^T\boldsymbol{D}^{-1}\boldsymbol{B})^{-1}\boldsymbol{B}^T\boldsymbol{D}^{-1}, \tag{9.8}$$

where the right-hand side of (9.8) involves only the inverses of $q \times q$ matrices, since $\boldsymbol{D}$ is a diagonal matrix. The determinant of $(\boldsymbol{BB}^T + \boldsymbol{D})$ can then be calculated as

$$| \boldsymbol{BB}^T + \boldsymbol{D} |=| \boldsymbol{D} | / | \boldsymbol{I}_q - \boldsymbol{B}^T(\boldsymbol{BB}^T + \boldsymbol{D})^{-1}\boldsymbol{B} | .$$

Unlike the PCA model, the factor analysis model (9.6) enjoys a powerful invariance property: changes in the scales of the feature variables in $\mathbf{y}_j$ appear only as scale changes in the appropriate rows of the matrix $\boldsymbol{B}$ of factor loadings.

9.3 Mixtures of factor analysers

A global nonlinear approach can be obtained by postulating a finite mixture of linear submodels for the distribution of the full observation vector $\mathbf{Y}_j$ given the (unobservable) factors $\boldsymbol{u}_j$; that is we can provide a local dimensionality reduction method by assuming that, conditional on its membership of the ith component of a mixture, the distribution of the observation $\mathbf{Y}_j$ can be modelled as

$$\mathbf{Y}_j = \boldsymbol{\mu}_i + \boldsymbol{B}_i\boldsymbol{U}_{ij} + \boldsymbol{e}_{ij} \qquad \text{with prob. } \pi_i \quad (i = 1, \ldots, g) \tag{9.9}$$

for $j = 1, \ldots, n$, where the factors $\boldsymbol{U}_{i1}, \ldots, \boldsymbol{U}_{in}$ are distributed independently $N(\boldsymbol{0}, \boldsymbol{I}_q)$, independently of the $\boldsymbol{e}_{ij}$, which are distributed independently $N(\boldsymbol{0}, \boldsymbol{D}_i)$, where $\boldsymbol{D}_i$ is a diagonal matrix $(i = 1, \ldots, g)$.

Thus the mixture of factor analysers model is given by (9.2), where the ith component-covariance matrix $\boldsymbol{\Sigma}_i$ has the form

$$\boldsymbol{\Sigma}_i = \boldsymbol{B}_i\boldsymbol{B}_i^T + \boldsymbol{D}_i \quad (i = 1, \ldots, g), \tag{9.10}$$

where $\boldsymbol{B}_i$ is a $p \times q$ matrix of factor loadings and $\boldsymbol{D}_i$ is a diagonal matrix $(i = 1, \ldots, g)$. The parameter vector $\boldsymbol{\Psi}$ now consists of the elements of the $\boldsymbol{\mu}_i$, the $\boldsymbol{B}_i$ and the $\boldsymbol{D}_i$, along with the mixing proportions π_i $(i = 1, \ldots, g - 1)$, on putting $\pi_g = 1 - \sum_{i=1}^{g-1} \pi_i$.

As $\frac{1}{2}q(q - 1)$ constraints are needed for $\boldsymbol{B}_i$ to be uniquely defined, the number of free parameters in (9.10) is

$$pq + p - \frac{1}{2}q(q - 1). \tag{9.11}$$

Thus with the representation (9.10), the reduction in the number of parameters for $\boldsymbol{\Sigma}_i$ is

$$\begin{aligned} r &= \frac{1}{2}p(p+1) - pq - p + \frac{1}{2}q(q-1) \\ &= \frac{1}{2}\{(p-q)^2 - (p+q)\}, \end{aligned} \tag{9.12}$$

assuming that q is chosen sufficiently smaller than p so that this difference is positive. The total number of parameters is

$$d_1 = (g-1) + 2gp + g\{pq - \tfrac{1}{2}q(q-1)\}. \tag{9.13}$$

We shall refer to this approach as MFA (mixtures of factor analysers).

9.4 Mixtures of common factor analysers (MCFA)

Even with this MFA approach, the number of parameters still might not be manageable, particularly if the number of dimensions p is large and/or the number of components (clusters) g is not small.

Baek and McLachlan (2008) and Baek *et al.* (2010) considered how this factor-analytic approach can be modified to provide a greater reduction in the number of parameters. They proposed to specify the normal mixture model (9.2) with the restrictions

$$\boldsymbol{\mu}_i = \boldsymbol{A}\boldsymbol{\xi}_i \quad (i = 1, \ldots, g) \tag{9.14}$$

and

$$\boldsymbol{\Sigma}_i = \boldsymbol{A}\boldsymbol{\Omega}_i\boldsymbol{A}^T + \boldsymbol{D} \quad (i = 1, \ldots, g), \tag{9.15}$$

where $\boldsymbol{A}$ is a $p \times q$ matrix, $\boldsymbol{\xi}_i$ is a q-dimensional vector, $\boldsymbol{\Omega}_i$ is a $q \times q$ positive definite symmetric matrix and $\boldsymbol{D}$ is a diagonal $p \times p$ matrix.

As to be made more precise in the next section, $\boldsymbol{A}$ is a matrix of loadings on q unobservable factors. The representation (9.14) and (9.15) is not unique, as it still holds if $\boldsymbol{A}$ were to be postmultiplied by any nonsingular matrix. Hence the number of free parameters in $\boldsymbol{A}$ is

$$pq - q^2. \tag{9.16}$$

Thus with the restrictions (9.14) and (9.15) on the component mean $\boldsymbol{\mu}_i$ and covariance matrix $\boldsymbol{\Sigma}_i$, respectively, the total number of parameters is reduced to

$$d_2 = (g-1) + p + q(p+g) + \tfrac{1}{2}gq(q+1) - q^2. \tag{9.17}$$

Baek and McLachlan (2008) termed this approach mixtures of common factor analysers (MCFA). This is because the matrix of factor loadings is common to the components before the component-specific rotation of the component factors to make them white noise. Note that the component-factor loadings are not common after this rotation, as in McNicholas and Murphy (2008). Baek and McLachlan (2008) showed for this approach how the EM algorithm can be implemented to fit this normal mixture model under the constraints (9.14) and (9.15). It can also be used to provide lower-dimensional plots of the data $\mathbf{y}_j$ $(j = 1, \ldots, n)$. It thus provides an alternative to canonical variates, which are calculated from the clusters under the assumption of equal component-covariance matrices.

In our implementation of this procedure, we follow Baek and McLachlan (2008) and postmultiply the solution $\hat{\boldsymbol{A}}$ for $\boldsymbol{A}$ by the nonsingular matrix as defined in the Appendix, which achieves the result

$$\hat{\boldsymbol{A}}^T\hat{\boldsymbol{A}} = \boldsymbol{I}_q, \tag{9.18}$$

where $\boldsymbol{I}_q$ denotes the $q \times q$ identity matrix; that is the p columns of $\hat{\boldsymbol{A}}$ are taken to be orthonormal. This solution is unique up to postmultiplication by an orthogonal matrix.

The MCFA approach, with its constraints (9.14) and (9.15) on the g component means and covariance matrices $\boldsymbol{\mu}_i$ and $\boldsymbol{\Sigma}_i$ $(i = 1, \ldots, g)$, can be viewed as a special case of the MFA approach. To see this we first note that the MFA approach with the factor-analytic representation (9.10) on $\boldsymbol{\Sigma}_i$ is equivalent to assuming that the distribution of the difference $\mathbf{Y}_j - \boldsymbol{\mu}_i$ can be modelled as

$$\mathbf{Y}_j - \boldsymbol{\mu}_i = \boldsymbol{B}_i\boldsymbol{U}_{ij} + \boldsymbol{e}_{ij} \quad \text{with prob. } \pi_i \quad (i = 1, \ldots, g) \tag{9.19}$$

for $j = 1, \ldots, n$, where the (unobservable) factors $\boldsymbol{U}_{i1}, \ldots, \boldsymbol{U}_{in}$ are distributed independently $N(\mathbf{0}, \boldsymbol{I}_q)$, independently of the $\boldsymbol{e}_{ij}$, which are distributed independently $N(\mathbf{0}, \boldsymbol{D}_i)$, where $\boldsymbol{D}_i$ is a diagonal matrix $(i = 1, \ldots, g)$.

This model may not lead to a sufficiently large enough reduction in the number of parameters, particularly if g is not small. Hence Baek and McLachlan (2008) and Baek *et al.* (2010) proposed the MCFA approach, whereby the ith component-conditional distribution of $\mathbf{Y}_j$ is modelled as

$$\mathbf{Y}_j = \boldsymbol{A}\boldsymbol{U}_{ij} + \boldsymbol{e}_{ij} \quad \text{with prob. } \pi_i \quad (i = 1, \ldots, g) \tag{9.20}$$

for $j = 1, \ldots, n$, where the (unobservable) factors $\boldsymbol{U}_{i1}, \ldots, \boldsymbol{U}_{in}$ are distributed independently $N(\boldsymbol{\xi}_i, \boldsymbol{\Omega}_i)$, independently of the $\boldsymbol{e}_{ij}$, which are distributed independently $N(\mathbf{0}, \boldsymbol{D})$, where $\boldsymbol{D}$ is a diagonal matrix $(i = 1, \ldots, g)$. Here $\boldsymbol{A}$ is a $p \times q$ matrix of factor loadings, which is taken to satisfy the relationship (9.18).

To see that the model underlying the MCFA approach as specified by (9.20) is a special case of the MFA approach as specified by (9.19), we note that we can rewrite (9.20) as

$$\begin{aligned}
\mathbf{Y}_j &= \boldsymbol{A}\boldsymbol{U}_{ij} + \boldsymbol{e}_{ij} \\
&= \boldsymbol{A}\boldsymbol{\xi}_i + \boldsymbol{A}(\boldsymbol{U}_{ij} - \boldsymbol{\xi}_i) + \boldsymbol{e}_{ij} \\
&= \boldsymbol{\mu}_i + \boldsymbol{A}\boldsymbol{K}_i\boldsymbol{K}_i^{-1}(\boldsymbol{U}_{ij} - \boldsymbol{\xi}_i) + \boldsymbol{e}_{ij} \\
&= \boldsymbol{\mu}_i + \boldsymbol{B}_i\boldsymbol{U}_{ij}^* + \boldsymbol{e}_{ij},
\end{aligned} \tag{9.21}$$

where

$$\mu_i = A\xi_i, \tag{9.22}$$
$$B_i = AK_i, \tag{9.23}$$
$$U_{ij}^* = K_i^{-1}(U_{ij} - \xi_i), \tag{9.24}$$

and where the U_{ij}^* are distributed independently $N(\mathbf{0},\ I_q)$. The covariance matrix of U_{ij}^* is equal to I_q, since K_i can be chosen so that

$$K_i^{-1}\Omega_i K_i^{-1^T} = I_q \quad (i = 1,\ \ldots,\ g). \tag{9.25}$$

On comparing (9.21) with (9.19), it can be seen that the MCFA model is a special case of the MFA model with the additional restrictions that

$$\mu_i = A\xi_i \quad (i = 1,\ \ldots,\ g), \tag{9.26}$$

$$B_i = AK_i \quad (i = 1,\ \ldots,\ g), \tag{9.27}$$

and

$$D_i = D \quad (i = 1,\ \ldots,\ g). \tag{9.28}$$

The latter restriction of equal diagonal covariance matrices for the component-specific error terms ($D_i = D$) is sometimes imposed with applications of the MFA approach to avoid potential singularities with small clusters (see McLachlan *et al.*, 2003).

Concerning the restriction (9.27) that the matrix of factor of loadings is equal to AK_i for each component, it can be viewed as adopting common factor loadings before the use of the transformation K_i to transform the factors so that they have unit variances and zero covariances, and hence the name mixtures of common factor analysers. It is also different to the MFA approach in that it considers the factor-analytic representation of the observations $\mathbf{Y}_j$ directly, rather than the error terms $\mathbf{Y}_j - \mu_i$.

As the MFA approach allows a more general representation of the component-covariance matrices and places no restrictions on the component means it is in this sense preferable to the MCFA approach if its application is feasible given the values of p and g. If the dimension p and/or the number of components g is too large, then the MCFA provides a more feasible approach at the expense of more distributional restrictions on the data, as demonstrated in Baek *et al.* (2010). The MCFA approach also has the advantage that the latent factors in its formulation are allowed to have different means and covariance matrices and are not white noise, as with the formulation of the MFA approach. Thus the (estimated) posterior means of the factors corresponding to the observed data can be used to portray the latter in low-dimensional spaces.

9.5 Some related approaches

The MCFA approach is similar in form to the approach proposed by Yoshida *et al.* (2004, 2006), who also imposed the additional restriction that the common diagonal covariance matrix $\boldsymbol{D}$ of the error terms is spherical,

$$\boldsymbol{D} = \sigma^2 \boldsymbol{I}_p, \tag{9.29}$$

and that the component-covariance matrices of the factors are diagonal. We shall call this approach MCUFSA (mixtures of common uncorrelated factors with spherical-error analysers). The total number of parameters with this approach is

$$d_3 = (g - 1) + pq + 1 + 2gq - \tfrac{1}{2}q(q + 1). \tag{9.30}$$

This restriction of sphericity of the errors and of diagonal covariance matrices in the component distributions of the factors can have an adverse effect on the clustering of high-dimensional datasets. The relaxation of these restrictions does considerably increase the complexity of the problem of fitting the model. However, as described in the Appendix, Baek and McLachlan (2008) and Baek *et al.* (2010) showed how it can be effected via the EM algorithm with the E- and M-steps being able to be carried out in closed form.

In Table 9.1, we have listed the number of parameters to be estimated for the models with the MFA, MCFA and MCUFSA approaches when $p = 50, 100$; $q = 2$; and $g = 4, 8$. For example, when we cluster $p = 50$ dimensional gene expression data into $g = 4$ groups using $q = 2$ factors, the MFA model requires 799 parameters to be estimated, while the MCUFSA needs only 117 parameters. Moreover, as the number of clusters grows from 4 to 8 the number of parameters for the MFA model grows almost twice as large as before, but that for MCUFSA remains almost the same (137 parameters). However, as MCUFSA needs less parameters to characterise

Table 9.1 The number of parameters in models for three factor-analytic approaches.

	p	g	q	Number of parameters
MFA	50	4	2	799
	50	8	2	1599
	100	4	2	1599
	100	8	2	3199
MCFA	50	4	2	169
	50	8	2	193
	100	4	2	319
	100	8	2	343
MCUFSA	50	4	2	117
	50	8	2	137
	100	4	2	217
	100	8	2	237

the structure of the clusters, it does not always provide a good fit. It may fail to fit the data adequately as it is assuming that the component-covariance matrices of the factors are diagonal and that the cluster-error distributions conditional on the factors are spherical. The model with the MFA approach has 170 and 194 parameters for $g = 4$ and 8, respectively, with $q = 2$ factors. The MCFA approach provides a good parsimonious compromise between the MFA and MCUFSA approaches.

In the context of the analysis of speech recognition data, Rosti and Gales (2004) considered a factor-analytic approach in which separate mixture distributions are adopted independently for the factors and for the error terms conditional on the factors. It thus contains the model (9.20) as a special case where the error distribution conditional on the factors is specified by a single normal distribution. However, they developed their procedure for only diagonal component-covariance matrices for the factors, whereas in the MCFA model these factor covariance matrices have no restrictions imposed on them. Some related work in this context of speech recognition includes those of Gales (1999), Gopinath *et al.* (1998), Kumar (1997) and Olsen and Gopinath (2002). More recently, Viroli (2010) has considered an approach similar to that of Rosti and Gales (2004).

In another approach with common factor loadings adopted recently by Galimberti *et al.* (2008) for the data after mean centring, the factors in (9.20) are taken to have a mixture distribution with the constraints that its mean is the null vector and its covariance matrix is the identity matrix. As their program applies only to mean-centred data, it cannot be used to visualise the original data.

In other recent work, Sanguinetti (2008) has considered a method of dimensionality reduction in a cluster analysis context. However, its underlying model assumes sphericity in the specification of the variances/covariances of the factors in each cluster. Our proposed method allows for oblique factors, which provides the extra flexibility needed to cluster more effectively high-dimensional datasets in practice. Xie *et al.* (2010) have considered a penalised version of mixture of factor analysers. The MCFA approach can also be formed by placing a mixture distribution on the features $\boldsymbol{U}_j$ in the model (9.6); see Galimberti *et al.* (2008), Viroli (2010) and Montanari and Viroli (2010).

9.6 Fitting of factor-analytic models

The fitting of mixtures of factor analysers as with the MFA approach has been considered in McLachlan *et al.* (2003), using a variant of the EM algorithm known as the alternating expectation-conditional maximisation algorithm (AECM). With the MCFA approach, we have fitted the same mixture model of factor analysers but with the additional restrictions (9.14) and (9.15) on the component means $\boldsymbol{\mu}_i$ and covariance matrices $\boldsymbol{\Sigma}_i$. We also have to impose the restriction (9.28) of common diagonal covariance matrices $\boldsymbol{D}$. The implementation of the EM algorithm for this model is described in the Appendix. In the EM framework, the component label $\mathbf{z}_j$ associated with the observation $\mathbf{y}_j$ is introduced as missing data, where $z_{ij} = (\mathbf{z}_j)_i$ is one or zero according to whether $\mathbf{y}_j$ belongs or does not belong to the ith component of the mixture ($i = 1, \ldots, g$; $j = 1, \ldots, n$). The unobservable factors $\boldsymbol{u}_{ij}$ are also introduced as missing data in the EM framework.

As part of the E-step, we require the conditional expectation of the component labels z_{ij} ($i = 1, \ldots, g$) given the observed data point $\mathbf{y}_j$ ($j = 1, \ldots, n$). It follows that

$$\begin{aligned} E_{\boldsymbol{\Psi}}\{Z_{ij} \mid \mathbf{y}_j\} &= \mathrm{pr}_{\boldsymbol{\Psi}}\{Z_{ij} = 1 \mid \mathbf{y}_j\} \\ &= \tau_i(\mathbf{y}_j; \boldsymbol{\Psi}) \qquad (i = 1, \ldots, g;\ j = 1, \ldots, n), \end{aligned}$$

where $\tau_i(\mathbf{y}_j; \boldsymbol{\Psi})$ is the posterior probability that $\mathbf{y}_j$ belongs to the ith component of the mixture. Here $E_{\boldsymbol{\Psi}}$ and $\mathrm{pr}_{\boldsymbol{\Psi}}$ denote expectation and probability, respectively, using $\boldsymbol{\Psi}$ as the parameter vector in the conditional distribution of the complete data given the observed data in the EM framework. From (9.20), it can be expressed under the MCFA model as

$$\tau_i(\mathbf{y}_j; \boldsymbol{\Psi}) = \frac{\pi_i \phi(\mathbf{y}_j; \boldsymbol{A\xi}_i, \boldsymbol{A\Omega}_i \boldsymbol{A}^{\mathrm{T}} + \boldsymbol{D})}{\sum_{h=1}^{g} \pi_h \phi(\mathbf{y}_j; \boldsymbol{A\xi}_h, \boldsymbol{A\Omega}_h \boldsymbol{A}^{\mathrm{T}} + \boldsymbol{D})} \tag{9.31}$$

for $i = 1, \ldots, g;\ j = 1, \ldots, n$.

We also require the conditional distribution of the unobservable (latent) factors $\boldsymbol{U}_{ij}$ given the observed data $\mathbf{y}_j$ ($j = 1, \ldots, n$). The conditional distribution of $\boldsymbol{U}_{ij}$ given $\mathbf{y}_j$ and its membership of the ith component of the mixture (that is $z_{ij} = 1$) is multivariate normal,

$$\boldsymbol{U}_{ij} \mid \mathbf{y}_j, z_{ij} = 1 \ \sim N(\boldsymbol{\xi}_{ij}, \boldsymbol{\Omega}_i^*), \tag{9.32}$$

where

$$\boldsymbol{\xi}_{ij} = \boldsymbol{\xi}_i + \boldsymbol{\gamma}_i^T (\mathbf{y}_j - \boldsymbol{A\xi}_i) \tag{9.33}$$

and

$$\boldsymbol{\Omega}_i^* = (\boldsymbol{I}_q - \boldsymbol{\gamma}_i^T \boldsymbol{A})\boldsymbol{\Omega}_i, \tag{9.34}$$

and where

$$\boldsymbol{\gamma}_i = (\boldsymbol{A\Omega}_i \boldsymbol{A}^T + \boldsymbol{D})^{-1} \boldsymbol{A\Omega}_i. \tag{9.35}$$

We can portray the observed data $\mathbf{y}_j$ in q-dimensional space by plotting the corresponding values of the $\hat{\boldsymbol{u}}_{ij}$, which are estimated conditional expectations of the factors $\boldsymbol{U}_{ij}$, corresponding to the observed data points $\mathbf{y}_j$. From (9.32) and (9.33),

$$\begin{aligned} E(\boldsymbol{U}_{ij} \mid \mathbf{y}_j, z_{ij} = 1) &= \boldsymbol{\xi}_{ij} \\ &= \boldsymbol{\xi}_i + \boldsymbol{\gamma}_i^T (\mathbf{y}_j - \boldsymbol{A\xi}_i). \end{aligned} \tag{9.36}$$

We let $\hat{\boldsymbol{u}}_{ij}$ denote the value of the right-hand side of (9.36) evaluated at the maximum likelihood estimates of $\boldsymbol{\xi}_i$, $\boldsymbol{\gamma}_i$ and $\boldsymbol{A}$. We can define the estimated value $\hat{\boldsymbol{u}}_j$ of the

jth factor corresponding to $\mathbf{y}_j$ as

$$\hat{\boldsymbol{u}}_j = \sum_{i=1}^{g} \tau_i(\mathbf{y}_j;\ \hat{\boldsymbol{\Psi}})\,\hat{\boldsymbol{u}}_{ij} \quad (j = 1, \ldots, n), \tag{9.37}$$

where, from (9.31), $\tau_i(\mathbf{y}_j;\ \hat{\boldsymbol{\Psi}})$ is the estimated posterior probability that $\mathbf{y}_j$ belongs to the ith component. An alternative estimate of the posterior expectation of the factor corresponding to the jth observation $\mathbf{y}_j$ is defined by replacing $\tau_i(\mathbf{y}_j;\ \hat{\boldsymbol{\Psi}})$ by $\hat{z}_{ij}$ in (9.37), where

$$\begin{aligned} \hat{z}_{ij} &= 1, \qquad \text{if } \hat{\tau}_i(\mathbf{y}_j;\ \hat{\boldsymbol{\Psi}}) \geq \hat{\tau}_h(\mathbf{y}_j;\ \hat{\boldsymbol{\Psi}}) \qquad (h = 1, \ldots, g; h \neq i), \\ &= 0, \qquad \text{otherwise.} \end{aligned} \tag{9.38}$$

9.7 Choice of the number of factors q

In practice consideration has to be given to the number of components g and the number of factors q in the mixture of factor analysers model. One obvious approach is to use the Bayesian Information Criterion (BIC) of Schwarz (1978). With BIC, $d \log n$ is added to twice the negative of the log-likelihood at the solution, where d denotes the number of free parameters in the model. The intent is to choose $g\,(q)$ to minimise the negative of this penalised form of the log-likelihood. In their examples and simulations where the true classification of the data was known, Baek *et al.* (2010) noted that the use of the BIC did not always lead to a choice of q that simultaneously minimises BIC and the overall apparent error rate, while maximising the adjusted Rand index of Hubert and Arabie (1985).

An alternative approach is to use the likelihood-ratio statistic for tests on g and q. For tests on g, it is well known that regularity conditions do not hold for the usual chi-square approximation to the asymptotic null distribution of the likelihood-ratio test statistic to be valid. However, they usually do hold for tests on the number of factors at a given level of g; see Geweke and Singleton (1980) who have identified situations where the regularity conditions do not hold.

9.8 Example

We report here the results of applying the MFA and MCFA approaches to a real dataset as undertaken by Baek *et al.* (2010). The dataset is based on the so-called Vietnam data which was considered in Smyth *et al.* (2006). The Vietnam dataset consists of the log transformed and standardised concentrations of 17 chemical elements to which four types of synthetic noise variables were added in Smyth *et al.* (2006) to study methods for clustering high-dimensional data. We used these data consisting of a total of 67 variables ($p = 67$; 17 chemical concentration variables plus 50 uniform noise variables). The concentrations were measured in hair samples from six classes ($g = 6$) of Vietnamese, and the total number of subjects were

Table 9.2 Comparison of MFA and MCFA approaches for implied clustering versus the true membership of Vietnam data.

Model	Factors	BIC	ARI	Jaccard	Error rate
	1	**46758**	0.5925	0.4974	0.2277
	2	48212	**0.6585**	**0.5600**	**0.1696**
MFA	3	49743	0.6322	0.5342	0.1830
	4	51351	0.5392	0.4510	0.2589
	5	52846	0.5700	0.4767	0.2679
	1	45171	0.3444	0.3447	0.4777
	2	44950	0.7288	0.6380	0.1384
MCFA	**3**	**44825**	**0.8063**	**0.7248**	**0.0893**
	4	44984	0.7081	0.6241	0.2277
	5	45151	0.6259	0.5385	0.2634

Source: Adapted from Baek *et al.* (2010).

$n = 224$. The noise variables were generated from the uniform distribution on the interval $[-2, 2]$.

Baek *et al.* (2010) implemented the MFA and MCFA approaches for $g = 6$ components with the number of factors q ranging from 1 to 5. For the MFA approach, they imposed the assumption of equal diagonal matrices $\boldsymbol{D}_i$ for the error term. The values they obtained for the ARI, the Jaccard index (Jaccard, 1901) and the overall apparent error rate are displayed in Table 9.2 for each value of q considered with MFA and MCFA approaches.

It can be seen that the lowest error rate and highest values of the ARI and Jaccard index are obtained by using $q = 3$ factors with the MCFA model, which coincides with the choice on the basis of the BIC. The best result with the MFA approach is obtained for $q = 2$ factors (the BIC suggests using $q = 1$). It can be seen that the error rate, ARI and Jaccard index for the MFA are not nearly as good as for MCFA.

9.9 Low-dimensional plots via MCFA approach

To illustrate the usefulness of the MCFA approach for portraying the results of a clustering in low-dimensional space, Baek *et al.* (2010) plotted for the Vietnam data the estimated posterior means of the factors $\hat{\boldsymbol{u}}_j$ as defined by (9.37) with the implied cluster labels shown (see Figure 9.1). In this plot, they used the second and third factors in the MCFA model with $q = 3$ factors. It can be seen that the clusters are represented in this plot with very little overlap. This is not the case in Figure 9.2, where the first two canonical variates are plotted. They were calculated using the implied clustering labels. It can be seen from Figure 9.2 that one cluster is essentially on top of another. The canonical variates are calculated on the basis of the assumption of equal cluster-covariance matrices, which does not apply here. The

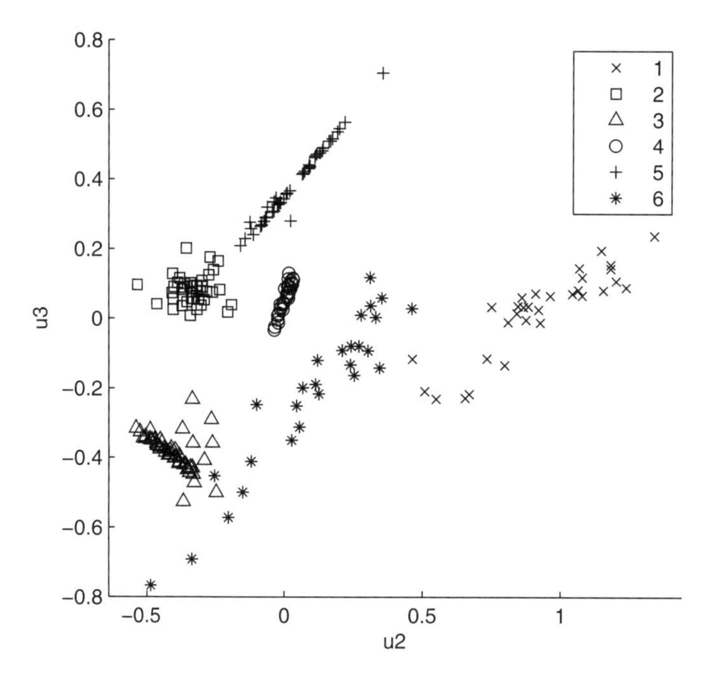

Figure 9.1 Plot of the (estimated) posterior mean factor scores via the MCFA approach with the six cluster labels shown for the Vietnam data. From Baek *et al.* (2010).

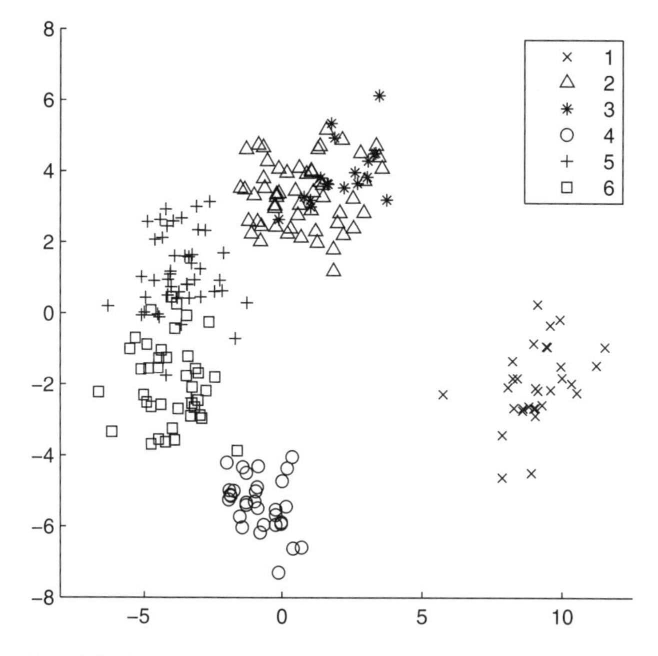

Figure 9.2 Plot of the first two canonical variates based on the implied clustering via MCFA approach with the six cluster labels shown for the Vietnam data. From Baek *et al.* (2010).

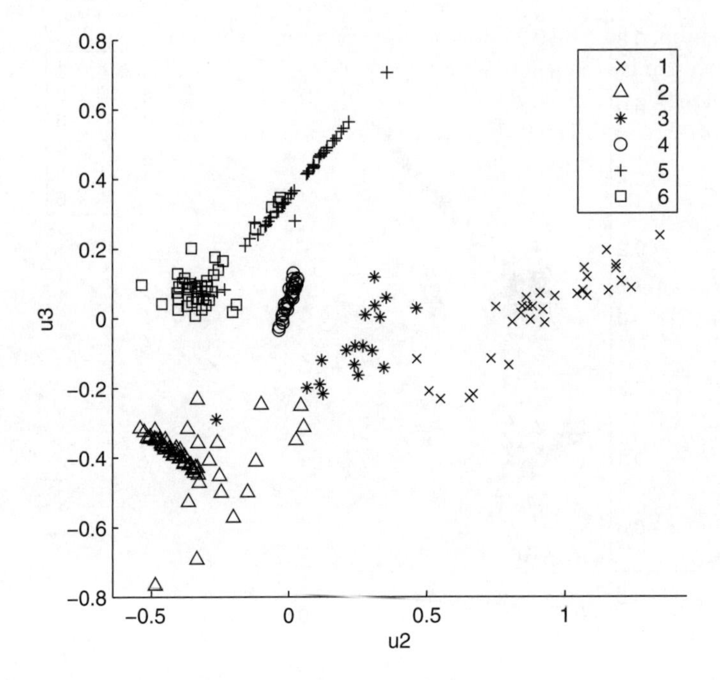

Figure 9.3 Plot of the (estimated) posterior mean factor scores via the MCFA approach with the true labels shown for the six classes in the Vietnam data. From Baek *et al.* (2010).

MCFA approach is not predicated on this assumption and so has more flexibility in representing the data in reduced dimensions.

We have also given in Figure 9.3 the plot corresponding to that in Figure 9.1 with the true cluster labels shown. It can be seen there is good agreement between the two plots. This is to be expected since the error rate of the MCFA model fitted with $q = 3$ factors is quite low (0.0893).

9.10 Multivariate *t*-factor analysers

The mixture of factor analysers model is sensitive to outliers since it adopts the multivariate normal family for the distributions of the errors and the latent factors. An obvious way to improve the robustness of this model for data that have longer tails than the normal or atypical observations is to consider using the multivariate *t*-family of elliptically symmetric distributions. It has an additional parameter called the degrees of freedom that controls the length of the tails of the distribution. Although the number of outliers needed for breakdown is almost the same as with the normal distribution, the outliers have to be much larger (Hennig, 2004).

Before we proceed to consider the mixture model proposed by McLachlan *et al.* (2007) that adopts the *t*-family for modelling the distribution of the component errors and also the latent factors, we give a brief account of the multivariate

t-distribution. The t-distribution for the ith component-conditional distribution of $\mathbf{Y}_j$ is obtained by embedding the normal $N_p(\boldsymbol{\mu}_i, \boldsymbol{\Sigma}_i)$ distribution in a wider class of elliptically symmetric distributions with an additional parameter ν_i called the degrees of freedom. This t-distribution can be characterised by letting W_j denote a random variable distributed as

$$W_j \sim \text{gamma}\,(\frac{1}{2}\nu_i, \frac{1}{2}\nu_i), \tag{9.39}$$

where the gamma (α, β) density function is equal to

$$f_{\text{G}}(w; \alpha, \beta) = \{\beta^{\alpha} w^{\alpha-1} / \Gamma(\alpha)\} \exp(-\beta w) I_{[0,\infty)}(w) \qquad (\alpha, \beta > 0), \tag{9.40}$$

and $I_A(w)$ denotes the indicator function that is 1 if w belongs to A and is zero otherwise. Then, if the conditional distribution of $\mathbf{Y}_j$ given $W_j = w_j$ is specified to be

$$\mathbf{Y}_j \mid w_j \sim N_p(\boldsymbol{\mu}_i, \boldsymbol{\Sigma}_i / w_j), \tag{9.41}$$

the unconditional distribution of $\mathbf{Y}_j$ has a (multivariate) t-distribution with mean $\boldsymbol{\mu}_i$, scale matrix $\boldsymbol{\Sigma}_i$ and degrees of freedom ν_i. The mean of this t-distribution is $\boldsymbol{\mu}_i$ and its covariance matrix is $\{\nu_i/(\nu_i - 2)\}\boldsymbol{\Sigma}_i$. We write

$$\mathbf{Y}_j \sim t_p(\boldsymbol{\mu}_i, \boldsymbol{\Sigma}_i, \nu_i), \tag{9.42}$$

and we let $f_{\text{t}}(\mathbf{y}_j; \boldsymbol{\mu}_i, \boldsymbol{\Sigma}_i, \nu_i)$ denote the corresponding density; see, for example, McLachlan and Peel (2000a, Chapter 7) and Kotz and Nadarajah (2004). As ν_i tends to infinity, the t-distribution approaches the normal distribution. Hence this parameter ν_i may be viewed as a robustness tuning parameter. It can be fixed in advance or it can be inferred from the data for each component.

The random vector $(\mathbf{Y}_j^{\text{T}}, \boldsymbol{U}_{ij}^{\text{T}})^{\text{T}}$ given its membership of the ith component of the mixture (that is $z_{ij} = 1$) has a multivariate normal distribution,

$$\begin{pmatrix} \mathbf{Y}_j \\ \boldsymbol{U}_{ij} \end{pmatrix} \mid z_{ij} = 1 \sim N_{p+q}(\boldsymbol{\mu}_i^*, \boldsymbol{\Sigma}_i^*) \quad (i = 1, \ldots, g), \tag{9.43}$$

where

$$\boldsymbol{\mu}_i^* = (\boldsymbol{\mu}_i^{\text{T}}, \mathbf{0}^{\text{T}})^{\text{T}} \tag{9.44}$$

and the covariance matrix $\boldsymbol{\Sigma}_i^*$ is given by

$$\boldsymbol{\Sigma}_i^* = \begin{pmatrix} \boldsymbol{B}_i\boldsymbol{B}_i^{\text{T}} + \boldsymbol{D}_i & \boldsymbol{B}_i \\ \boldsymbol{B}_i^{\text{T}} & \boldsymbol{I}_q \end{pmatrix}. \tag{9.45}$$

McLachlan *et al.* (2007) formulated t-analysers model by replacing the multivariate normal distribution in (9.43) for the ith component-conditional distribution of $\mathbf{Y}_j$ by the multivariate t-distribution with mean vector vector $\boldsymbol{\mu}_i$, scale matrix $\boldsymbol{\xi}_i$ and ν_i degrees with the factor analytic restriction (9.47) on the component-scale matrices $\boldsymbol{\Sigma}_i$. Thus their postulated mixture model of t-factor analysers assumes that $\mathbf{y}_1, \ldots, \mathbf{y}_n$ is an observed random sample from the t-mixture density

$$f(\mathbf{y}_j; \boldsymbol{\Psi}) = \sum_{i=1}^{g} \pi_i f_t(\mathbf{y}_j; \boldsymbol{\mu}_i, \boldsymbol{\Sigma}_i, \nu_i), \tag{9.46}$$

where

$$\boldsymbol{\Sigma}_i = \boldsymbol{B}_i \boldsymbol{B}_i^{\mathrm{T}} + \boldsymbol{D}_i \quad (i = 1, \ldots, g) \tag{9.47}$$

and where now the vector of unknown parameters $\boldsymbol{\Psi}$ consists of the degrees of freedom ν_i in addition to the mixing proportions π_i and the elements of the $\boldsymbol{\mu}_i$, $\boldsymbol{B}_i$ and $\boldsymbol{D}_i$ $(i = 1, \ldots, g)$. As in the mixture of factor analysers model, $\boldsymbol{B}_i$ is a $p \times q$ matrix and $\boldsymbol{D}_i$ is a diagonal matrix.

In order to fit this model (9.46) with the restriction (9.47), it is computationally convenient to exploit its link with factor analysis. Accordingly, corresponding to (9.19), we assume that

$$\mathbf{Y}_j = \boldsymbol{\mu}_i + \boldsymbol{B}_i \boldsymbol{U}_{ij} + \boldsymbol{e}_{ij} \qquad \text{with prob. } \pi_i \quad (i = 1, \ldots, g) \tag{9.48}$$

for $j = 1, \ldots, n$, where the joint distribution of the factor $\boldsymbol{U}_{ij}$ and of the error $\boldsymbol{e}_{ij}$ needs to be specified so that it is consistent with the t-mixture formulation (9.46) for the marginal distribution of $\mathbf{Y}_j$.

From (9.43), we have for the usual factor analysis model that conditional on membership of the ith component of the mixture the joint distribution of $\mathbf{Y}_j$ and its associated factor (vector) $\boldsymbol{U}_{ij}$ is multivariate normal,

$$\begin{pmatrix} \mathbf{Y}_j \\ \boldsymbol{U}_{ij} \end{pmatrix} \mid z_{ij} = 1 \sim N_{p+q}(\boldsymbol{\mu}_i^*, \boldsymbol{\Sigma}_i^*) \quad (i = 1, \ldots, g). \tag{9.49}$$

where the mean $\boldsymbol{\mu}_i^*$ and the covariance matrix $\boldsymbol{\xi}_i$ are defined by (9.44) and (9.45). We now replace the normal distribution by the t-distribution in (9.49) to postulate that

$$\begin{pmatrix} \mathbf{Y}_j \\ \boldsymbol{U}_{ij} \end{pmatrix} \mid z_{ij} = 1 \sim t_{p+q}(\boldsymbol{\mu}_i^*, \boldsymbol{\Sigma}_i^*, \nu_i) \quad (i = 1, \ldots, g). \tag{9.50}$$

This specification of the joint distribution of $\mathbf{Y}_j$ and its associated factors in (9.48) will imply the t-mixture model (9.46) for the marginal distribution of $\mathbf{Y}_j$ with the restriction (9.47) on its component-scale matrices.

Using the characterisation of the t-distribution discussed in Section 9.10, it follows that we can express (9.49) alternatively as

$$\begin{pmatrix} \mathbf{Y}_j \\ \boldsymbol{U}_{ij} \end{pmatrix} \mid w_j, z_{ij} = 1 \sim N_{p+q}(\boldsymbol{\mu}_i^*, \boldsymbol{\Sigma}_i^*/w_j), \tag{9.51}$$

where w_{ij} is a value of the weight variable W_j taken to have the gamma distribution (9.40). It can be established from (9.51) that

$$\boldsymbol{U}_{ij} \mid w_j, z_{ij} = 1 \sim N_q(\boldsymbol{0}, \boldsymbol{I}_q/w_j) \tag{9.52}$$

and

$$\boldsymbol{e}_{ij} \mid w_j, z_{ij} = 1 \sim N_p(\boldsymbol{0}, \boldsymbol{D}_i/w_j), \tag{9.53}$$

and hence that

$$\boldsymbol{U}_{ij} \mid z_{ij} = 1 \sim t_q(\boldsymbol{0}, \boldsymbol{I}_q, \nu_i) \tag{9.54}$$

and

$$\boldsymbol{e}_{ij} \mid z_{ij} = 1 \sim t_p(\boldsymbol{0}, \boldsymbol{D}_i, \nu_i). \tag{9.55}$$

Thus with this formulation, the error terms $\boldsymbol{e}_{ij}$ and the factors $\boldsymbol{U}_{ij}$ are distributed according to the t-distribution with the same degrees of freedom. However, the factors and error terms are no longer independently distributed as in the normal-based model for factor analysis, but they are uncorrelated. To see this, we have from (9.51) that conditional on w_j, $\boldsymbol{U}_{ij}$ and $\boldsymbol{e}_{ij}$ are uncorrelated and, hence, unconditionally uncorrelated.

More recently, Baek and McLachlan (2010) have extended the MCFA model to mixtures of common t-factor analysers.

9.11 Discussion

In practice, much attention is being given to the use of normal mixture models in density estimation and clustering. However, for high-dimensional datasets, the component-covariance matrices are highly parameterised and some form of reduction in the number of parameters is needed, particularly when the number of observations n is not large relative to the number of dimensions p. One way to proceed is to use the three-step approach of McLachlan *et al.* (2002), whereby on the first step the number of variables is reduced by their individual screening, followed on the second step by a clustering of the retained variables into groups and finally on the third step by a clustering of the observation vectors via the fitting of a mixture model to the representatives of the groups of variables formed on the second step. Another way of proceeding is to work with mixtures of factor analysers (MFA) as studied in McLachlan and Peel (2000a, 2000b). This approach achieves a reduction in the number of parameters through its factor-analytic representation of the component-covariance matrices. However, it may not provide a sufficient reduction in the number of parameters, particularly when the number g of clusters (components) to be imposed on the data is not small. In such instances the number of parameters can be reduced appreciably by using a factor-analytic representation of the component-covariance matrices with common factor loadings. The factor loadings are common before the component-specific rotation of the factors to have white noise; that is, before this rotation, the factors are allowed to have different component-specific distributions. This sharing of the factor loadings enables the model to be used to cluster high-dimensional data into many clusters and to provide low-dimensional plots of the clusters so obtained. The latter plots are given in terms of the (estimated) posterior means of the factors corresponding to the observed data. These projections are not useful with the MFA approach as in its formulation the factors are taken to be white noise with no cluster-specific discriminatory features for the factors.

The MFA approach does allow a more general representation of the component variances/covariances and places no restrictions on the component means. Thus it is more flexible in its modelling of the data. The MCFA provides a comparable approach that can be applied in situations where the dimension p and the number of clusters g can be quite large.

In practice, we can use BIC or the likelihood-ratio statistic to provide a guide to the choice of the number of factors q and the number of components g to be used. On the latter choice it is well known that regularity conditions do not hold for the usual chi-square approximation to the asymptotic null distribution of the likelihood-ratio test statistic to be valid. However, they usually should hold for tests on the number of factors at a given level of g; see Geweke and Singleton (1980). In their examples and simulation experiments, Baek *et al.* (2010) used the BIC, although it did not always lead to the best choice of the number of factors q. Hence further investigation would appear to be needed on the use of BIC and other criteria for choosing the number of factors q for a given number of components g.

Appendix

The model (9.20) underlying the MCFA approach can be fitted via the EM algorithm to estimate the vector $\boldsymbol{\Psi}$ of unknown parameters. It consists of the mixing proportions π_i, the factor component-mean vectors $\boldsymbol{\xi}_i$, the distinct elements of the factor component-covariance matrices $\boldsymbol{\Omega}_i$, the projection matrix $\boldsymbol{A}$ based on sharing of factor loadings and the common diagonal matrix $\boldsymbol{D}$ of the residuals given the factor scores within a component of the mixture. We now demonstrate the application of the EM algorithm as applied by Baek *et al.* (2010) to this problem.

In order to apply the EM algorithm to this problem, we introduce the component-indicator labels z_{ij}, where z_{ij} is one or zero according to whether $\mathbf{y}_j$ belongs or does not belong to the ith component of the model. We let $\mathbf{z}_j$ be the component-label vector, $\mathbf{z}_j = (z_{1j}, \ldots, z_{gj})^{\mathrm{T}}$. The $\mathbf{z}_j$ are treated as missing data, along with the (unobservable) latent factors $\boldsymbol{u}_{ij}$ within this EM framework. The complete-data log-likelihood is then given by

$$\log L_c(\boldsymbol{\Psi}) = \sum_{i=1}^{g}\sum_{j=1}^{n} z_{ij}\left[\log \pi_i + \log \phi(\mathbf{y}_j; \boldsymbol{A}\boldsymbol{u}_{ij}, \boldsymbol{D}) + \log \phi(\boldsymbol{u}_{ij}; \boldsymbol{\xi}_i, \boldsymbol{\Omega}_i)\right]. \tag{9.56}$$

• **E-step**: On the E-step, we require the conditional expectation of the complete-data log-likelihood, $\log L_c(\boldsymbol{\Psi})$, given the observed data $\mathbf{y} = (\mathbf{y}_1^T, \ldots, \mathbf{y}_n^T)^T$, using the current fit for $\boldsymbol{\Psi}$. Let $\boldsymbol{\Psi}^{(k)}$ be the value of $\boldsymbol{\Psi}$ after the kth iteration of the EM algorithm. Then, more specifically, on the $(k+1)$th iteration the E-step requires the computation of the conditional expectation of $\log L_c(\boldsymbol{\Psi})$ given $\mathbf{y}$, using $\boldsymbol{\Psi}^{(k)}$ for $\boldsymbol{\Psi}$, which is denoted by $Q(\boldsymbol{\Psi}; \boldsymbol{\Psi}^{(k)})$.

We let

$$\tau_{ij}^{(k)} = \tau_i(\mathbf{y}_j; \boldsymbol{\Psi}^{(k)}), \tag{9.57}$$

where $\tau_i(\mathbf{y}_j; \boldsymbol{\Psi})$ is defined by (9.31). Also, we let $E_{\boldsymbol{\Psi}^{(k)}}$ refer to the expectation operator, using $\boldsymbol{\Psi}^{(k)}$ for $\boldsymbol{\Psi}$. Then the so-called Q-function, $Q(\boldsymbol{\Psi}; \boldsymbol{\Psi}^{(k)})$, can be written as

$$Q(\boldsymbol{\Psi}; \boldsymbol{\Psi}^{(k)}) = \sum_{i=1}^{g} \sum_{j=1}^{n} \tau_{ij}^{(k)} \left[\log \pi_i + w_{1ij}^{(k)} + w_{1ij}^{(k)} \right], \tag{9.58}$$

where

$$w_{1ij}^{(k)} = E_{\boldsymbol{\Psi}^{(k)}} \left[\log \phi(\mathbf{y}_j; \boldsymbol{A}\boldsymbol{u}_{ij}, \boldsymbol{D}) \mid \mathbf{y}_j, z_{ij} = 1 \right] \tag{9.59}$$

and

$$w_{2ij}^{(k)} = E_{\boldsymbol{\Psi}^{(k)}} \left[\log \phi(\boldsymbol{u}_{ij}; \boldsymbol{\xi}_i, \boldsymbol{\Omega}_i) \mid \mathbf{y}_j, z_{ij} = 1 \right]. \tag{9.60}$$

• **M-step**: On the $(k+1)$th iteration of the EM algorithm, the M-step consists of calculating the updated estimates $\pi_i^{(k+1)}$, $\boldsymbol{\xi}_i^{(k+1)}$, $\boldsymbol{\Omega}_i^{(k+1)}$, $\boldsymbol{A}^{(k+1)}$ and $\boldsymbol{D}^{(k+1)}$ by solving the equation

$$\partial Q(\boldsymbol{\Psi}; \boldsymbol{\Psi}^{(k)}) / \partial \boldsymbol{\Psi} = \mathbf{0}. \tag{9.61}$$

The updated estimates of the mixing proportions π_i are given as in the case of the normal mixture model by

$$\pi_i^{(k+1)} = \sum_{j=1}^{n} \tau_{ij}^{(k)} / n \quad (i = 1, \ldots, g). \tag{9.62}$$

Concerning the other parameters, it can be shown using vector and matrix differentiation that

$$\partial Q(\boldsymbol{\Psi}; \boldsymbol{\Psi}^{(k)}) / \partial \boldsymbol{\xi}_i = \boldsymbol{\Omega}_i^{-1} \sum_{j=1}^{n} \tau_{ij}^{(k)} E_{\boldsymbol{\Psi}^{(k)}} \left[(\boldsymbol{u}_{ij} - \boldsymbol{\xi}_i) \mid \mathbf{y}_j \right], \tag{9.63}$$

$$\partial Q(\boldsymbol{\Psi}; \boldsymbol{\Psi}^{(k)}) / \partial \boldsymbol{\Omega}_i^{-1} = \sum_{j=1}^{n} \tau_{ij}^{(k)} \frac{1}{2} \{ \boldsymbol{\Omega}_i - E_{\boldsymbol{\Psi}^{(k)}} \left[(\boldsymbol{u}_{ij} - \boldsymbol{\xi}_i)(\boldsymbol{u}_{ij} - \boldsymbol{\xi}_i)^T \mid \mathbf{y}_j \right] \}, \tag{9.64}$$

$$\partial Q(\boldsymbol{\Psi};\, \boldsymbol{\Psi}^{(k)})/\partial \boldsymbol{D}^{-1} = \sum_{i=1}^{g}\sum_{j=1}^{n} \tau_{ij}^{(k)} \frac{1}{2}\{\boldsymbol{D} - E_{\boldsymbol{\Psi}^{(k)}}\left[(\mathbf{y}_j - \boldsymbol{A}\boldsymbol{u}_{ij})(\mathbf{y}_j - \boldsymbol{u}_{ij})^{\mathrm{T}} \mid \mathbf{y}_j\right]\}, \tag{9.65}$$

$$\partial Q(\boldsymbol{\Psi};\, \boldsymbol{\Psi}^{(k)})/\partial \boldsymbol{A} = \sum_{i=1}^{g}\sum_{j=1}^{n} \tau_{ij}^{(k)} \left\{ \boldsymbol{D}^{-1} \left[\mathbf{y}_j E_{\boldsymbol{\Psi}^{(k)}}(\boldsymbol{u}_{ij}^{\mathrm{T}} \mid \mathbf{y}_j) - \boldsymbol{A} E_{\boldsymbol{\Psi}^{(k)}}(\boldsymbol{u}_{ij}\boldsymbol{u}_{ij}^{\mathrm{T}} \mid \mathbf{y}_j) \right] \right\}. \tag{9.66}$$

On equating (9.63) to the zero vector, it follows that $\boldsymbol{\xi}_i^{(k+1)}$ can be expressed as

$$\boldsymbol{\xi}_i^{(k+1)} = \boldsymbol{\xi}_i^{(k)} + \frac{\sum_{j=1}^{n} \tau_{ij}^{(k)} \boldsymbol{\gamma}_i^{(k)\mathrm{T}} \mathbf{y}_{ij}^{(k)}}{\sum_{j=1}^{n} \tau_{ij}^{(k)}}, \tag{9.67}$$

where

$$\mathbf{y}_{ij}^{(k)} = \mathbf{y}_j - \boldsymbol{A}^{(k)}\boldsymbol{\xi}_i^{(k)} \tag{9.68}$$

and

$$\boldsymbol{\gamma}_i^{(k)} = \left(\boldsymbol{A}^{(k)}\boldsymbol{\Omega}_i^{(k)}\boldsymbol{A}^{(k)^{\mathrm{T}}} + \boldsymbol{D}^{(k)} \right)^{-1} \boldsymbol{A}^{(k)}\boldsymbol{\Omega}_i^{(k)}. \tag{9.69}$$

On equating (9.64) to the null matrix, it follows that

$$\boldsymbol{\Omega}_i^{(k+1)} = \frac{\sum_{j=1}^{n} \tau_{ij}^{(k)} \boldsymbol{\gamma}_i^{(k)^{\mathrm{T}}} \mathbf{y}_{ij}^{(k)} \mathbf{y}_{ij}^{(k)^{\mathrm{T}}} \boldsymbol{\gamma}^{(k)}}{\sum_{j=1}^{n} \tau_{ij}^{(k)}} + (\boldsymbol{I}_q - \boldsymbol{\gamma}_i^{(k)^{\mathrm{T}}} \boldsymbol{A}^{(k)})\boldsymbol{\Omega}_i^{(k)}. \tag{9.70}$$

On equating (9.65) to the zero vector, we obtain

$$\boldsymbol{D}^{(k+1)} = \mathrm{diag}(\boldsymbol{D}_1^{(k)} + \boldsymbol{D}_2^{(k)}), \tag{9.71}$$

where

$$\boldsymbol{D}_1^{(k)} = \frac{\sum_{i=1}^{g}\sum_{j=1}^{n} \tau_{ij}^{(k)} \boldsymbol{D}^{(k)}(\boldsymbol{I}_p - \boldsymbol{\beta}_i^{(k)})}{\sum_{i=1}^{g}\sum_{j=1}^{n} \tau_{ij}^{(k)}} \tag{9.72}$$

and

$$\boldsymbol{D}_2^{(k)} = \frac{\sum_{i=1}^{g}\sum_{j=1}^{n} \tau_{ij}^{(k)} \boldsymbol{\beta}_i^{(k)^{\mathrm{T}}} \mathbf{y}_{ij}^{(k)} \mathbf{y}_{ij}^{(k)^{\mathrm{T}}} \boldsymbol{\beta}_i^{(k)}}{\sum_{i=1}^{g}\sum_{j=1}^{n} \tau_{ij}^{(k)}}, \tag{9.73}$$

and where

$$\boldsymbol{\beta}_i^{(k)} = (\boldsymbol{A}^{(k)}\boldsymbol{\Omega}_i^{(k)}\boldsymbol{A}^{(k)^{\mathrm{T}}} + \boldsymbol{D}^{(k)})^{-1}\boldsymbol{D}^{(k)}. \tag{9.74}$$

On equating (9.66) to the null matrix, we obtain

$$\boldsymbol{A}^{(k+1)} = \left(\sum_{i=1}^{g} \boldsymbol{A}_{1i}^{(k)}\right)\left(\sum_{i=1}^{g} \boldsymbol{A}_{2i}^{(k)}\right)^{-1}, \tag{9.75}$$

where

$$\boldsymbol{A}_{1i}^{(k+1)} = \sum_{j=1}^{n} \tau_{ij}^{(k)} \left(\mathbf{y}_j \boldsymbol{\xi}_i^{(k)^{\mathrm{T}}} + \mathbf{y}_{ij}^{(k)^{\mathrm{T}}} \boldsymbol{\gamma}_i^{(k)}\right), \tag{9.76}$$

$$\boldsymbol{A}_{2i}^{(k+1)} = \sum_{j=1}^{n} \tau_{ij}^{(k)} \left[\left(\boldsymbol{I}_q - \boldsymbol{\gamma}_i^{(k)^{T}} \boldsymbol{A}^{(k)}\right) \boldsymbol{\Omega}_i^{(k)} + \boldsymbol{r}_i^{(k)} \boldsymbol{r}_i^{(k)^{T}}\right] \tag{9.77}$$

and

$$\boldsymbol{r}_i^{(k)} = \boldsymbol{\xi}_i^{(k)} + \boldsymbol{\gamma}_i^{(k)^{\mathrm{T}}} \mathbf{y}_{ij}^{(k)}. \tag{9.78}$$

We have to specify an initial value for the vector $\boldsymbol{\Psi}$ of unknown parameters in the application of the EM algorithm. A random start is obtained by first randomly assigning the data into g groups. Let n_i, $\bar{\mathbf{y}}_i$, and $\boldsymbol{S}_i$ be the number of observations, the sample mean and the sample covariance matrix, respectively, of the ith group of the data so obtained $(i = 1, \ldots, g)$. We then proceed as follows:

- Set $\pi_i^{(0)} = n_i/n$.
- Generate random numbers from the standard normal distribution $N(0, 1)$ to obtain values for the (j, k)th element of $\boldsymbol{A}^*$ $(j = 1, \ldots, p; k = 1, \ldots, q)$.
- Define $\boldsymbol{A}^{(0)}$ by $\boldsymbol{A}^*$.
- On noting that the transformed data $\boldsymbol{D}^{-1/2}\mathbf{Y}_j$ satisfies the probabilistic PCA model of Tipping and Bishop (1999) with $\sigma_i^2 = 1$, it follows that for a given $\boldsymbol{D}^{(0)}$ and $\boldsymbol{A}^{(0)}$, we can specify $\boldsymbol{\Omega}_i^{(0)}$ as

$$\boldsymbol{\Omega}_i^{(0)} = \boldsymbol{A}^{(0)^{\mathrm{T}}} \boldsymbol{D}^{(0)^{1/2}} \boldsymbol{H}_i(\boldsymbol{\Lambda}_i - \tilde{\sigma}_i^2 I_q)\boldsymbol{H}_i^{\mathrm{T}} \boldsymbol{D}^{(0)^{1/2}} \boldsymbol{A}^{(0)},$$

where $\tilde{\sigma}_i^2 = \sum_{h=q+1}^{p} \lambda_{ih}/(p-q)$. The q columns of the matrix $\boldsymbol{H}_i$ are the eigenvectors corresponding to the eigenvalues $\lambda_{i1} \geq \lambda_{i2} \geq \cdots \geq \lambda_{iq}$ of

$$\boldsymbol{D}^{(0)^{-1/2}} \boldsymbol{S}_i \boldsymbol{D}^{(0)^{-1/2}}, \tag{9.79}$$

where $\boldsymbol{S}_i$ is the covariance matrix of the $\mathbf{y}_j$ in the ith group and $\boldsymbol{\Lambda}_i$ is the diagonal matrix with diagonal elements equal to $\lambda_{i1}, \ldots, \lambda_{iq}$. Concerning the choice of $\boldsymbol{D}^{(0)}$, we can take $\boldsymbol{D}^{(0)}$ to be the diagonal matrix formed from the diagonal elements of the (pooled) within-cluster sample covariance matrix of the $\mathbf{y}_j$. The initial value for $\boldsymbol{\xi}_i$ is $\boldsymbol{\xi}_i^{(0)} = \boldsymbol{A}^{(0)^{T}} \bar{\mathbf{y}}_i$.

Some clustering procedure such as k-means can be used to provide nonrandom partitions of the data, which can be used to obtain another set of initial values for the parameters. In our analyses we used both initialisation methods.

As noted previously, the solution $\hat{A}$ for the matrix of factor loadings is unique only up to postmultiplication by a nonsingular matrix. We chose to postmultiply by the nonsingular matrix for which the solution is orthonormal; that is

$$\hat{A}^{\mathrm{T}}\hat{A} = I_q. \tag{9.80}$$

To achieve this with $\hat{A}$ computed as above, we note that we can use the Cholesky decomposition to find the upper triangular matrix C of order q so that

$$\hat{A}^{\mathrm{T}}\hat{A} = C^{\mathrm{T}}C. \tag{9.81}$$

Then it follows that if we replace $\hat{A}$ by

$$\hat{A}C^{-1}, \tag{9.82}$$

then it will satisfy the requirement (9.80). With the adoption of the estimate (9.82) for $\hat{A}$, we need to adjust the updated estimates $\hat{\xi}_i$ and $\hat{\Omega}_i$ to be

$$C\hat{\xi}_i \tag{9.83}$$

and

$$C\hat{\Omega}_i C^{\mathrm{T}}, \tag{9.84}$$

where $\hat{\xi}_i$ and $\hat{\Omega}_i$ are given by the limiting values of (9.67) and (9.70), respectively. Note that if we premultiply both sides of 9.22 with the parameters replaced by their estimates, we have, on using (9.80) that

$$\hat{A}_i^{\mathrm{T}}\hat{\mu}_i^{\mathrm{T}} = \hat{\xi}_i \quad (i = 1, \ldots, p). \tag{9.85}$$

An R version of our program is available at `http://www.maths.uq.edu.au/~gjm/`.

References

Baek, J. and McLachlan, G. J. (2010) Mixtures of common t-factor analyzers for clustering high-dimensional microarray data (Submitted for publication).

Baek, J. and McLachlan, G. J. (2008) Mixtures of factor analyzers with common factor loadings for the clustering and visualisation of high-dimensional data, Technical Report NI08018-SCH, Preprint Series of the Isaac Newton Institute for Mathematical Sciences, Cambridge.

Baek, J. McLachlan, G. J. and Flack L. K. (2010) Mixtures of factor analyzers with common factor loadings: applications to the clustering and visualisation of high-dimensional data. *IEEE Transactions on Pattern Analysis and Machine Intelligence*, **32**, 1298–1309.

Banfield, J. D. and Raftery, A. E. (1993) Model-based Gaussian and non-Gaussian clustering. *Biometrics*, **49**, 803–821.

Chang, W. C. (1983) On using principal components before separating a mixture of two multivariate normal distributions. *Applied Statistics*, **32**, 267–275.

Dempster, A. P., Laird, N. M. and Rubin, D. B. (1977) Maximum likelihood from incomplete data via the EM algorithm (with discussion). *Journal of the Royal Statistical Society, Series B*, **39**, 1–38.

Gales, M. (1999) Semi-tied covariance matrices for hidden Markov models. *IEEE Transactions on Speech and Audio Processing*, **7**, 272–281.

Galimberti, G., Montanari, A. and Viroli, C. (2008) Latent classes of objects and variable selection. In *Proceedings of COMPSTAT 2008* (ed. P. Brito), pp. 373–383. Springer-Verlag.

Geweke, J. F. and Singleton, K. J. (1980) Interpreting the likelihood ratio statistic in factor models when sample size is small. *Journal of the American Statistical Association*, **75**, 133–137.

Ghahramani, Z. and Hinton, G. E. (1997) The EM algorithm for factor analyzers. Technical Report CRG-TR-96-1, The University of Toronto, Toronto.

Gopinath, R., Ramabhadran, B. and Dharanipragada, S. (1998) Factor analysis invariant to linear transformations of data. In *Proceedings of the International Conference on Speech and Language Processing*, pp. 397–400.

Hennig, C. (2004) Breakdown points for maximum likelihood–estimators of location–scale mixtures. *Annals of Statistics*, **32**, 1313–1340.

Hinton, G. E., Dayan, P. and Revow, M. (1997) Modeling the manifolds of images of handwritten digits. *IEEE Transactions on Neural Networks*, **8**, 65–73.

Hubert, L. and Arabie, P. (1985) Comparing partitions. *Journal of Classification*, **2**, 193–218.

Jaccard, P. (1901) Distribution de la florine alpine dans la Bassin de Dranses et dans quelques régions voisines. *Bulletin de la Société Vaudoise des Sciences Naturelles*, **37**, 241–272.

Kotz, S. and Nadarajah, S. (2004) *Multivariate t–Distributions and Their Applications*. Cambridge University Press.

Kumar, N. (1997) Investigation of silicon-auditory models and generalization of linear discriminant analysis for improved speech recognition. PhD Thesis, Johns Hopkins University, Maryland.

Lawley, D. N. and Maxwell, A. E. (1971) *Factor Analysis as a Statistical Method*. Butterworths, London.

McLachlan, G. J. and Krishnan, T. (2008) *The EM Algorithm and Extensions,* 2nd edn. John Wiley & Sons, Ltd.

McLachlan, G. J. and Peel, D. (2000a) *Finite Mixture Models*. John Wiley & Sons, Ltd.

McLachlan, G. J. and Peel, D. (2000b) Mixtures of factor analyzers. In *Proceedings of the Seventeenth International Conference on Machine Learning* (ed. P. Langley), pp. 599–606.

McLachlan, G. J., Bean, R. W. and Peel, D. (2002) Mixture model-based approach to the clustering of microarray expression data. *Bioinformatics*, **18**, 413–422.

McLachlan, G. J., Peel, D. and Bean, R. W. (2003) Modelling high-dimensional data by mixtures of factor analyzers. *Computational Statistics and Data Analysis*, **41**, 379–388.

McLachlan, G. J., Bean, R. W. and Ben-Tovim Jones, L. (2007) Extension of the mixture of factor analyzers model to incorporate the multivariate *t* distribution. *Computational Statistics and Data Analysis*, **51**, 5327–5338.

McNicholas, P. D. and Murphy, T. B. (2008) Parsimonious Gaussian mixture models. *Statistics and Computing*, **18**, 285–296.

Montanari, A. and Viroli, C. (2010) Heteroscedastic factor mixture analysis. *Statistical Modelling* (in press).

Olsen, P. and Gopinath, R. (2002) Modeling inverse covariance matrices by basis expansion. In *Proceedings of the International Conference on Acoustics, Speech, and Signal Processing*, Vol. 1, pp. 945–948.

Sanguinetti, G. (2008) Dimensionality reduction of clustered data sets. *IEEE Transactions on Pattern Analysis and Machine Intelligence*, **30**, 535–540.

Schwarz, G. (1978) Estimating the dimension of a model. *Annals of Statistics*, **6**, 461–464.

Smyth, C., Coomans, C. and Everingham, Y. (2006) Clustering noisy data in a reduced dimension space via multivariate regression trees. *Pattern Recognition*, **39**, 424–431.

Tipping, M. E. and Bishop, C. M. (1997) Mixtures of probabilistic principal component analysers. Technical Report NCRG/97/003, Neural Computing Research Group, Aston University, Birmingham.

Tipping, M. E. and Bishop, C. M. (1999) Mixtures of probabilistic principal component analysers. *Neural Computation*, **11**, 443–482.

Rosti, A.-V. I. and Gales, M. J. F. (2001) Generalised linear Gaussian models. Technical Report CUED/F-INFENG/TR.420, Cambridge University Engineering Department.

Rosti, A.-V. I. and Gales, M. J. F. (2004) Factor analysis hidden Markov models for speech recognition. *Computer Speech and Language*, **18**, 181–200.

Viroli, C. (2010) Dimensionally reduced model-based clustering through mixtures of factor mixture analyzers *Journal of Classification* (in press).

Xie, B., Pan, W. and Shen, X. (2010) Penalized mixtures of factor analyzers with application to clustering high-dimensional microarray data. *Bioinformatics*, **26**, 501–508.

Yoshida, R., Higuchi, T. and Imoto, S. (2004) A mixed factors model for dimension reduction and extraction of a group structure in gene expression data. In *Proceedings of the 2004 IEEE Computational Systems Bioinformatics Conference*, pp.161–172.

Yoshida, R., Higuchi, T. Imoto, S. and Miyano, S. (2006) ArrayCluster: an analytic tool for clustering, data visualization and model finder on gene expression profiles. *Bioinformatics*, **22**, 1538–1539.

10

Dealing with label switching under model uncertainty

Sylvia Frühwirth-Schnatter

10.1 Introduction

This chapter considers identification of a finite mixture distribution with K components,

$$p(\mathbf{y}|\boldsymbol{\theta}_1, \ldots, \boldsymbol{\theta}_K, \boldsymbol{\eta}) = \sum_{k=1}^{K} \eta_k f_{\mathcal{T}}(\mathbf{y}|\boldsymbol{\theta}_k), \qquad (10.1)$$

where $\mathbf{y}$ is the realisation of a univariate or multivariate, discrete- or continuous-valued random variable and the component densities $f_{\mathcal{T}}(\mathbf{y}|\boldsymbol{\theta}_k)$ arise from the same distribution family $\mathcal{T}(\boldsymbol{\theta})$ indexed by a parameter $\boldsymbol{\theta}$ taking values in Θ. Given a sample $\mathbf{y} = (\mathbf{y}_1, \ldots, \mathbf{y}_N)$, identification of (10.1) concerns estimating K, the component-specific parameters $\boldsymbol{\theta}_1, \ldots, \boldsymbol{\theta}_K$ as well as the weight distribution $\boldsymbol{\eta} = (\eta_1, \ldots, \eta_K)$. A comprehensive review of finite mixture models and their estimation is provided by the monographs of Titterington *et al.* (1985), McLachlan and Peel (2000) and Frühwirth-Schnatter (2006). This chapter is based on Bayesian inference using MCMC methods, which date back to Diebolt and Robert (1994); see Frühwirth-Schnatter (2006, Chapter 3) and Marin *et al.* (2005) for a review.

Mixtures: Estimation and Applications, First Edition. Edited by Kerrie L. Mengersen, Christian P. Robert and D. Michael Titterington.

Identification requires handling the label switching problem caused by invariance of the representation (10.1) with respect to reordering the components:

$$p(\mathbf{y}|\boldsymbol{\theta}_1, \ldots, \boldsymbol{\theta}_K, \boldsymbol{\eta}) = \sum_{k=1}^{K} \eta_k f_{\mathcal{T}}(\mathbf{y}|\boldsymbol{\theta}_k) = \sum_{k=1}^{K} \eta_{\rho(k)} f_{\mathcal{T}}(\mathbf{y}|\boldsymbol{\theta}_{\rho(k)}),$$

where ρ is an arbitrary permutation of $\{1, \ldots, K\}$. The resulting multimodality and perfect symmetry of the posterior distribution $p(\boldsymbol{\theta}_1, \ldots, \boldsymbol{\theta}_K, \boldsymbol{\eta}|\mathbf{y})$ makes it difficult to perform component-specific inference.

Many useful methods have been developed to force a unique labelling on draws from this posterior distribution when the number of components is known (Celeux, 1998; Celeux *et al.*, 2000; Frühwirth-Schnatter, 2001b; Jasra *et al.*, 2005; Marin *et al.*, 2005; Nobile and Fearnside, 2007; Sperrin *et al.*, 2010; Spezia, 2009; Stephens, 2000b). To identify a unique labelling, most papers work in the usually high-dimensional space Θ^K. In Section 10.2, a simple relabelling technique suggested in Frühwirth-Schnatter (2006, page 97) is discussed, which is based on k-means clustering in the point-process representation of the MCMC draws and operates in the parameter space Θ or an even smaller subspace $\tilde{\Theta} \subset \Theta$ rather than in the high-dimensional space Θ^K.

The rest of this chapter deals with identifying a unique labelling under model uncertainty. Section 10.3 sheds some light on the role the prior $p(\boldsymbol{\eta})$ of the weight distribution plays when the true number K_{tr} of components is unknown. It is shown that the very popular uniform prior on $\boldsymbol{\eta}$ is usually a poor choice for overfitting models. A prior decision has to be made through the choice of $p(\boldsymbol{\eta})$, whether for overfitting mixture models empty components or identical, nonempty components should be introduced. As a consequence of this choice, either the number of nonempty components K_0 or the total number of components K is a better estimator of K_{tr}.

Section 10.4 discusses identification of finite mixture models that strongly overfit heterogeneity in the component-specific parameters. While standard priors lead to strong underfitting of K_{tr} in such a situation, well-known variable-selection shrinkage priors are applied to handle overfitting heterogeneity. Such priors are able to discriminate between coefficients that are more or less homogeneous and coefficients that are heterogeneous and avoid underfitting of K_{tr} by reducing automatically the prior variance of homogeneous components.

10.2 Labelling through clustering in the point-process representation

10.2.1 The point-process representation of a finite mixture model

Stephens (2000a) showed that a finite mixture distribution has a representation as a marked point process and may be seen as a distribution of the points $\{\boldsymbol{\theta}_1, \ldots, \boldsymbol{\theta}_K\}$

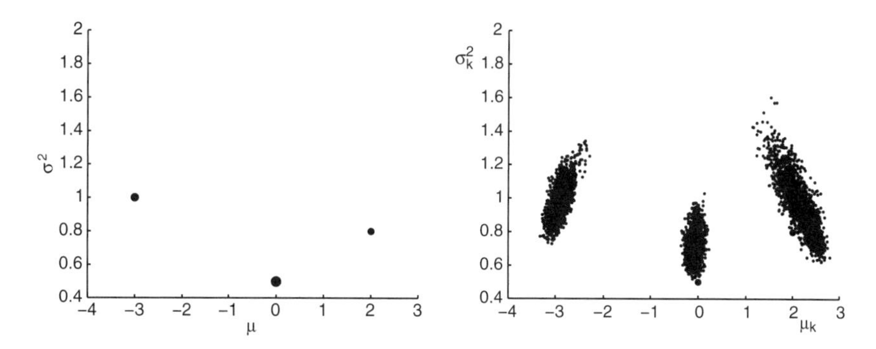

Figure 10.1 Point-process representation of a univariate normal mixture distribution with $K_{tr} = 3$ components (left-hand side) and point-process representation of the MCMC draws (right-hand side) obtained by fitting a normal mixture distribution with $K = 3$ components to 500 simulated observations.

over the parameter space Θ with marks proportional to the weights; see the left-hand side of Figure 10.1. Evidently, the point-process representation is invariant to relabelling the components of the finite mixture distribution.

Frühwirth-Schnatter (2006, Subsection 3.7.1) introduced the point-process representation of a sequence of posterior draws $\boldsymbol{\theta}_1^{(m)}, \ldots, \boldsymbol{\theta}_K^{(m)}, \boldsymbol{\eta}^{(m)}, m = 1, \ldots, M$, obtained through Bayesian inference, where an individual point-process representation is produced for each posterior draw. If the number K of fitted components matches the true number of components K_{tr}, then the posterior draws cluster in this point-process representation around the 'true' points $\{\boldsymbol{\theta}_1^{\text{true}}, \ldots, \boldsymbol{\theta}_K^{\text{true}}\}$. This is illustrated in the right-hand side of Figure 10.1 where posterior sampling is carried out using MCMC methods based on data augmentation and Gibbs sampling; see for example, Frühwirth-Schnatter (2006, Section 3.5), under the priors $\boldsymbol{\eta} \sim \mathcal{D}(4, \ldots, 4)$, $\mu_k \sim \mathcal{N}(m, R^2)$ and $\sigma_k^2 \sim \mathcal{G}^{-1}(c_0, C_0)$ with $C_0 \sim \mathcal{G}(g_0, G_0)$ (Richardson and Green, 1997). The hyperparameters m and R are chosen as the midpoint and the length of the observation interval, $c_0 = 2$, $g_0 = 0.5$ and $G_0 = 100 g_0 / R^2$.

The point-process representation of the MCMC draws allows posterior differences in the component-specific parameters to be studied despite potential label switching, which makes it very useful for identification. Frühwirth-Schnatter (2001b) suggested identifying a finite mixture model by formulating identifiability constraints based on a visual inspection of bivariate scatterplots of the point-process representation. While this method works quite well in lower dimensions, it is difficult or even impossible to extend it to higher dimensions, because the point-process representation is difficult to visualise and simple identifiability constraints usually do not exist.

Example 1: Multivariate normal mixtures

Consider, for illustration, a mixture of $K_{tr} = 4$ multivariate normal distributions of dimension $r = 6$ with $\eta_1 = \cdots = \eta_4 = 1/4$, component-specific means

$$\begin{pmatrix} \boldsymbol{\mu}_1 & \boldsymbol{\mu}_2 & \boldsymbol{\mu}_3 & \boldsymbol{\mu}_4 \end{pmatrix} = \begin{pmatrix} -2 & -2 & -2 & 0 \\ 3 & 0 & -3 & 3 \\ 4 & 4 & 4 & 4 \\ 0 & 0 & 0 & 0 \\ 0 & 2 & 0 & 0 \\ 1 & 0 & 1 & 0 \end{pmatrix},$$

and component-specific covariance matrices $\Sigma_1 = 0.5\mathbf{I}_6$, $\Sigma_2 = 4\mathbf{I}_6 + 0.2\mathbf{e}_6$, $\Sigma_3 = 4\mathbf{I}_6 - 0.2\mathbf{e}_6$ and $\Sigma_4 = \mathbf{I}_6$, where $\mathbf{I}_r$ is the $(r \times r)$ identity matrix and $\mathbf{e}_r$ is an $(r \times r)$ matrix of ones. The component-specific parameter $\boldsymbol{\theta}_k = (\boldsymbol{\mu}_k, \Sigma_k)$ is of dimension $r + r(r+1)/2 = 27$. A simple identifiability constraint like $\mu_{1,l} < \cdots < \mu_{4,l}$ does not exist for this example, because in each dimension l the component-specific means $\mu_{1,l}, \ldots, \mu_{4,l}$ take on at most three distinct values.

To visualise the point-process representation of this mixture distribution pairwise scatterplots of selected elements of $\boldsymbol{\theta}_k$ are considered in Figure 10.2, where $\mu_{k,j}$ is plotted versus $\mu_{k,l}$ for all combinations (j, l). The number of points visible in these two-dimensional projections is equal to the number of distinct values of the vector $(\mu_{k,j}, \mu_{k,l})$ across $k = 1, \ldots, K$ and is smaller than $K_{tr} = 4$ in most cases; see, for example, the plot of $\mu_{\cdot,1}$ versus $\mu_{\cdot,3}$.

A sample of $N = 1000$ observations $\mathbf{y} = (\mathbf{y}_1, \ldots, \mathbf{y}_N)$ is simulated from this model and a multivariate normal mixture with $K = 4$ components is fitted using Bayesian inference based on the Dirichlet prior $\mathcal{D}(16.5, \ldots, 16.5)$ for $\boldsymbol{\eta}$, which will be motivated in Subsection 10.3.1, and the standard prior for $\boldsymbol{\mu}_k$ and Σ_k (Stephens, 1997) where $\mu_{k,l} \sim \mathcal{N}\left(m_l, R_l^2\right)$, independently for each $k = 1, \ldots, K$ and $l = 1, \ldots, r$, and $\Sigma_k \sim \mathcal{W}_r^{-1}(c_0, \mathbf{C}_0)$ with $\mathbf{C}_0 \sim \mathcal{W}_r(g_0, \mathbf{G}_0)$. The hyperparameters m_l and R_l are chosen as the midpoint and the length of the observation interval of the lth feature, while $c_0 = 2.5 + (r-1)/2$, $g_0 = 0.5 + (r-1)/2$, and $\mathbf{G}_0 = 100 g_0/c_0 \text{Diag}\left(1/R_1^2 \ldots 1/R_r^2\right)$.

Figure 10.3 shows the point-process representation of 5000 MCMC draws simulated from the corresponding posterior after a burn-in of 1000 draws. As for the univariate normal mixture considered in Figure 10.1, the MCMC draws cluster around the true points visible in Figure 10.2 despite the higher dimensions. It is expected that two MCMC draws corresponding to the same component-specific parameter cluster around the same point in the point-process representation even if label switching took place between the draws. This property is used in Subsection 10.2.2 to identify the mixture in an automatic fashion.

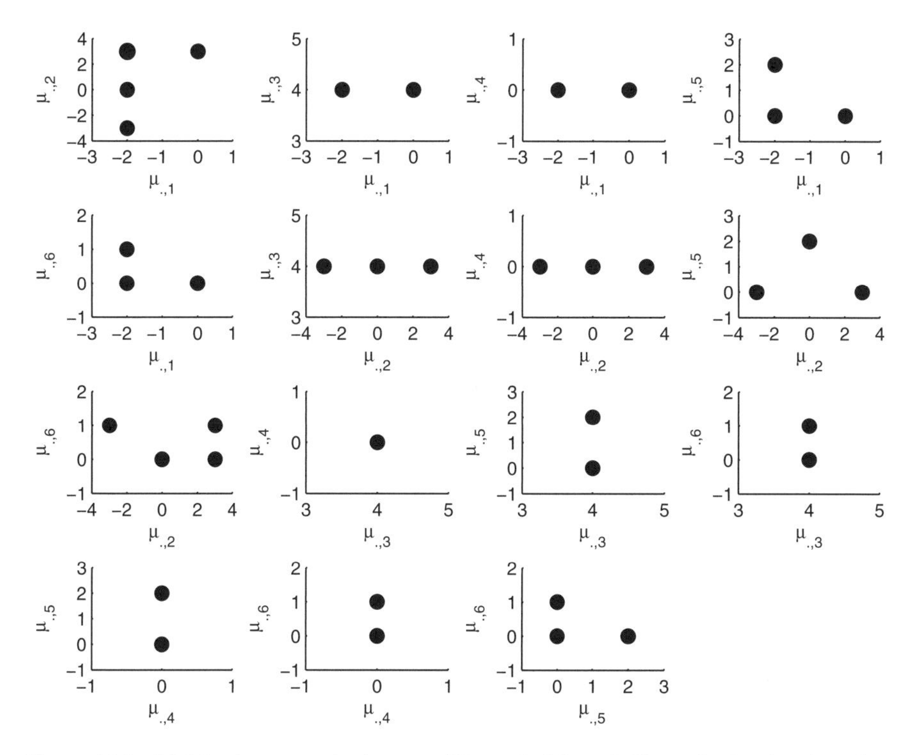

Figure 10.2 Multivariate normal mixtures of Example 1 ($r = 6$, $K_{tr} = 4$), showing the point-process representation of the true mixture distribution using projections on to $(\mu_{\cdot,j}, \mu_{\cdot,l})$ for all combinations (j, l).

10.2.2 Identification through clustering in the point-process representation

Frühwirth-Schnatter (2006, page 97) suggested using k-means clustering in the point-process representation of the MCMC draws to identify a finite mixture model. The method is related to that of Celeux (1998), but it identifies K clusters in the space Θ instead of $K!$ clusters in the space Θ^K. Hence it is much easier to implement in particular for high-dimensional problems.

Clustering in the point-process representation works as follows:

(a) Apply k-means clustering with K clusters to all KM posterior draws of the component-specific parameter $\boldsymbol{\theta}_k^{(m)}$, $k = 1, \ldots, K$, $m = 1, \ldots, M$. This delivers a classification index $I_k^{(m)}$ taking values in $\{1, \ldots, K\}$ for each of the KM posterior draws. It is advisable to standardise $\theta_{k,l}^{(m)}$ for each l across all KM posterior draws prior to clustering, in particular if the elements of $\boldsymbol{\theta}_k$ have a different scale.

(b) For each $m = 1, \ldots, M$, construct the classification sequence $\rho_m = (I_1^{(m)}, \ldots, I_K^{(m)})$ and check if ρ_m is a permutation of $\{1, \ldots, K\}$. Note that

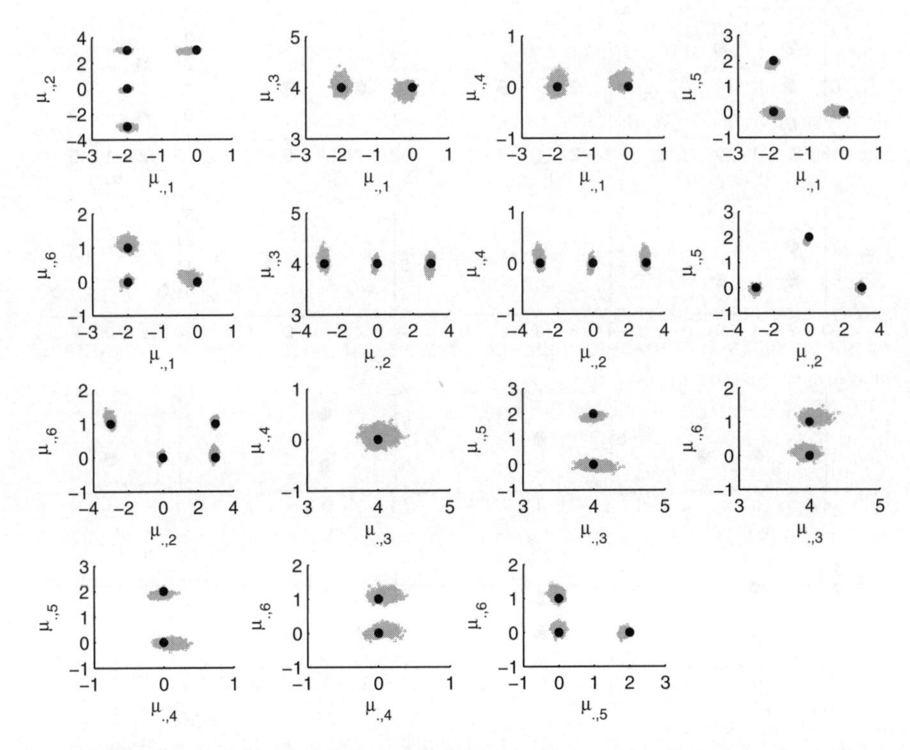

Figure 10.3 Multivariate normal mixtures of Example 1 ($r = 6$, $K_{tr} = 4$), showing the point-process representation of 5000 MCMC draws obtained by fitting a mixture with $K = 4$ components; the point-process representation of the true mixture is marked by $\bullet$.

condition (3.57) given in Frühwirth-Schnatter (2006, page 97) is not sufficient; instead one could simply sort the elements of ρ_m and check if the resulting sequence is equal to $\{1, \ldots, K\}$. Determine the number M_ρ of classification sequences ρ_m that are not a permutation.

(c) For all $M - M_\rho$ MCMC draws where ρ_m is a permutation, a unique labelling is achieved by reordering the draws through the inverse ρ_m^{-1} of ρ_m:

(c1) $\eta_1^{(m)}, \ldots, \eta_K^{(m)}$ is substituted by $\eta_{\rho_m^{-1}(1)}^{(m)}, \ldots, \eta_{\rho_m^{-1}(K)}^{(m)}$;

(c2) $\boldsymbol{\theta}_1^{(m)}, \ldots, \boldsymbol{\theta}_K^{(m)}$ is substituted by $\boldsymbol{\theta}_{\rho_m^{-1}(1)}^{(m)}, \ldots, \boldsymbol{\theta}_{\rho_m^{-1}(K)}^{(m)}$.

The same permutation ρ_m may be used to relabel the MCMC draws $\mathbf{S}^{(m)} = (S_1^{(m)}, \ldots, S_N^{(m)})$ of the hidden allocations:

(c3) $S_1^{(m)}, \ldots, S_N^{(m)}$ is substituted by $\rho_m(S_1^{(m)}), \ldots, \rho_m(S_N^{(m)})$.

In step (a), the clustering of $\boldsymbol{\theta}_k^{(m)}$ is unrelated to the clustering of $\boldsymbol{\theta}_j^{(m)}$ for the same MCMC draw, and hence draws from different components may be allocated to the same cluster. This is in contrast to other relabelling methods that force $\boldsymbol{\theta}_k^{(m)}$ and $\boldsymbol{\theta}_j^{(m)}$ to belong to different clusters, in order to guarantee that in step (b) ρ_m is a permutation for all draws.

It is not essential to apply k-means clustering in step (a) to the entire component-specific parameter $\boldsymbol{\theta}_k$ to obtain sensible classification sequences ρ_m for relabelling. In particular, for higher-dimensional problems it is possible to further reduce complexity by considering only a subset of the elements of $\boldsymbol{\theta}_k$ instead of the entire vector. The resulting classification sequences are used in step (c) to identify the entire vector $\boldsymbol{\theta}_k$. Pamminger and Frühwirth-Schnatter (2010), for instance, consider clustering categorial time-series panel data using finite mixtures of Markov chain models and discuss an application to a panel of 10 000 individual wage mobility data from the Austrian labour market. The component-specific parameter $\boldsymbol{\theta}_k$ is a transition matrix with 30 distinct elements. Instead of clustering all elements of $\boldsymbol{\theta}_k$, they use only the six persistence probabilities, which are expected to be different across groups based on economic knowledge.

Furthermore, it is possible to apply k-means clustering in step (a) to some functional of $\boldsymbol{\theta}_k$. Frühwirth-Schnatter and Pyne (2010), for instance, consider mixtures of multivariate skew normal and skew-t distributions and discuss an application to flow cytometry. For identification they apply k-means clustering to the component-specific mean $\mathrm{E}(\mathbf{y}|\boldsymbol{\theta}_k)$, which is a nonlinear function of $\boldsymbol{\theta}_k$.

Finally, Frühwirth-Schnatter (2008) shows that the method is easily extended to more general finite mixture models like Markov switching models. Hahn *et al.* (2010), for instance, consider continuous-time multivariate Markov switching models and discuss an application to modelling multivariate financial time series. For labelling through k-means clustering they use the component-specific mean $\boldsymbol{\mu}_k$ and the component-specific volatilities $\mathrm{Diag}\,(\Sigma_k)^{1/2}$.

All classification sequences $\rho_m, m = 1 \ldots, M$, derived by k-means clustering are expected to be permutations if the point-process representation of the MCMC draws contains K well-separated simulation clusters. If a small fraction of nonpermutations is present, then the posterior draws may be stratified according to whether or not they are identified. Only the subsequence of identified draws is used in step (c) for component-specific inference known to be sensitive to label switching, like parameter estimation and classification.

However, if the fraction of nonpermutations is high for the chosen number of components K, any inference that is known to be sensitive to label switching should be avoided for that particular K. A high fraction of nonpermutations typically happen in cases where the mixture is overfitting the number of components; see Subsection 10.3.2.

Example 1 (*continued*): Multivariate normal mixtures

For illustration, labelling through k-means clustering in the point-process representation is applied to the simulated data introduced in Subsection 10.2.1. Clustering in step (a) is based on the six elements $(\mu_{k,1}, \ldots, \mu_{k,6})$ of $\boldsymbol{\mu}_k$. However, one could add measures describing Σ_k, such as the six component-specific volatilities $\mathrm{Diag}\,(\Sigma_k)^{1/2}$ or the determinant $|\Sigma_k|$.

The point-process representation of the 5000 MCMC draws shown in Figure 10.3 contains $MK = 20\ 000$ realisations of $\boldsymbol{\mu}_k^{(m)}$. Applying k-means

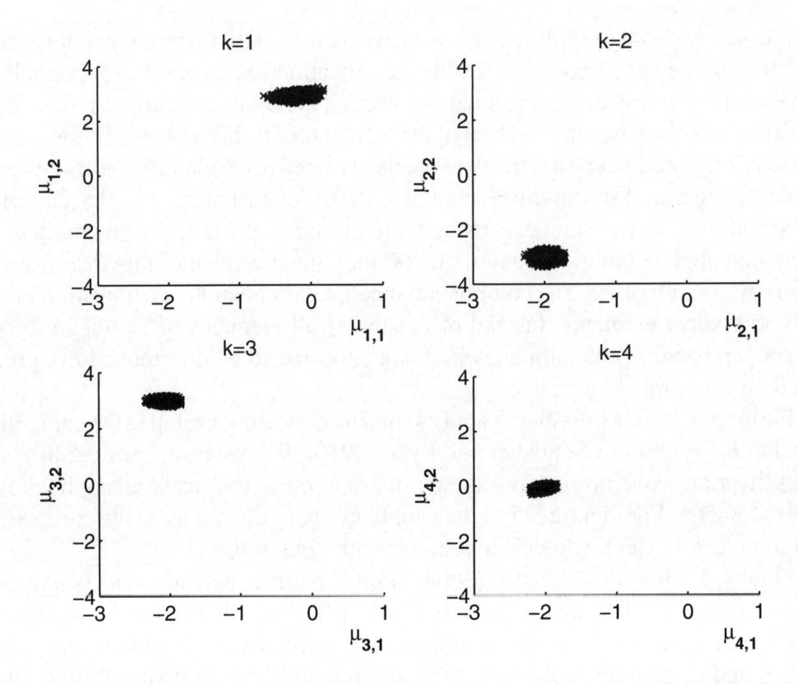

Figure 10.4 Multivariate normal mixtures of Example 1, showing identified MCMC draws of $(\mu_{k,1}, \mu_{k,2})$ for $k = 1, \ldots, 4$.

clustering with four clusters delivers a classification index $I_k^{(m)}$ for each of these 20 000 realisations. Based on these indices, 5000 classification sequences $\rho_m = (I_1^{(m)}, \ldots, I_4^{(m)})$ are constructed for $m = 1, \ldots, 5000$, each of which corresponds to the mth MCMC draw. It turns out that all but $M_\rho = 23$ classification sequences are permutations of $\{1, \ldots, 4\}$. These permutations ρ_m are used to identify the mixture through relabelling the corresponding 4973 MCMC draws, as described in step (c). Figure 10.4 shows scatterplots of the 4973 identified MCMC draws of $(\mu_{k,1}, \mu_{k,2})$ for each $k = 1, \ldots, 4$.

10.3 Identifying mixtures when the number of components is unknown

Identification through clustering in the point-process representation is possible even when the true number K_{tr} of nonempty, nonidentical components is unknown. However, one has to deal appropriately with the irregular likelihood function associated with overfitting mixtures when the number of components K is larger than K_{tr}. It turns out that the prior on the weight distribution $\boldsymbol{\eta} = (\eta_1, \ldots, \eta_K)$ plays an important role in this respect.

10.3.1 The role of Dirichlet priors in overfitting mixtures

The usual prior on the weight distribution is a Dirichlet prior, i.e. $\boldsymbol{\eta} \sim \mathcal{D}(e_1, \ldots, e_K)$. Very often an exchangeable prior is assumed with $e_k \equiv e_0$, $k = 1, \ldots, K$. A very popular choice, found in many papers, is $\boldsymbol{\eta} \sim \mathcal{D}(1, \ldots, 1)$, i.e. the uniform distribution on the unit simplex. Bernardo and Girón (1988) investigate reference priors for overfitting mixtures where the component-specific parameters are known and show that the $\mathcal{D}(1, \ldots, 1)$ prior approximates the reference prior for $K = 2$ as well as for $K > 2$, whenever all component mixture densities converge to the same distribution.

However, under the uniform prior, the posterior distribution is nothing but the normalised likelihood function, which is highly irregular for overfitting mixtures because it reflects two different ways of reducing the overfitting mixture to a mixture with the true number of components; see Frühwirth-Schnatter (2006, Section 1.3.2). Assume, for instance, that the mixture is overfitting by one degree, meaning that $K = K_{tr} + 1$. One perspective of the likelihood evolution is to 'let two component-specific parameters be identical' where the difference $\boldsymbol{\theta}_k - \boldsymbol{\theta}_j$ between two component-specific parameters is shrunken towards 0. In this case, only the sum $\eta_k + \eta_j$ of the corresponding weights is identified. A second perspective is to 'leave one group empty' where one of the weights η_k is shrunken towards 0. In this case, the corresponding component-specific parameter $\boldsymbol{\theta}_k$ is not identified.

Rousseau and Mengersen (2010) study the asymptotic behaviour of the posterior distribution in overfitting mixture models. They show that the hyperparameter $(e_1, \ldots, e_K)$ strongly influences the way the posterior density handles overfitting. If $(e_1, \ldots, e_K)$ satisfies the condition $\max_{k=1,\ldots,K} e_k < d/2$, where $d = \dim \boldsymbol{\theta}_k$, then the posterior density is asymptotically concentrated on regions handling overfitting by leaving $K - K_{tr}$ groups empty. On the other hand, if $\min_{k=1,\ldots,K} e_k > d/2$, then at least one superfluous component has a positive weight and the posterior is asymptotically concentrated on regions where at least two component-specific parameters are more or less identical. Hence, choosing the right prior on the weight distribution helps the posterior distribution to focus on precisely one of the two ways to handle overfitting mixtures and facilitates posterior inference.

It is interesting to discuss various popular prior choices for $\boldsymbol{\eta}$ in the light of the results obtained by Rousseau and Mengersen (2010). It follows immediately that the uniform prior is not a good choice for overfitting univariate normal mixtures, because $e_0 = d/2$. On the other hand, this prior remains sensible for multivariate normal mixtures, at least asymptotically.

Frühwirth-Schnatter (2006, Section 4.2.2) recommends bounding the posterior away from regions allowing empty groups to deal with overfitting mixtures and claims that choosing $e_0 = 4$ serves this goal in general. However, Rousseau and Mengersen (2010) imply that this is true only for low-dimensional mixtures where $d < 8$, such as bivariate normal mixtures or switching regression models with at most seven covariates. With increasing d, the hyperparameter e_0 has to be increased in parallel to achieve this effect.

Ishwaran *et al.* (2001) recommend choosing $e_0 = \alpha/K_{\max}$, where $K_{\max}$ is the maximum number of components, in order to approximate a Dirichlet process prior (Ferguson, 1973) by the Dirichlet prior $\mathcal{D}(e_0, \dots, e_0)$. Green and Richardson (2001) show that this prior converges to a Dirichlet process prior with concentration parameter α as $K_{\max}$ increases. Simulation studies in Ishwaran *et al.* (2001) show that the behaviour of this prior with $\alpha = 1$ is quite good, at least if $K_{\max}$ is sufficiently large. Hence e_0 typically is smaller than $d/2$ and handles overfitting mixtures by favouring empty groups a posteriori.

In this context, it is also illuminating to consider the stick-breaking representation of the Dirichlet distribution ($k = 1, \dots, K-1$):

$$\begin{aligned} \eta_k|\eta_1, \dots, \eta_{k-1} &= \left(1 - \sum_{j=1}^{k-1} \eta_j\right) \phi_k, \\ \phi_k &\sim \mathcal{B}\left(e_k, \sum_{j=k+1}^{K} e_j\right), \end{aligned} \tag{10.2}$$

where $\phi_1, \dots, \phi_{K-1}$ are independent and $\phi_K = 1$. Equation (10.2) relies on the factor ϕ_k to define how much of the remaining weight is assigned to group k, given the sum of the weights $\eta_1, \dots, \eta_{k-1}$ of the groups $1, \dots, k-1$.

Figure 10.5 displays the priors $\phi_k \sim \mathcal{B}[e_0, (K-k)e_0], k = 1, \dots, K-1$, resulting from exchangeable priors for various hyperparameters e_0. Choosing $e_0 = 0.1$ leads to a prior where ϕ_k is pulled either towards 0 or 1. If ϕ_k is close to 0, then according to (10.2) hardly any weight is assigned to η_k, saving the remaining sum of weights for the remaining groups $k+1, \dots, K$. If ϕ_k is close to 1, then η_k takes most of the remaining sum of weights causing the remaining groups $k+1, \dots, K$ to be almost empty. Hence choosing $e_0 = 0.1$ favours empty groups, as expected from Rousseau and Mengersen (2010).

On the other hand, choosing $e_0 = 4$ and $e_0 = 16.5$ as for the mixtures considered in Subsection 10.2.1 bounds ϕ_k away from 0 and 1, favouring nonempty groups a priori. Choosing $e_0 = 1$ leads to a prior, which is rather indifferent in this respect.

10.3.2 The meaning of K for overfitting mixtures

Considering the results obtained in Subsection 10.3.1 it should be evident that no unique labelling exists when the model overfits the number of components. Either $\boldsymbol{\theta}_k$ is not identified, because η_k is 0, or at least two component-specific parameters $\boldsymbol{\theta}_k$ and $\boldsymbol{\theta}_j$ are the same.

Hence, most papers recommend the use of some model selection method to estimate K_{tr} and to identify a unique labelling for a model with K fixed at $\hat{K}_{tr}$. Popular methods are BIC (Keribin, 2000), marginal likelihoods (Frühwirth-Schnatter, 2004), the integrated classification likelihood (Biernacki *et al.*, 2000) or RJMCMC (Dellaportas and Papageorgiou, 2006; Richardson and Green, 1997).

To obtain meaningful inference for mixtures with an unknown number of components, it turns out to be important to distinguish in the fitted mixture between the total number of components K and the number of nonempty components K_0.

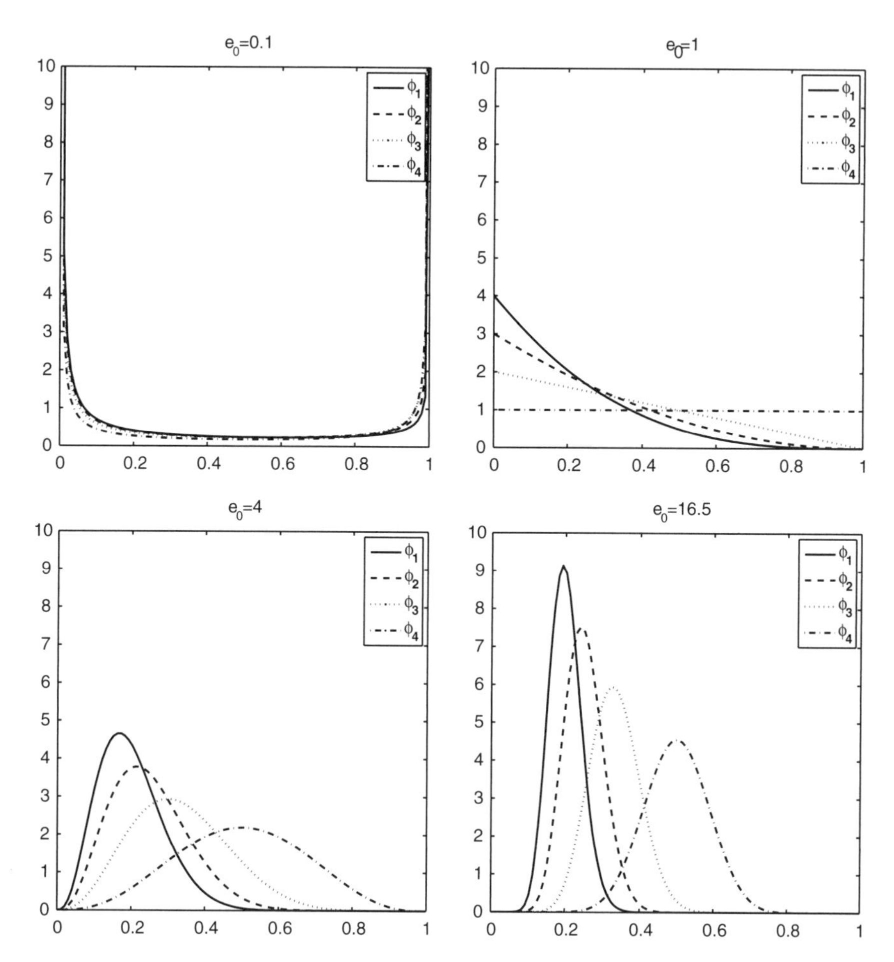

Figure 10.5 Prior on ϕ_k for $K = 5$; $e_0 = 0.1$ (top left-hand side), $e_0 = 1$ (top right-hand side), $e_0 = 4$ (bottom left-hand side) and $e_0 = 16.5$ (bottom right-hand side).

Since the total number K may be increased by adding empty components without changing the likelihood function, Nobile (2004) suggests using in general the number of nonempty components K_0 as an estimator for K. However, whether or not the total number of components K or the number of nonempty components K_0 in the posterior fitted mixture is closer to K_{tr} depends again on the prior of the weight distribution.

In the light of Rousseau and Mengersen (2010), it is to be expected that the number of nonempty groups K_0 is a poor estimator of K_{tr} for a prior where $e_0 > d/2$, because most of the posterior draws from an overfitting mixture come from regions where at least two nonempty components are more or less equal. For instance, if the mixture is overfitting by one degree, then the number K_0 of nonempty groups is equal to the total number of components K in the fitted mixture, and, consequently, K_0 will overestimate the number of nonempty, nonidentical components K_{tr}. For a prior favouring the 'let component-specific parameters be identical' perspective it is more sensible to use the marginal likelihood $p(\mathbf{y}|K)$ or the posterior $p(K|\mathbf{y})$ obtained by RJMCMC to estimate K_{tr}.

On the other hand, for a prior favouring the 'leave some groups empty' perspective, some of the groups are very likely a posteriori empty for an overfitting mixture and the total number of components K will overestimate K_{tr}. As a consequence, the marginal likelihood $p(\mathbf{y}|K)$ and the posterior probabilities $p(K|\mathbf{y})$ in RJMCMC will overestimate K_{tr} under such a prior (Nobile, 2004). In the light of Rousseau and Mengersen (2010), it is to be expected that a prior where $e_0 < d/2$ is a poor choice for selecting the number of components using the marginal likelihood or RJMCMC, because most of the posterior draws come from regions allowing empty groups. This inflates the marginal likelihood as shown by Nobile (2004) and explains why RJMCMC in combination with the uniform prior often leads to overfitting K, in particular for univariate mixtures of normals.

For a prior favouring the 'leave some groups empty' perspective, the number K_0 of nonempty groups is a much better estimator of K_{tr}. To estimate K_0, consider the posterior draws of $N_k(\mathbf{S})$, where $N_k(\mathbf{S})$ is the number of observations allocated to group k. For each MCMC draw of the allocation vector $\mathbf{S}$ define the corresponding number of empty groups $K_n(\mathbf{S})$ as

$$K_n(\mathbf{S}) = K - \sum_{k=1}^{K} \mathbb{I}_{N_k(\mathbf{S})=0}, \tag{10.3}$$

where $\mathbb{I}$ is the indicator function, which in the current setting takes the value 1 for any empty group, i.e. if and only if $N_k(\mathbf{S}) = 0$. Note that $K_n(\mathbf{S})$ is invariant to labelling the components. An estimator $\hat{K}_0$ of K_0 is then given by the value $K_n(\mathbf{S})$ visited most often by the MCMC procedure. The corresponding number of visits is denoted by M_0.

10.3.3 The point-process representation of overfitting mixtures

The point-process representation of the true finite mixture distribution is invariant under addition of an empty component, because the corresponding mark is 0 and the point is not visible. When one component is split into two, then the position of the points remains unchanged. Hence it is to be expected that the point-process representation of the MCMC draws obtained from Bayesian inference behaves similarly for an overfitting mixture. The precise behaviour, however, depends again on the prior of the weight distribution.

For a prior favouring the 'leave some groups empty' perspective, $K - K_{tr}$ groups have very small weights η_k for many MCMC draws. Within data augmentation no observation is allocated to these groups and the corresponding component-specific parameters $\boldsymbol{\theta}_k$ are sampled from the prior $p(\boldsymbol{\theta}_1, \ldots, \boldsymbol{\theta}_K)$. In the point-process representation of the MCMC draws, all draws from a nonempty component cluster around the points corresponding to the true mixture and are overlaid by a misty cloud of draws from the prior, corresponding to the empty components. This is illustrated in Figure 10.6 where a mixture with $K = 5$ components is fitted under

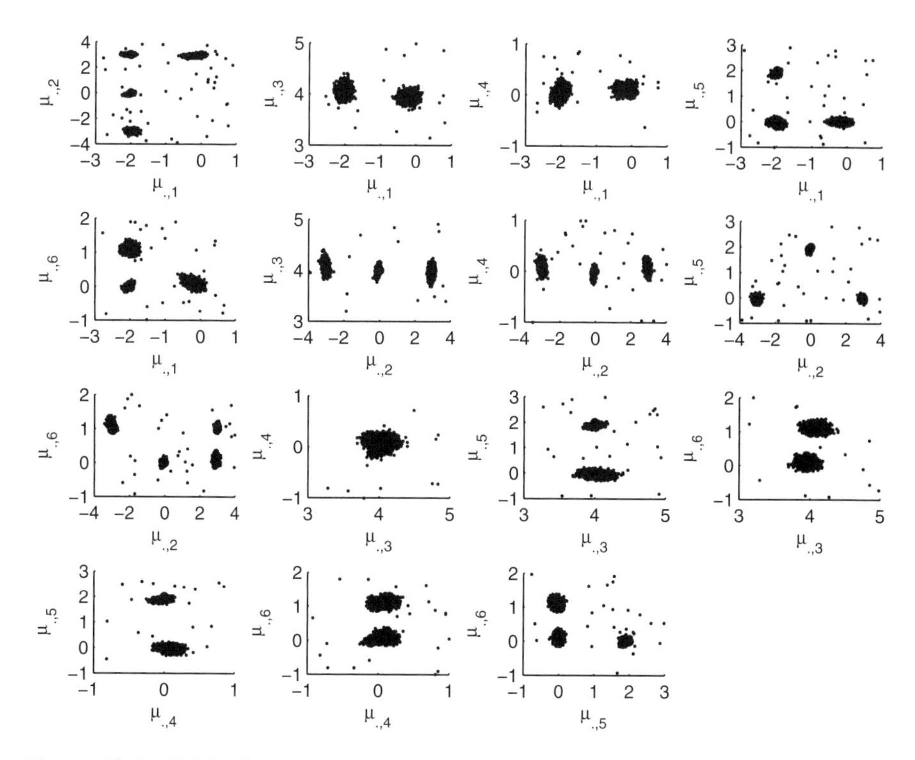

Figure 10.6 Multivariate normal mixtures of Example 1 ($r = 6$, $K_{tr} = 4$), showing the point-process representation of 5000 MCMC draws obtained by fitting a mixture with $K = 5$ components using the standard prior for the component-specific parameters and $e_0 = 0.1$.

$e_0 = 0.1$ to the data simulated in Subsection 10.2.1 from a multivariate normal mixture with $K_{tr} = 4$ components.

When k-means clustering with K clusters is applied to the point-process representation of an overfitting mixture using this prior, it is not clear what is going to happen. If the prior mode is close to the posterior mode, there usually is a high proportion of nonpermutations, indicating that the mixture is overfittting. On the other hand, in cases where the mode of the prior is quite distinct from the posterior mode, it might happen that a considerable fraction of the resulting classification sequences ρ_m are actually permutations of $\{1, \ldots, K\}$ with $K - K_{tr}$ groups corresponding to draws from the prior. However, posterior inference based on such draws is not likely to be useful.

An alternative is to estimate the number $\hat{K}_0$ of nonempty groups as described in Subsection 10.3.2 and to consider only the subsequence of the MCMC draws where $K_n(\mathbf{S}) = \hat{K}_0$ for further inference. The size of this subsequence is equal to M_0. For each of these MCMC draws, no observation is allocated to exactly $K - \hat{K}_0$ groups. To identify a mixture with $\hat{K}_0$ nonempty components, the empty groups are removed for each MCMC draw, leaving M_0 draws from a mixture with $\hat{K}_0$ nonempty components. Labelling through k-means clustering with K_0 clusters as described in Subsection 10.2 is applied to these reduced MCMC draws. The number $M_{0,\rho}$ of classification sequences that are not permutations is determined. All $M_0 - M_{0,\rho}$ draws where the classification sequence ρ_m is a permutation of $\{1, \ldots, \hat{K}_0\}$ are identified and considered for further posterior inference.

The situation is quite different for a prior favouring the 'let component-specific parameters be identical' strategy for overfitting mixtures. As far as the point-process representation of an overfitting mixture using this prior is concerned, at least two component-specific parameters $\boldsymbol{\theta}_k$ and $\boldsymbol{\theta}_j$ are strongly pulled together for many MCMC draws, meaning that both $\boldsymbol{\theta}_k$ and $\boldsymbol{\theta}_j$ cluster around a single point corresponding to the point-process representation of the true mixture. This is illustrated in Figure 10.7, where a mixture with five components is fitted as in Figure 10.6, but with $e_0 = 16.5$.

When k-means clustering with K clusters as described in Subsection 10.2 is applied to such a point-process representation, then a high proportion of nonpermutations results and it is not possible to identify the original mixture model from the MCMC draws of the overfitting mixture. The best strategy is first select an optimal number K under this prior using model choice based on marginal likelihoods or RJMCMC and to identify the mixture for the optimal K. In addition, the proportion of nonpermutations using a prior avoiding empty groups provides an informal way

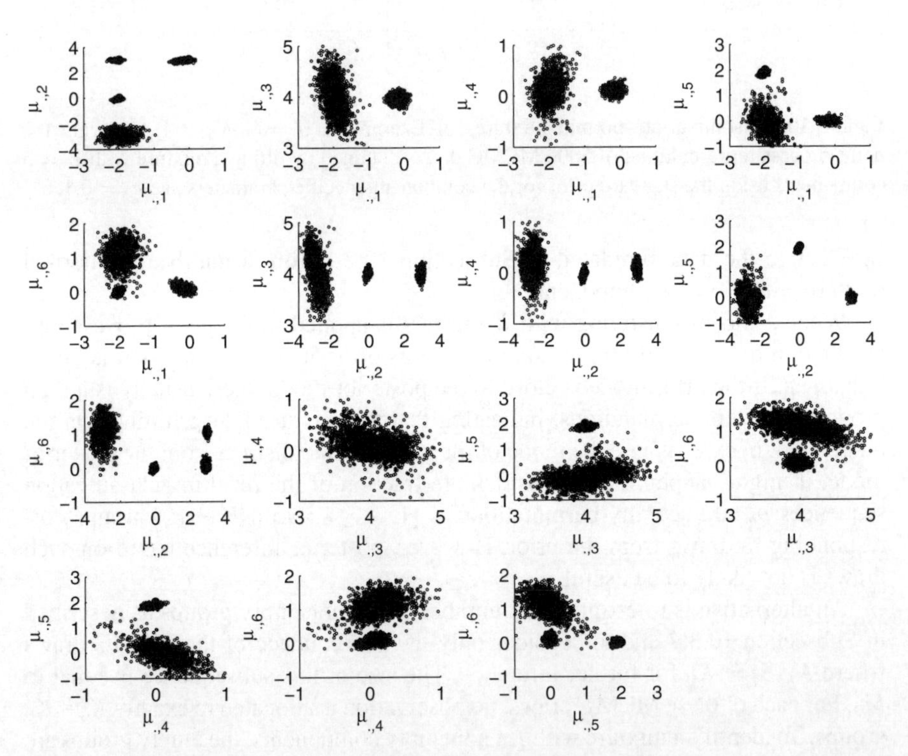

Figure 10.7 Multivariate normal mixtures of Example 1 ($r = 6$, $K_{tr} = 4$), showing the point-process representation of 5000 MCMC draws obtained by fitting a mixture with $K = 5$ components using the standard prior for the component-specific parameters and $e_0 = 16.5$.

of selecting K, because a high proportion of nonpermutations for a K-component mixture typically is a sign that K is overfitting K_{tr}.

10.3.4 Examples

Normal mixtures with increasing numbers K of components are fitted to various simulated datasets combining the standard priors for the component-specific parameters mentioned earlier with various priors for the weight distribution, namely the Dirichlet prior with scale $e_0 = 0.1$, which favours for overfitting mixtures empty groups, the one with $e_0 = (r + r(r+1)/2)/2 + 3$, which favours similar, nonempty groups, and the standard uniform prior corresponding to $e_0 = 1$.

For each combination of K and e_0, 5000 draws are simulated from the posterior distribution after a burn-in of 1000 draws using simple MCMC based on data augmentation and Gibbs sampling. An estimator $\hat{K}_0$ of the number of nonempty components is derived as explained after Equation (10.3) and the number M_0 of MCMC draws having exactly $\hat{K}_0$ nonempty groups is determined. Furthermore, k-means clusterings using K clusters is performed using (μ_k, σ_k) for univariate mixtures and the components of $\boldsymbol{\mu}_k$ for multivariate mixtures, and the number M_ρ of classification sequences that are not permutations of $\{1, \ldots, K\}$ is reported.

The marginal likelihood $p(\mathbf{y}|K, e_0)$ is computed using bridge sampling as in Frühwirth-Schnatter (2004). The importance density is a mixture distribution constructed from all $K!$ permutations of randomly selected complete data posteriors $p(\boldsymbol{\theta}_1, \ldots, \boldsymbol{\theta}_K, \boldsymbol{\eta}|\mathbf{S}^{(s)}, \mathbf{y})$, $s = 1, \ldots, S$, where S decreases with K to make the computations feasible; see also Frühwirth-Schnatter (2006, Section 5.3).

Finally, for each combination of K and e_0, BIC-MCMC is determined from the largest log-likelihood value $\log p(\mathbf{y}|\boldsymbol{\theta}_1, \ldots, \boldsymbol{\theta}_K, \boldsymbol{\eta})$ observed across the MCMC draws. Whereas the classical BIC criterion is independent of the prior, BIC-MCMC depends on the hyperparameter e_0, because different regions of the parameter space are visited during MCMC steps.

Example 2: Univariate normal mixtures

Table 10.1 reports results for $N = 1000$ observations generated from a mixture of $K_{tr} = 2$ univariate normal distributions with $\mu_1 = -4$, $\mu_2 = 3$, $\sigma_1^2 = \sigma_2^2 = 0.5$ and $\eta_1 = 0.2$. The marginal likelihood $p(\mathbf{y}|K, e_0)$ selects with high posterior probability a mixture with $K = 2$ nonempty components using the prior associated with $e_0 = 4$. Using k-means clustering with two clusters in the corresponding MCMC point-process representation yields classification sequences that are permutations for all cases but $M_\rho = 53$ draws. Posterior inference is then based on the resulting 4947 identified MCMC draws.

The marginal likelihood selects a mixture with $K = 3$ components using the prior associated with $e_0 = 0.1$. Whereas the total number K of components overfits the true number of components, the number of nonempty components $\hat{K}_0 = 2$ yields the true value. This mixture is not identifiable in the point-process representation, because a high fraction of MCMC draws are not a permutation of $\{1, 2, 3\}$ ($M_\rho = 4398$). However, a two-component mixture is identifiable from the $M_0 = 4223$ MCMC draws having exactly two nonempty groups. After the one

Table 10.1 Univariate normal mixtures of Example 2, showing the influence of various priors of the weight distribution (e_0) on model choice with respect to the number of components, with the model selected by each criterion indicated in bold.

e_0	K	$\log p(\mathbf{y}\|K, e_0)$	BIC-MCMC	$\hat{K}_0$	M_0	M_ρ	$M_{0,\rho}$
	1	−2471.7	4938.1				
0.1	2	−1576.6	**3144.5**	2	5000	61	61
	3	**−1576.5**	3161.1	2	4223	4398	18
	4	−1623.4	3277.9	2	4070	4841	117
	5	−1623.8	3297.9	2	3740	4841	104
1	2	−1583.9	**3161.3**	2	5000	0	0
	3	**−1583.5**	3781.8	3	4997	5000	3852
	4	−1585.9	3186.8	4	4897	5000	4293
	5	−1592.7	3207.8	5	4652	5000	4501
4	2	**−1575.9**	**3144.5**	2	5000	53	53
	3	−1578	3164	3	5000	2342	2342
	4	−1629.2	3275.9	4	5000	4660	4660
	5	−1636.2	3296.6	5	5000	4660	4660

empty group is removed for each of these draws, k-means clustering with two clusters is applied. All classification indices except $M_{\rho,0} = 18$ are permutations of $\{1, 2\}$. Posterior inference is then based on the resulting 4205 identified draws.

For the uniform prior corresponding to $e_0 = 1$, the marginal likelihood is practically the same for mixtures with $K = 2$ and $K = 3$ components. Taking the mixture with the (slightly) larger marginal likelihood leads to overfitting the true number of components. Finally, BIC-MCMC selects the correct number of components in this example for all priors.

Example 1 (continued): Multivariate normal mixtures

Table 10.2 reports the results for the six-dimensional data simulated in Subsection 10.2.1 from a normal mixture with $K_{tr} = 4$ components. BIC-MCMC selects $K = 3$ components and underfits the true number of components, regardless of the prior.

The marginal likelihood selects a mixture with four nonempty components under the prior $e_0 = 16.5$, which yields the true number of components. Furthermore, this mixture is identifiable from the point-process representation, because all MCMC draws apart from $M_\rho = 23$ draws are permutations. The corresponding 4978 identified MCMC draws have already been plotted in Figure 10.4.

For the prior associated with $e_0 = 0.1$, the marginal likelihood selects a mixture with $K = 5$ components. However, only $\hat{K}_0 = 4$ groups are nonempty. Again, under this prior, the total number of components overfits the true number of components, while the number of nonempty components in the selected mixture yields the true value. This mixture is not identifiable in the point-process representation, because a high fraction of MCMC draws is not a permutation of $\{1, \ldots, 5\}$ ($M_\rho = 4130$). However, a four-component mixture is identifiable from the $M_0 = 5000$ MCMC

Table 10.2 Multivariate normal mixtures of Example 1 ($r = 6$, $K_{tr} = 4$), showing the influence of various priors of the weight distribution (e_0) on model choice with respect to the number of components, with the model selected by each criterion indicated in bold.

e_0	K	$\log p(\mathbf{y}\mid K, e_0)$	BIC-MCMC	$\hat{K}_0$	M_0	M_ρ	$M_{0,\rho}$
0.1	2	−9745.7	19534.1	2	5000	0	0
	3	−9509.9	**19135.8**	3	5000	0	0
	4	−9509.5	19323.6	3	4747	3952	0
	5	**−9483.4**	19369.1	4	5000	4130	0
	6	−9483.8	19563.4	4	4918	4619	566
1	2	−9744	19529.7	2	5000	0	0
	3	−9506.7	**19133.3**	3	5000	0	0
	4	−9490.2	19170.2	4	3003	1655	2588
	5	**−9482.9**	19371.8	4	4988	3855	4595
	6	−9487.8	19557.8	4	4288	4669	3988
16.5	2	−9742.6	19535.9	2	5000	0	0
	3	−9505.6	**19135.9**	3	5000	0	0
	4	**−9474.9**	19177.9	4	5000	23	23
	5	−9503.6	19339.5	5	5000	3631	3631
	6	−9538.5	19490.8	6	5000	3688	3688

draws having exactly four nonempty groups. After the one empty group is removed for each of these draws, k-means clustering with four clusters is applied. All classification indices are permutations of $\{1, \ldots, 4\}$ ($M_{0,\rho} = 0$). Posterior inference is then based on the resulting 5000 identified draws.

For the uniform prior, the marginal likelihood selects a mixture with five components, where only four of them are nonempty. However, it is not possible to identify the mixture from the MCMC draws by removing empty groups as described above, because the majority of these draws ($M_{0,\rho} = 4595$) is not a permutation of $\{1, \ldots, 4\}$.

Example 3: Multivariate normal mixtures

$N = 1000$ eight-dimensional observations were generated from a mixture of four multivariate normal distributions with $\eta_1 = \cdots = \eta_4 = 1/4$,

$$\begin{pmatrix} \boldsymbol{\mu}_1 & \boldsymbol{\mu}_2 & \boldsymbol{\mu}_3 & \boldsymbol{\mu}_4 \end{pmatrix} = \begin{pmatrix} -2 & 2 & -2 & 2 \\ 0 & 1 & 0 & 0 \\ 4 & 3 & 1 & 3 \\ 1 & 0 & 2 & 0 \\ -2 & -2 & 2 & 2 \\ 2 & 0 & -2 & 0 \\ 1 & 1 & 2 & 1 \\ -1 & 0 & 1 & 0 \end{pmatrix},$$

Table 10.3 Multivariate normal mixtures of Example 3 ($r = 8$, $K_{tr} = 4$), showing the influence of various priors of the weight distribution (e_0) on model choice with respect to the number of components, with the model selected by each criterion indicated in bold.

e_0	K	$\log p(\mathbf{y}\|K, e_0)$	BIC-MCMC	$\hat{K}_0$	M_0	M_ρ	$M_{\rho,0}$
0.1	3	−13272.2	26683.2	3	5000	0	0
	4	−13172.4	**26635.1**	4	5000	0	0
	5	−13171.7	26953.6	4	5000	3975	0
	6	**−13171.6**	27262.9	4	5000	4683	0
1	3	−13268.9	26679.4	3	5000	0	0
	4	**−13167.4**	**26633**	4	5000	0	0
	5	−13171.3	26942.7	4	5000	3987	0
	6	−13175.4	27262	4	5000	4712	0
25	3	−13268.8	26678.9	3	5000	0	0
	4	**−13162.4**	**26634.9**	4	5000	0	0
	5	−13202	26905.4	5	5000	89	89
	6	−13259.9	27186.1	6	5000	275	275

$\Sigma_1 = 1.2\mathbf{I}_8$, $\Sigma_2 = \mathbf{I}_8 + 0.4\mathbf{e}_8$, $\Sigma_3 = 2\mathbf{I}_8 - 0.2\mathbf{e}_8$ and Σ_4 being a correlation matrix with $\Sigma_{4,jl} = 0.8^{|j-l|}$. All results are reported in Table 10.3.

The mixture components are extremely well-separated, and hence it is to be expected that the likelihood function is quite informative concerning clustering of the observations. As a consequence, little information is lost when estimation is based on the mixture likelihood function rather than the complete data likelihood function where the allocations $\mathbf{S}$ are known. Table 10.3 shows that the prior of $p(\boldsymbol{\eta})$ loses to a certain degree its impact on selecting the right number of components for such an informative likelihood function. For instance, BIC-MCMC selects the right number of components regardless of the prior.

The marginal likelihood selects mixtures with $K = 4$ nonempty components ($\hat{K}_0 = 4$) for $e_0 = 1$ and $e_0 = 25$ and identification through k-means clustering in the point-process representation is possible for all MCMC draws in both cases.

For the prior calibration $e_0 = 0.1$, the marginal likelihood selects a mixture with six components. However, the marginal likelihood is virtually the same for a mixture with five components. Taking either $K = 5$ or $K = 6$ as an estimator for K_{tr} again overfits the true number of components, while the number of nonempty components $\hat{K}_0 = 4$ yields the true number of components for both mixtures. For both mixtures, all $M_0 = 5000$ MCMC draws have four nonempty groups. After the empty groups are removed from each of these MCMC draws k-means clustering with four clusters is applied to identify a four-component mixture. All classification indices turn out to be permutations of $\{1, \ldots, 4\}$. Posterior inference is then based on the resulting 5000 identified MCMC draws.

10.4 Overfitting heterogeneity of component-specific parameters

10.4.1 Overfitting heterogeneity

The standard finite mixture model assumes that a priori all elements in the component-specific parameter $\boldsymbol{\theta}_k$ are different. In particular, in higher dimensions it might happen that some components of $\boldsymbol{\theta}_k$ are homogeneous in certain dimensions, i.e $\theta_{1,j} = \cdots = \theta_{K,j}$ for a subset of indices $j \in \{1, \ldots, r\}$. Fitting a mixture with unconstrained parameters $\boldsymbol{\theta}_1, \ldots, \boldsymbol{\theta}_K$ leads to overfitting heterogeneity.

For instance, overfitting heterogeneity is present in the multivariate mixture example considered in Subsection 10.2.1. The corresponding point-process representation of the true mixture distribution shown in Figure 10.2 indicates overfitting heterogeneity in a natural way. For instance, the projection reduces to a single point for any pair $(\mu_{\cdot,j}, \mu_{\cdot,l})$ where both $\mu_{k,j}$ and $\mu_{k,l}$ are homogeneous. For any pair $(\mu_{\cdot,j}, \mu_{\cdot,l})$ where one element is homogeneous, the number of points is equal to the number of distinct values of the heterogeneous element and all points lie on a line parallel to the axis of the heterogeneous element.

Similarly, the point-process representation of the MCMC draws indicate homogeneity of $\mu_{k,j}$ and $\mu_{k,l}$, because in this case only a single simulation cluster is visible in the scatter plot of the draws of $(\mu_{\cdot,j}, \mu_{\cdot,l})$, as shown in Figure 10.3. This information may be used to identify in an exploratory manner which elements of $\boldsymbol{\theta}_k$ are homogenous; see, for instance, Frühwirth-Schnatter (2001a, 2001b).

For the example considered above, the marginal likelihood selected the true number of components (in Subsection 10.3.4), despite overfitting heterogeneity. However, when using an increasing degree of overfitting heterogeneity, selecting K based on the marginal likelihood tends to underfit the true number of components, because the marginal likelihood penalises the assumption of heterogeneity in $\theta_{k,j}$ whenever $\theta_{k,j}$ is homogeneous, even when the true number of components has been chosen.

Example 4: Multivariate normal mixtures

Consider, for instance, $N = 1000$ eight-dimensional observations generated by a mixture of $K_{tr} = 4$ multivariate normal distributions with $\eta_1 = \cdots = \eta_4 = 1/4$,

$$\begin{pmatrix} \boldsymbol{\mu}_1 & \boldsymbol{\mu}_2 & \boldsymbol{\mu}_3 & \boldsymbol{\mu}_4 \end{pmatrix} = \begin{pmatrix} -1 & -1 & 1 & 1 \\ 0 & 0 & 0 & 0 \\ 4 & 4 & 4 & 4 \\ 0 & 0 & 0 & 0 \\ -1 & 1 & -1 & 1 \\ 0 & 0 & 0 & 0 \\ 1 & 1 & 1 & 1 \\ 0 & 0 & 0 & 0 \end{pmatrix},$$

$\Sigma_1 = 0.5\mathbf{I}_r$, $\Sigma_2 = 4\mathbf{I}_r + 0.2\mathbf{e}_r$, $\Sigma_3 = 4\mathbf{I}_r - 0.2\mathbf{e}_r$ and $\Sigma_4 = \mathbf{I}_r$. This mixture is similar to the one considered in Example 1. However, the dimension of r is higher and the mean is different only for features one and five, while it is the same in all other dimensions.

Fitting an unconstrained mixture with $K = 4$ components considerably overfits the number of parameters, because in six dimensions four parameters are introduced where a single parameter would have been sufficient. This leads to difficulties with identifying the mixture with the right number of components; see Table 10.4 where normal mixtures with increasing numbers K of components are fitted as

Table 10.4 Multivariate normal mixtures of Example 4 ($r = 8$, $K_{tr} = 4$), showing the comparison of various priors of the weight distribution (e_0) and various priors for the component means, with the model selected by each criterion indicated in bold.

Prior $\mu_{k,j}$	e_0	K	$\log p(\mathbf{y}\|K, e_0)$	BIC-MCMC	$\hat{K}_0$	M_0	M_ρ	$M_{0,\rho}$
Standard	0.1	2	**−13077.7**	**26269.9**	2	5000	0	0
	0.1	3	−13090.9	26431.7	3	5000	0	0
	0.1	4	−13108.4	26628.3	4	5000	0	0
	0.1	5	−13105	26937.2	4	5000	5000	0
	0.1	6	−13107	27252.8	4	5000	5000	0
	25	2	**−13081.2**	**26271.9**	2	5000	0	0
	25	3	−13083.5	26425.1	3	5000	0	0
	25	4	−13098.7	26624	4	5000	0	0
	25	5	−13131	26863.5	5	5000	91	91
	25	6	−13170.9	27118.6	6	5000	2521	2521
Laplace	0.1	2	−13044.7	**26269.1**	2	5000	0	0
	0.1	3	−13039.2	26429.2	3	5000	0	0
	0.1	4	−13036.9	26721.8	3	4161	4477	3794
	0.1	5	−13036.3	26927.5	4	3909	4655	3782
	0.1	6	**−13032.4**	27179.8	4	3341	5000	3341
	25	2	−13048.1	**26271.8**	2	5000	0	0
	25	3	−13031.7	26416.3	3	5000	0	0
	25	4	**−13027**	26630.4	4	5000	0	0
	25	5	−13043.3	26858.4	5	5000	77	77
	25	6	−13063.3	27095	6	5000	1628	1628
Normal-gamma	0.1	2	−13008.2	**26266**	2	5000	0	0
	0.1	3	−12980.6	26424.9	3	5000	0	0
	0.1	4	−12956.7	26611.8	4	5000	0	0
	0.1	5	**−12941.7**	26874.2	5	3960	2482	1729
	0.1	6	−12951.2	27158	5	2577	4997	2577
	25	2	−13011.8	26270.4	2	5000	0	0
	25	3	−12974.6	26422.6	3	5000	0	0
	25	4	**−12946.4**	26626.7	4	5000	0	0
	25	5	−12949.2	26851.5	5	5000	218	218
	25	6	−12970.4	27131.1	6	5000	1478	1478

in Subsection 10.3.4 using the standard prior described above. Regardless of e_0, the marginal likelihood picks a mixture with two nonempty components ($\hat{K}_0 = 2$), and thus strongly underfits the true number of components. On the other hand, k-means clustering in the point-process representation clearly indicates that it is possible to go beyond $K = 2$ without losing identifiability, because the number of nonpermutations M_ρ is equal to 0 up to $K = 4$.

This discrepancy is handled in Subsection 10.4.2 by substituting the standard prior on the component-specific means by a shrinkage prior.

10.4.2 Using shrinkage priors on the component-specific location parameters

If it is expected that the mixture model is overfitting the heterogeneity of the components, one could use shrinkage priors well-known from variable selection in regression models (Fahrmeir *et al.*, 2010; Park and Casella, 2008) instead of standard priors.

In this section two such shrinkage priors are discussed for the component-specific mean in a multivariate normal mixture, namely the normal-gamma prior (Griffin and Brown, 2010) and the Bayesian Lasso prior (Park and Casella, 2008). Alternative shrinkage priors for univariate and multivariate normal mixtures have been considered recently by Yau and Holmes (2010).

Subsequently, priors are centred at m_l, being equal to the mean of the data in dimension l rather than at the midpoint as for the standard prior.

The normal-gamma prior

The normal-gamma prior has been studied recently by Griffin and Brown (2010) in the context of regression models. In the context of mixtures, this means using a hierarchical prior for the mean, where the prior variance follows a gamma distribution:

$$\mu_{k,l}|B_l \sim \mathcal{N}\left(m_l, R_l^2 B_l\right), \qquad B_l \sim \mathcal{G}(\nu, \nu), \tag{10.4}$$

where $l = 1, \ldots, r$. Note that $\text{Var}(\mu_{k,l}) = R_l^2 \text{E}(B_l)$ is equal to R_l^2 for all $k = 1, \ldots, 4$ as for the standard prior. In each dimension l the scaling factor B_l acts as a local shrinkage factor, which is able to pull the group means $\mu_{k,l}$ towards m_l by assuming values considerably smaller than one. For values ν that are smaller than one, an important shrinkage effect may be achieved in this way.

It is very easy to extend the MCMC setting for this prior:

(a) run the usual MCMC steps, i.e. sampling the component-specific parameters, the weight distribution and the allocations, conditional on the hyperparameters $B_1, \ldots, B_r$;
(b) sample the hyperparameters $B_1, \ldots, B_r$ conditional on $\boldsymbol{\mu}_1, \ldots, \boldsymbol{\mu}_K$: sample each B_l independently from the corresponding generalised inverse Gaussian

posterior distribution:

$$B_l|\mu_{1,l}, \ldots, \mu_{K,l} \sim \mathcal{GIG}(p_K, a_l, b_l),$$
$$p_K = \nu - \frac{K}{2}, \qquad a_l = 2\nu, \qquad b_l = \sum_{k=1}^{K} (\mu_{k,l} - m_l)^2 / R_l^2. \tag{10.5}$$

Alternatively, $B_l^{-1}|\mu_{1,l}, \ldots, \mu_{K,l}$ may be sampled from $\mathcal{GIG}(-p_K, b_l, a_l)$.

It should be noted that the marginal prior $p(\mu_{1,l}, \ldots, \mu_{K,l})$ is available in closed form:

$$p(\mu_{1,l}, \ldots, \mu_{K,l}) = \frac{\nu^\nu}{(2\pi)^{K/2}\Gamma(\nu)} 2K_{p_K}(\sqrt{a_l b_l}) \left(\frac{b_l}{a_l}\right)^{p_K/2},$$

where $K_p(z)$ is the modified Bessel function of the second kind and p_K, a_l and b_l are the posterior moments defined in (10.5). This is useful for the computation of the marginal likelihood $p(\mathbf{y}|K)$ using this prior.

The Bayesian Lasso prior

The second prior is obtained by substituting the normal prior for the mean by a Laplace prior leading to the Bayesian Lasso finite mixture model:

$$\mu_{k,l}|B_l \sim \text{Lap}(m_l, B_l), \qquad B_l \sim \mathcal{G}^{-1}(s_l, S_l), \tag{10.6}$$

where $l = 1, \ldots, r$. With $s_l > 2$ and $S_l = (s_l - 1)(s_l - 2)R_l$, the prior expected variance $\text{Var}(\mu_{k,l}) = \text{E}(B_l)$ is equal to R_l^2 for all $k = 1, \ldots, 4$ as for the standard prior. The prior (10.6) allows for the following hierarchical formulation:

$$\mu_{k,l}|\psi_{k,l} \sim \mathcal{N}(m_l, \psi_{k,l}),$$
$$\psi_{k,l}|B_l \sim \mathcal{E}\left[1/(2B_l^2)\right].$$

This representation is useful because it leads to an easy extension of the MCMC scheme:

(a) run the usual MCMC steps, i.e. sampling the component-specific parameters, the weight distribution and the allocations, conditional on the prior variances $\boldsymbol{\psi} = (\psi_{k,1}, l = 1, \ldots, r, k = 1, \ldots, K)$;

(b) sample the hyperparameters $B_1, \ldots, B_r$ and $\boldsymbol{\psi}$ conditional on $\boldsymbol{\mu}_1, \ldots, \boldsymbol{\mu}_K$: sample each B_l independently from the corresponding inverted gamma posterior distribution:

$$B_l|\mu_{1,l}, \ldots, \mu_{K,l} \sim \mathcal{G}^{-1}\left(s_l + K, S_l + \sum_{k=1}^{K} |\mu_{k,l} - m_l|\right),$$

and sample each $\psi_{k,l}^{-1}$ independently from the corresponding inverse Gaussian posterior distribution:

$$\psi_{k,l}^{-1}|\mu_{k,l}, B_l \sim \text{InvGau}\left(\frac{1}{B_l|\mu_{k,l} - m_l|}, \frac{1}{B_l^2}\right).$$

Once more, the marginal prior $p(\mu_{1,l}, \ldots, \mu_{K,l})$ is available in closed form:

$$p(\mu_{1,l}, \ldots, \mu_{K,l}) = \frac{\Gamma(s_l + K)S_l^{s_l}}{\Gamma(s_l)(S_l + \sum_{k=1}^{K} |\mu_{k,l} - m_l|)^{s_l+K}}, \tag{10.7}$$

which is useful for the computation of the marginal likelihood $p(\mathbf{y}|K)$ using this prior.

Example 4 (continued): Multivariate normal mixtures

Table 10.4 summarises the estimation of normal mixtures with increasing numbers K of components as in Subsection 10.4.1 using the Laplace prior with $s_l = 3$ and using the normal-gamma prior with $\nu = 0.5$.

When $e_0 = 25$, the marginal likelihood selects the true number of components for both the Laplace prior and the normal-gamma prior. Hence these shrinkage priors in combination with $e_0 = 25$ manage overfitting heterogeneity nicely as opposed to the standard prior. For both shrinkage priors, k-means clustering in the point-process representation of the MCMC draws obtained for $K = 4$ using the elements of $\boldsymbol{\mu}_k$ leads to classification sequences ρ_m, all of which are permutations ($M_\rho = 0$). Figure 10.8 shows scatterplots of the 5000 identified MCMC draws of $(\mu_{k,1}, \mu_{k,5})$ for $k = 1, \ldots, 4$ for the normal-gamma prior using the hyperparameter $e_0 = 25$. The corresponding scatterplot for the Bayesian Lasso prior is not reported because it looks very similar.

In addition, the posterior draws of the scaling factors $B_l, l = 1, \ldots, 8$, obtained from the normal-gamma prior, help to identify in which direction l the elements $\mu_{k,l}$ of the component-specific mean $\boldsymbol{\mu}_k$ are homogeneous. Figure 10.9 shows box-plots of B_l using the hyperparameter $e_0 = 25$, but the figure looks very similar for $e_0 = 0.1$. The posterior distribution of B_l is strongly pulled towards 0 for all homogeneous elements, i.e. in all dimensions l except one and five, and is bounded away from zero for all heterogeneous elements. Hence the normal-gamma clearly separates between homogeneous and heterogeneous elements.

For the prior calibration $e_0 = 0.1$, the marginal likelihood selects a mixture with five nonempty components for the normal-gamma prior, and hence the number of true components is overfitted. For the Laplace prior, a mixture with six components is selected, two of which are empty ($\hat{K}_0 = 4$). $M_0 = 3341$ MCMC draws have four nonempty groups. After the two empty groups are removed from each of these MCMC draws, as described in Subsection 10.3.3, k-means clustering with four clusters is applied to identify a four-component mixture. However, it turns out that a high fraction of $M_{0,\rho} = 3341$ sequences are not permutations of $\{1, \ldots, 4\}$. Hence, using $e_0 = 0.1$ is not possible to identify the mixture.

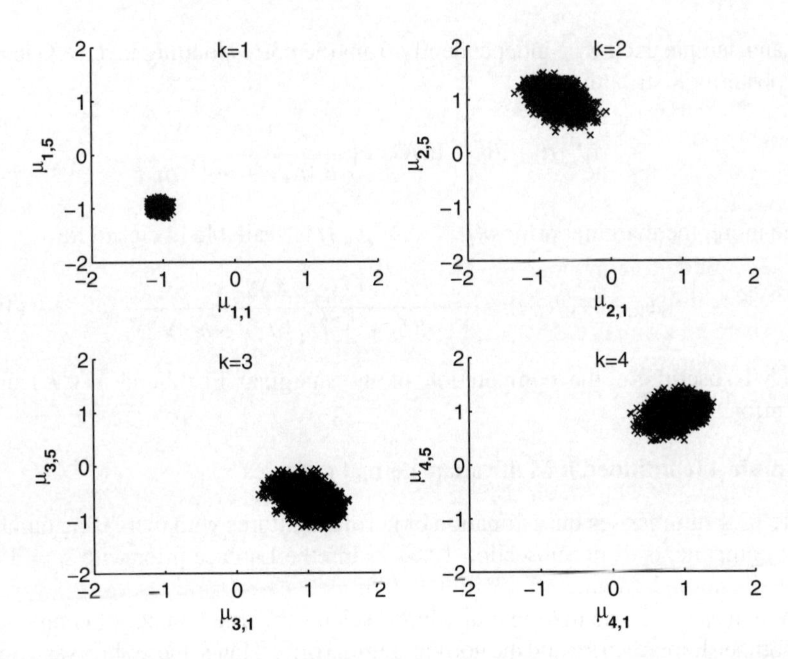

Figure 10.8 Multivariate normal mixtures of Example 4 ($r = 8$, $K_{tr} = 4$), showing identified MCMC draws of $(\mu_{k,1}, \mu_{k,5})$, $k = 1, \ldots, 4$, for a mixture with four components, based on a normal-gamma prior and $e_0 = 25$.

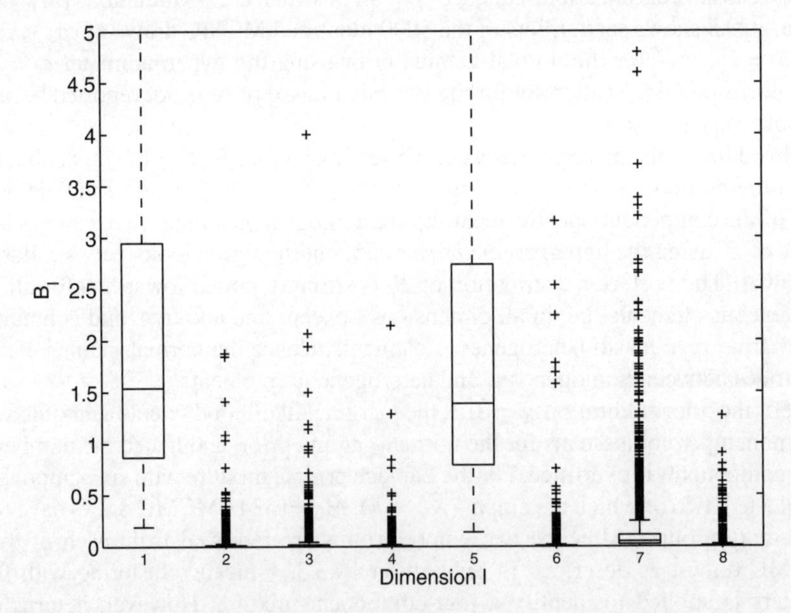

Figure 10.9 Multivariate normal mixtures of Example 4 ($r = 8$, $K_{tr} = 4$), showing the boxplot of the MCMC draws of the shrinkage factor B_l for the normal-gamma prior using $e_0 = 25$ for $l = 1, \ldots, 8$.

10.5 Concluding remarks

This contribution investigated how to identify a finite mixture model under uncertainty with respect to model specification. The marginal likelihood is clearly useful for selecting the total number of K components in a finite mixture model, if the model is not strongly overfitting heterogeneity in the component-specific parameters. However, the prior on the weight distribution plays an important role in inferring the true number K_{tr} of components from the selected mixture. For a prior where the hyperparameter e_0 in the Dirichlet prior $\mathcal{D}(e_0, \dots, e_0)$ is larger than $d/2$, where $d = \dim \boldsymbol{\theta}$, e.g. $e_0 = d/2 + 3$, the total number K of components of the selected mixture is actually a sensible estimator of K_{tr}. The mixture is identified by applying k-means clustering with K clusters to the point-process representation of the MCMC draws.

On the other hand, if e_0 is considerably smaller than one, e.g. $e_0 = 0.1$, then the estimated number $\hat{K}_0$ of nonempty components yields a much better estimator of K_{tr} than K. With such a prior, the mixture is identified by postprocessing all MCMC draws with exactly $\hat{K}_0$ nonempty groups. First, the empty groups are removed and then k-means clustering with $\hat{K}_0$ clusters is applied to the point-process representation of these MCMC draws.

Furthermore, the identification of finite mixture models that are strongly overfitting heterogeneity in the component-specific parameters has been considered. While standard priors lead to strong underfitting of K_{tr} in such a situation, shrinkage priors well-known from variable selection were applied to finite mixtures. These priors avoid underfitting of K_{tr} by reducing automatically the prior variance of homogeneous components.

Several simulated datasets demonstrated the usefulness of this approach towards identification. However, several important issues remained untouched and would deserve careful investigation. First of all, shrinkage priors have to be extended to handle overfitting heterogeneity not only in the mean but also in the component-specific covariance matrix.

Second, the issue of how genuine multimodality where minor modes are present impacts the point-process representation of the MCMC draws needs to be investigated. It is not clear how labelling by k-means clustering could be extended to deal with genuine multimodality; see Grün and Leisch (2009) for an alternative approach.

A final issue is model uncertainty with respect to the distribution family underlying the mixture. Criteria like BIC are known to overfit the number of components in the case of misspecifying this distribution family and it would be interesting to find priors similar to shrinkage priors that avoid this effect.

References

Bernardo, J. M. and Girón, F. G. (1988) A Bayesian analysis of simple mixture problems. In *Bayesian Statistics 3* (eds J. M. Bernardo, M. H. DeGroot, D. V. Lindley and A. F. M. Smith) pp. 67–78, Clarendon.

Biernacki, C., Celeux, G. and Govaert, G. (2000) Assessing a mixture model for clustering with the integrated completed likelihood. *IEEE Transactions on Pattern Analysis and Machine Intelligence*, **22**, 719–725.

Celeux, G. (1998) Bayesian inference for mixture: the label switching problem. In *COMPSTAT 98* (eds P. J. Green and R. Rayne), pp. 227–232. Physica, Heidelberg.

Celeux, G., Hurn, M. and Robert, C. P. (2000) Computational and inferential difficulties with mixture posterior distributions. *Journal of the American Statistical Association*, **95**, 957–970.

Dellaportas, P. and Papageorgiou, I. (2006) Multivariate mixtures of normals with unknown number of components. *Statistics and Computing*, **16**, 57–68.

Diebolt, J. and Robert, C. P. (1994) Estimation of finite mixture distributions through Bayesian sampling. *Journal of the Royal Statistical Society, Series B*, **56**, 363–375.

Fahrmeir, L., Kneib, T. and Konrath, S. (2010) Bayesian regularisation in structured additive regression: a unifying perspective on shrinkage, smoothing and predictor selection. *Statistics and Computing*, **20**, 203–219.

Ferguson, T. S. (1973) A Bayesian analysis of some nonparametric problems. *Annals of Statistics*, **1**, 209–230.

Frühwirth-Schnatter, S. (2001a) Fully Bayesian analysis of switching Gaussian state space models. *Annals of the Institute of Statistical Mathematics*, **53**, 31–49.

Frühwirth-Schnatter, S. (2001b) Markov chain Monte Carlo estimation of classical and dynamic switching and mixture models. *Journal of the American Statistical Association*, **96**, 194–209.

Frühwirth-Schnatter, S. (2004) Estimating marginal likelihoods for mixture and Markov switching models using bridge sampling techniques. *The Econometrics Journal*, **7**, 143–167.

Frühwirth-Schnatter, S. (2006) *Finite Mixture and Markov Switching Models*. Springer.

Frühwirth-Schnatter, S. (2008) Comment on article by T. Rydén on 'EM versus Markov chain Monte Carlo for estimation of hidden Markov models'. *Bayesian Analysis*, **3**, 689–698.

Frühwirth-Schnatter, S. and Pyne, S. (2010) Bayesian Inference for finite mixtures of univariate and multivariate skew normal and skew-*t* distributions. *Biostatistics*, **11**, 317 – 336.

Green, P. J. and Richardson, S. (2001) Modelling heterogeneity with and without the Dirichlet process. *Scandinavian Journal of Statistics*, **28**, 355–375.

Griffin, J. E. and Brown, P. J. (2010) Inference with normal-gamma prior distributions in regression problems. *Bayesian Analysis*, **5**, 171–188.

Grün, B. and Leisch, F. (2009) Dealing with label switching in mixture models under genuine multimodality. *Journal of Multivariate Analysis*, **100**, 851–861.

Hahn, M., Frühwirth-Schnatter, S. and Sass, J. (2010) Markov chain Monte Carlo methods for parameter estimation in multidimensional continuous time Markov switching models. *Journal of Financial Econometrics*, **8**, 88–121.

Ishwaran, H., James, L. F. and Sun, J. (2001) Bayesian model selection in finite mixtures by marginal density decompositions. *Journal of the American Statistical Association*, **96**, 1316–1332.

Jasra, A., Holmes, C. C. and Stephens, D. A. (2005) Markov chain Monte Carlo methods and the label switching problem in Bayesian mixture modelling. *Statistical Science*, **20**, 50–67.

Keribin, C. (2000) Consistent estimation of the order of mixture models. *Sankhya A*, **62**, 49–66.

Marin, J. M., Mengersen, K. and Robert, C. P. (2005) Bayesian modelling and inference on mixtures of distributions. In *Bayesian Thinking, Modelling and Computation* (eds D. Dey and C. Rao) Vol. 25 of *Handbook of Statistics*, pp. 459–507, North-Holland.

McLachlan, G. J. and Peel, D. (2000) *Finite Mixture Models*. John Wiley & Sons, Ltd.

Nobile, A. (2004) On the posterior distribution of the number of components in a finite mixture. *Annals of Statistics*, **32**, 2044–2073.

Nobile, A. and Fearnside, A. (2007) Bayesian finite mixtures with an unknown number of components: the allocation sampler. *Statistics and Computing*, **17**, 147–162.

Pamminger, C. and Frühwirth-Schnatter, S. (2010) Bayesian clustering of categorical time series. *Bayesian Analysis*, **5**, 345–368.

Park, T. and Casella, G. (2008) The Bayesian Lasso. *Journal of the American Statistical Association*, **103**, 681–686.

Richardson, S. and Green, P. J. (1997) On Bayesian analysis of mixtures with an unknown number of components. *Journal of the Royal Statistical Society, Series B*, **59**, 731–792.

Rousseau, J. and Mengersen, K. (2010) Asymptotic behaviour of the posterior distribution in overfitted mixture models. Technical Report, ENSEA-CREST.

Sperrin, M., Jaki, T. and Wit, E. (2010) Probabilistic relabelling strategies for the label switching problem in Bayesian mixture models. *Statistics and Computing*, **20**, 357–366.

Spezia, L. (2009) Reversible jump and the label switching problem in hidden Markov models. *Journal of Statistical Planning and Inference*, **139**, 2305–2315.

Stephens, M. (1997) Bayesian methods for mixtures of normal distributions. DPhil Thesis, University of Oxford.

Stephens, M. (2000a) Bayesian analysis of mixture models with an unknown number of components – an alternative to reversible jump methods. *Annals of Statistics*, **28**, 40–74.

Stephens, M. (2000b) Dealing with label switching in mixture models. *Journal of the Royal Statistical Society, Series B*, **62**, 795–809.

Titterington, D. M., Smith, A. F. M. and Makov, U. E. (1985) *Statistical Analysis of Finite Mixture Distributions*. John Wiley & Sons, Ltd.

Yau, C. and Holmes, C. (2010) Hierarchical Bayesian nonparametric mixture models for clustering with variable relevance determination. Technical Report, Department of Statistics, University of Oxford.

11

Exact Bayesian analysis of mixtures

Christian P. Robert and Kerrie L. Mengersen

11.1 Introduction

In this chapter, we show how a complete and exact Bayesian analysis of a parametric mixture model is possible in cases when components of the mixture are taken from exponential families and when conjugate priors are used. This restricted set-up allows us to show the relevance of the Bayesian approach as well as to exhibit the limitations of a complete analysis, namely that it is impossible to conduct this analysis when the sample size is too large, when the data are not from an exponential family, or when priors that are more complex than conjugate priors are used.

We want to stress from the beginning that this chapter is anything but a formal exercise: to understand how the Bayesian analysis of a mixture model unravels and automatically exploits the missing data structure of the model is crucial for grasping the details of simulation methods (not covered in this chapter; see, for example Lee *et al.*, 2009; Robert and Casella, 2004) that take full advantage of the missing structures. It also allows for a comparison between exact and approximate techniques when the former are available. While the relevant references are pointed out in due time, we note here that the chapter builds upon the foundational paper of Fearnhead (2005).

We thus assume that a sample $\mathbf{x} = (x_1, \ldots, x_n)$ from the mixture model

$$\sum_{i=1}^{k} p_i \, h(x) \exp\left[\theta_i \cdot R(x) - \Psi(\theta_i)\right] \tag{11.1}$$

Mixtures: Estimation and Applications, First Edition. Edited by Kerrie L. Mengersen, Christian P. Robert and D. Michael Titterington.

is available, where $\theta \cdot R(x)$ denotes the scalar product between the vectors θ and $R(x)$. We are selecting on purpose the natural representation of an exponential family (see, for example, Robert, 2001, Chapter 3), in order to facilitate the subsequent derivation of the posterior distribution.

When the components of the mixture are Poisson $\mathcal{P}(\lambda_i)$ distributions, if we define $\theta_i = \log \lambda_i$, the Poisson distribution indeed is written as a natural exponential family:

$$f(x|\theta_i) = (1/x!) \exp\left(\theta_i x - e^{\theta_i}\right).$$

For a mixture of multinomial distributions $\mathcal{M}(m; q_{i1}, \ldots, q_{iv})$, the natural representation is given by

$$f(x|\theta_i) = (m!/x_1! \cdots x_v!) \exp(x_1 \log q_{i1} + \cdots + x_v \log q_{iv})$$

and the overall (natural) parameter is thus $\theta_i = (\log q_{i1}, \ldots, \log q_{iv})$.

In the normal $\mathcal{N}(\mu_i, \sigma_i^2)$ case, the derivation is more delicate when both parameters are unknown since

$$f(x|\theta_i) = \frac{1}{\sqrt{2\pi\sigma_i^2}} \exp\left(-\frac{\mu_i^2}{2\sigma_i^2} + \frac{\mu_i x}{\sigma_i^2} + \frac{-x^2}{2\sigma_i^2}\right).$$

In this particular setting, the natural parameterisation is in $\theta_i = (\mu_i/\sigma_i^2, 1/\sigma_i^2)$ while the statistic $R(x) = (x, -x^2/2)$ is two-dimensional. The moment cumulant function is then $\Psi(\theta) = \theta_1^2/\theta_2$.

11.2 Formal derivation of the posterior distribution

11.2.1 Locally conjugate priors

As described in the standard literature on mixture estimation (Dempster *et al.*, 1977; Frühwirth-Schnatter, 2006; MacLachlan and Peel, 2000), the missing variable decomposition of a mixture likelihood associates each observation in the sample with one of the k components of the mixture (11.1), i.e.

$$x_i | z_i \sim f(x_i | \theta_{z_i}).$$

Given the component allocations $\mathbf{z}$, we end up with a cluster of (sub)samples from different distributions from the same exponential family. Priors customarily used for the analysis of these exponential families can therefore be extended to the mixtures as well.

While conjugate priors do not formally exist for mixtures of exponential families, we will define *locally conjugate priors* as priors that are conjugate for the completed distribution, that is for the likelihood associated with both the observations and the missing data $\mathbf{z}$. This amounts to taking regular conjugate priors for the parameters of the different components and a conjugate Dirichlet prior on the weights of the mixture,

$$(p_1, \ldots, p_k) \sim \mathcal{D}(\alpha_1, \ldots, \alpha_k).$$

When we consider the complete likelihood

$$\begin{aligned} L^c(\theta, p|\mathbf{x}, \mathbf{z}) &= \prod_{i=1}^{n} p_{z_i} \exp\left[\theta_{z_i} \cdot R(x_i) - \Psi(\theta_{z_i})\right] \\ &= \prod_{j=1}^{k} p_j^{n_j} \exp\left[\theta_j \cdot \sum_{z_i=j} R(x_i) - n_j \Psi(\theta_j)\right] \\ &= \prod_{j=1}^{k} p_j^{n_j} \exp\left[\theta_j \cdot S_j - n_j \Psi(\theta_j)\right], \end{aligned}$$

it is easily seen that we remain within an exponential family since there exists a sufficient statistic with fixed dimension, $(n_1, S_1, \ldots, n_k, S_k)$. If we use a Dirichlet prior,

$$\pi(p_1, \ldots, p_k) = \frac{\Gamma(\alpha_1 + \cdots + \alpha_k)}{\Gamma(\alpha_1) \cdots \Gamma(\alpha_k)} p_1^{\alpha_1 - 1} \cdots p_k^{\alpha_k - 1},$$

on the vector of the weights $(p_1, \ldots, p_k)$ defined on the simplex of $\mathbb{R}^k$, and (generic) conjugate priors on the θ_js,

$$\pi_j(\theta_j] \propto \exp\left[\theta_j \cdot s_{0j} - \lambda_j \Psi(\theta_j)\right],$$

the posterior associated with the complete likelihood $L^c(\theta, p|\mathbf{x}, \mathbf{z})$ is then of the same family as the prior:

$$\begin{aligned} \pi(\theta, p|\mathbf{x}, \mathbf{z}) &\propto \pi(\theta, p) \times L^c(\theta, p|\mathbf{x}, \mathbf{z}) \\ &\propto \prod_{j=1}^{k} p_j^{\alpha_j - 1} \exp\left[\theta_j \cdot s_{0j} - \lambda_j \Psi(\theta_j)\right] \times p_j^{n_j} \exp\left[\theta_j \cdot S_j - n_j \Psi(\theta_j)\right] \\ &= \prod_{j=1}^{k} p_j^{\alpha_j + n_j - 1} \exp\left[\theta_j \cdot (s_{0j} + S_j) - (\lambda_j + n_j) \Psi(\theta_j)\right]; \end{aligned}$$

the parameters of the prior are transformed from α_j to $\alpha_j + n_j$, from s_{0j} to $s_{0j} + S_j$ and from λ_j into $\lambda_j + n_j$.

For instance, in the case of the Poisson mixture, the conjugate priors are gamma $\mathcal{G}(a_j, b_j)$, with corresponding posteriors (for the complete likelihood), gamma $\mathcal{G}(a_j + S_j, b_j + n_j)$ distributions, in which S_j denotes the sum of the observations in the jth group.

For a mixture of multinomial distributions, $\mathcal{M}(m; q_{j1}, \ldots, q_{jv})$, the conjugate priors are Dirichlet $\mathcal{D}_v(\beta_{j1}, \ldots, \beta_{jv})$ distributions, with corresponding posteriors $\mathcal{D}_v(\beta_{j1} + s_{j1}, \ldots, \beta_{jv} + s_{jv})$, s_{ju} denoting the number of observations from component j in group u ($1 \le u \le v$), with $\sum_u s_{jv} = n_j m$.

In the normal mixture case, the standard conjugate priors are products of normal and inverse gamma distributions, i.e.

$$\mu_j|\sigma_j \sim \mathcal{N}(\xi_j, \sigma_j^2/c_j) \quad \text{and} \quad \sigma_j^{-2} \sim \mathcal{G}(a_j/2, b_j/2)\,.$$

Indeed, the corresponding posterior is

$$\mu_j|\sigma_j \sim \mathcal{N}((c_j\xi_j + n_j\overline{x}_j, \sigma_j^2/(c_j + n_j))$$

and

$$\sigma_j^{-2} \sim \mathcal{G}(\{a_j + n_j\}/2, \{b_j + n_j\hat{\sigma}_j^2 + (\overline{x}_j - \xi_j)^2/(c_j^{-1} + n_j^{-1})\})\,,$$

where $n_j\overline{x}_j$ is the sum of the observations allocated to component j and $n_j\hat{\sigma}_j^2$ is the sum of the squares of the differences from $\overline{x}_j$ for the same group (with the convention that $n_j\hat{\sigma}_j^2 = 0$ when $n_j = 0$).

11.2.2 True posterior distributions

These straightforward derivations do not correspond to the observed likelihood, but to the completed likelihood. While this may be enough for some simulation methods like Gibbs sampling (see, for example, Diebolt and Robert, 1990, 1994), we need further developments for obtaining the true posterior distribution.

If we now consider the observed likelihood, it is natural to expand this likelihood as a sum of completed likelihoods over all possible configurations of the partition space of allocations, that is a sum over k^n terms. Except in the very few cases that are processed below, including Poisson and multinomial mixtures (see Section 11.2.3), this sum does not simplify into a smaller number of terms because no summary statistics exist. From a Bayesian point of view, the complexity of the model is therefore truly of magnitude $O(k^n)$.

The observed likelihood is thus

$$\sum_{\mathbf{z}} \prod_{j=1}^{k} p_j^{n_j} \exp\left[\theta_j \cdot S_j - n_j \Psi(\theta_j)\right]$$

(with the dependence of (n_j, S_j) upon $\mathbf{z}$ omitted for notational purposes) and the associated posterior is, up to a constant,

$$\begin{aligned}\sum_{\mathbf{z}} \prod_{j=1}^{k} p_j^{n_j+\alpha_j-1} &\exp\left[\theta_j \cdot (s_{0j} + S_j) - (n_j + \lambda_j)\Psi(\theta_j)\right] \\ &= \sum_{\mathbf{z}} \omega(\mathbf{z})\, \pi(\theta, \mathbf{p}|\mathbf{x}, \mathbf{z}),\end{aligned}$$

where $\omega(\mathbf{z})$ is the normalising constant missing in

$$\prod_{j=1}^{k} p_j^{n_j+\alpha_j-1} \exp\left[\theta_j \cdot (s_{0j} + S_j) - (n_j + \lambda_j)\Psi(\theta_j)\right]$$

i.e.

$$\omega(\mathbf{z}) \propto \frac{\prod_{j=1}^{k} \Gamma(n_j + \alpha_j)}{\Gamma(\sum_{j=1}^{k}(n_j + \alpha_j))} \times \prod_{j=1}^{k} K(s_{0j} + S_j, n_j + \lambda_j),$$

if $K(\xi, \delta)$ is the normalising constant of $\exp[\theta_j \cdot \xi - \delta\Psi(\theta_j)]$, i.e.

$$K(\xi, \delta) = \int \exp[\theta_j \cdot \xi - \delta\Psi(\theta_j)]\, \mathrm{d}\theta.$$

The posterior $\sum_{\mathbf{z}} \omega(\mathbf{z})\, \pi(\theta, \mathbf{p}|\mathbf{x}, \mathbf{z})$ is therefore a mixture of conjugate posteriors where the parameters of the components as well as the weights can be computed in closed form. The availability of the posterior does not mean that alternative estimates like MAP and MMAP estimates can be computed easily. However, this is a useful closed-form result in the sense that moments can be computed exactly: for instance, if there is no label-switching problem (Jasra *et al.*, 2005; Stephens, 2000) and if the posterior mean is producing meaningful estimates, we have that

$$\mathbb{E}\left[\nabla\Psi(\theta_j)|\mathbf{x}\right] = \sum_{\mathbf{z}} \omega(\mathbf{z}) \frac{s_{0j} + S_j}{n_j + \lambda_j},$$

since, for each allocation vector $\mathbf{z}$, we are in an exponential family set-up where the posterior mean of the expectation $\Psi(\theta)$ of $R(x)$ is available in closed form. (Obviously, the posterior mean only makes sense as an estimate for very discriminative priors; see Jasra *et al.*, 2005.) Similarly, estimates of the weights p_j are given by

$$\mathbb{E}\left[p_j|\mathbf{x}\right] = \sum_{\mathbf{z}} \omega(\mathbf{z}) \frac{n_j + \alpha_j}{n + \alpha.},$$

where $\alpha. = \sum_j \alpha_j$. Therefore, the only computational effort required is the summation over all partitions.

This decomposition further allows for a closed-form expression of the marginal distributions of the various parameters of the mixture. For instance, the (marginal) posterior distribution of θ_i is given by

$$\sum_{\mathbf{z}} \omega(\mathbf{z}) \frac{\exp\left[\theta_j \cdot (s_{0j} + S_j) - (n_j + \lambda_j)\Psi(\theta_j)\right]}{K(s_{0j} + S_j, n_j + \lambda_j)}.$$

(Note that, when the hyperparameters α_j, s_{0j} and n_j are independent of j, this posterior distribution is independent of j.) Similarly, the posterior distribution of the vector $(p_1, \ldots, p_k)$ is equal to

$$\sum_{\mathbf{z}} \omega(\mathbf{z})\, \mathcal{D}(n_1 + \alpha_1, \ldots, n_k + \alpha_k)\,.$$

If k is small and n is large, and when all hyperparameters are equal, the posterior should then have k spikes or peaks, due to the label-switching/lack of identifiability phenomenon.

We will now proceed through standard examples.

11.2.3 Poisson mixture

In the case of a two-component Poisson mixture,

$$x_1, \ldots, x_n \overset{\text{iid}}{\sim} p\,\mathcal{P}(\lambda_1) + (1-p)\,\mathcal{P}(\lambda_2)\,,$$

let us assume a uniform prior on p (i.e. $\alpha_1 = \alpha_2 = 1$) and exponential priors $\mathcal{E}xp(1)$ and $\mathcal{E}xp(1/10)$ on λ_1 and λ_2, respectively. (The scales are chosen to be fairly different for the purpose of illustration. In a realistic setting, it would be sensible either to set those scales in terms of the scale of the problem, if known, or to estimate the global scale following the procedure of Mengersen and Robert, 1996.)

The normalising constant is then equal to

$$\begin{aligned} K(\xi, \delta) &= \int_{-\infty}^{\infty} \exp\left[\theta_j \xi - \delta \log(\theta_j)\right]\, \mathrm{d}\theta \\ &= \int_0^{\infty} \lambda_j^{\xi-1} \exp(-\delta\lambda_j)\, \mathrm{d}\lambda_j \\ &= \delta^{-\xi}\,\Gamma(\xi)\,, \end{aligned}$$

with $s_{01} = 1$ and $s_{02} = 10$, and the corresponding posterior is (up to the normalisation of the weights)

$$\begin{aligned} &\sum_{\mathbf{z}} \frac{\prod_{j=1}^{2} \Gamma(n_j+1)\Gamma(1+S_j)/(s_{0j}+n_j)^{S_j+1}}{\Gamma(2+\sum_{j=1}^{2} n_j)}\, \pi(\theta, \mathbf{p}|\mathbf{x}, \mathbf{z}) \\ &= \sum_{\mathbf{z}} \frac{\prod_{j=1}^{2} n_j!\, S_j!/(s_{0j}+n_j)^{S_j+1}}{(N+1)!}\, \pi(\theta, \mathbf{p}|\mathbf{x}, \mathbf{z}) \\ &\propto \sum_{\mathbf{z}} \prod_{j=1}^{2} n_j!\, S_j!/(s_{0j}+n_j)^{S_j+1}\, \pi(\theta, \mathbf{p}|\mathbf{x}, \mathbf{z})\,, \end{aligned}$$

with $\pi(\theta, \mathbf{p}|\mathbf{x}, \mathbf{z}))$ corresponding to a beta $\mathcal{B}e(1+n_j, 1+N-n_j)$ distribution on p and to a gamma $\mathcal{G}a(S_j+1, s_{0j}+n_j)$ distribution on λ_j $(j = 1, 2)$.

An important feature of this example is that the sum does not need to involve all of the 2^n terms, simply because the individual terms in the previous sum factorise in (n_1, n_2, S_1, S_2), which then acts like a local sufficient statistic. Since $n_2 = n - n_1$ and $S_2 = \sum x_i - S_1$, the posterior only requires as many distinct terms as there are distinct values of the pair (n_1, S_1) in the completed sample. For instance, if the sample is $(0, 0, 0, 1, 2, 2, 4)$, the distinct values of the pair (n_1, S_1) are $(0, 0), (1, 0), (1, 1), (1, 2), (1, 4), (2, 0), (2, 1), (2, 2), (2, 3), (2, 4), (2, 5), (2, 6), \ldots, (6, 5), (6, 7), (6, 8), (7, 9)$. There are therefore 41 distinct terms in the posterior, rather than $2^8 = 256$.

The problem of computing the number (or cardinality) $\mu_n(n_1, S_1)$ of terms in the k^n sums with the same statistic (n_1, S_1) has been tackled by Fearnhead (2005) in that he proposes a recursive formula for computing $\mu(n_1, S_1)$ in an efficient way, as expressed below for a k component mixture:

Theorem (Fearnhead, 2005): If $\mathbf{e}_j$ denotes the vector of length k made up of zeros everywhere except at component j where it is equal to one, if

$$\mathbf{n} = (n_1, \ldots, n_k), \quad \mathbf{n} - \mathbf{e}_j = (n_1, \ldots, n_j - 1, \ldots, n_k) \quad \text{and} \quad y\mathbf{e}_j = (0, \ldots, y, \ldots, 0),$$

then

$$\mu_1(\mathbf{e}_j, y_1\mathbf{e}_j) = 1 \quad \text{and} \quad \mu_n(\mathbf{n}, \mathbf{s}) = \sum_{j=1}^{k} \mu_{n-1}(\mathbf{n} - \mathbf{e}_j, \mathbf{s} - y_n\mathbf{e}_j).$$

Therefore, once the $\mu_n(\mathbf{n}, \mathbf{s})$ are all computed, the posterior can be written as

$$\sum_{(n_1,S_1)} \mu_n(n_1, S_1) \prod_{j=1}^{2} \left[n_j!\, S_j!/(s_{0j} + n_j)^{S_j+1} \right] \pi(\theta, \mathbf{p}|\mathbf{x}, n_1, S_1),$$

up to a constant, since the complete likelihood posterior only depends on the sufficient statistic (n_1, S_1).

Now, the closed-form expression allows for a straightforward representation of the marginals. For instance, the marginal in λ_1 is given by

$$\sum_{\mathbf{z}} \left(\prod_{j=1}^{2} n_j!\, S_j!/(s_{0j} + n_j)^{S_j+1} \right) (n_1 + 1)^{S_1+1} \lambda^{S_1} \exp[-(n_1 + 1)\lambda_1]/n_1!$$

$$= \sum_{(n_1,S_1)} \mu_n(n_1, S_1) \prod_{j=1}^{2} n_j!\, S_j!/(s_{0j} + n_j)^{S_j+1}$$
$$\times (n_1 + 1)^{S_1+1} \lambda_1^{S_1} \exp[-(n_1 + 1)\lambda_1]/n_1!$$

up to a constant, while the marginal in λ_2 is

$$\sum_{(n_1,S_1)} \mu_n(n_1, S_1) \prod_{j=1}^{2} \left[n_j!\, S_j!/(s_{0j} + n_j)^{S_j+1} \right] (n_2 + 10)^{S_2+1} \lambda_2^{S_2} \exp[-(n_2 + 10)\lambda_2]/n_2!$$

again up to a constant, and the marginal in p is

$$\sum_{(n_1,S_1)} \mu_n(n_1, S_1) \frac{\prod_{j=1}^{2} n_j!\, S_j!/(s_{0j} + n_j)^{S_j+1}}{(N+1)!} \frac{(N+1)!}{n_1!(N-n_1)!} p^{n_1}(1-p)^{N-n_1}$$

$$= \sum_{u=0}^{N} \sum_{S_1;n_1=u} \mu_n(u, S_1) \frac{S_1!(S-S_1)!\, p^u(1-p)^{N-u}}{(u+1)^{S_1+1}(n-u+10)^{S-S_1+1}},$$

still up to a constant, if S denotes the sum of all observations.

As pointed out above, another interesting outcome of this closed-form representation is that marginal likelihoods (or evidences) can also be computed in closed form. The marginal distribution of $\mathbf{x}$ is directly related to the unnormalised weights $\omega(\mathbf{z}) = \omega(n_1, S_1)$ in that

$$m(\mathbf{x}) = \sum_{\mathbf{z}} \omega(\mathbf{z}) = \sum_{(n_1,S_1)} \mu_n(n_1, S_1)\omega(n_1, S_1)$$

$$\propto \sum_{(n_1,S_1)} \mu_n(n_1, S_1) \prod_{j=1}^{2} n_j!\, S_j!/(s_{0j} + n_j)^{S_j+1},$$

up to the product of factorials $1/y_1! \cdots y_n!$ (but this is irrelevant in the computation of the Bayes factor).

In practice, the derivation of the cardinalities $\mu_n(n_1, S_1)$ can be done recursively as in Fearnhead (2005): include each observation y_k by updating all the $\mu_{k-1}(n_1, S_1, k-1-n_1, S_2)$s in both $\mu_k(n_1+1, S_1+y_k, n_2, S_2)$ and $\mu_k(n_1, S_1, n_2+1, S_2+y_k)$, and then check for duplicates. (A naïve R implementation is available from the authors upon request.)

Now, even with this considerable reduction in the complexity of the posterior distribution (to be compared with k^n), the number of terms in the posterior still grows very fast both with n and with the number of components k, as shown through a few simulated examples in Table 11.1. (The missing items in the table simply took too much time or too much memory on the local mainframe when using our R program. Fearnhead (2005) used a specific C program to overcome this difficulty with larger sample sizes.) The computational pressure also increases with the range of the data; that is for a given value of (k, n), the number of rows in `cardin` is much larger when the observations are larger, as shown for instance in the first three rows of Table 11.1. A simulated Poisson $\mathcal{P}(\lambda)$ sample of size 10 is primarily made up of zeros when $\lambda = 0.1$, but mostly takes different values when $\lambda = 10$. The impact on the number of sufficient statistics can be easily assessed when $k = 4$. (Note that the simulated dataset corresponding to $(n, \lambda) = (10, 0.1)$ in Table 11.1 corresponds to a sample only made up of zeros, which explains the $n + 1 = 11$ values of the sufficient statistic $(n_1, S_1) = (n_1, 0)$ when $k = 2$.)

An interesting comment one can make about this decomposition of the posterior distribution is that it may happen that, as already noted in Casella *et al.* (2004), a small number of values of the local sufficient statistic (n_1, S_1) carry most of

Table 11.1 Number of pairs (n_1, S_1) for simulated datasets from a Poisson $\mathcal{P}(\lambda)$ and different numbers of components.

(n, λ)	$k = 2$	$k = 3$	$k = 4$
(10, 0.1)	11	66	286
(10, 1)	52	885	8160
(10, 10)	166	7077	120,908
(20, 0.1)	57	231	1771
(20, 1)	260	20,607	566,512
(20, 10)	565	100,713	—
(30, 0.1)	87	4060	81,000
(30, 1)	520	82,758	—
(30, 10)	1413	637,020	—
(40, 0.1)	216	13,986	—
(40, 1)	789	271,296	—
(40, 10)	2627	—	—

Missing terms are due to excessive computational or storage requirements.

the posterior weight. Table 11.2 provides some occurrences of this feature, as for instance in the case $(n, \lambda) = (20, 10)$.

We now turn to a minnow dataset made of 50 observations, for which we need a minimal description. As seen in Figure 11.1, the datapoints take large values, which is a drawback from a computational point of view since the number of statistics to be registered is much larger than when all data points are small. For this reason, we can only process the mixture model with $k = 2$ components.

Table 11.2 Number of sufficient statistics (n_i, S_i) corresponding to the 99 % largest posterior weights/total number of pairs for datasets simulated either from a Poisson $\mathcal{P}(\lambda)$ or from a mixture of two Poisson $\mathcal{P}(\lambda_i)$, and different numbers of components.

(n, λ)	$k = 2$	$k = 3$	$k = 4$
(10, 1)	20/44	209/675	1219/5760
(10, 10)	58/126	1292/4641	13247/78060
(20, 0.1)	38/40	346/630	1766/6160
(20, 1)	160/196	4533/12819	80925/419824
(10, 0.1, 10, 2)	99/314	5597/28206	—
(10, 1, 10, 5)	21/625	13981/117579	—
(15, 1, 15, 3)	50/829	62144/211197	—
(20, 10)	1/580	259/103998	—
(30, 0.1)	198/466	20854/70194	30052/44950
(30, 1)	202/512	18048/80470	—
(30, 5)	1/1079	58820/366684	—

Missing terms are due to excessive computational or storage requirements.

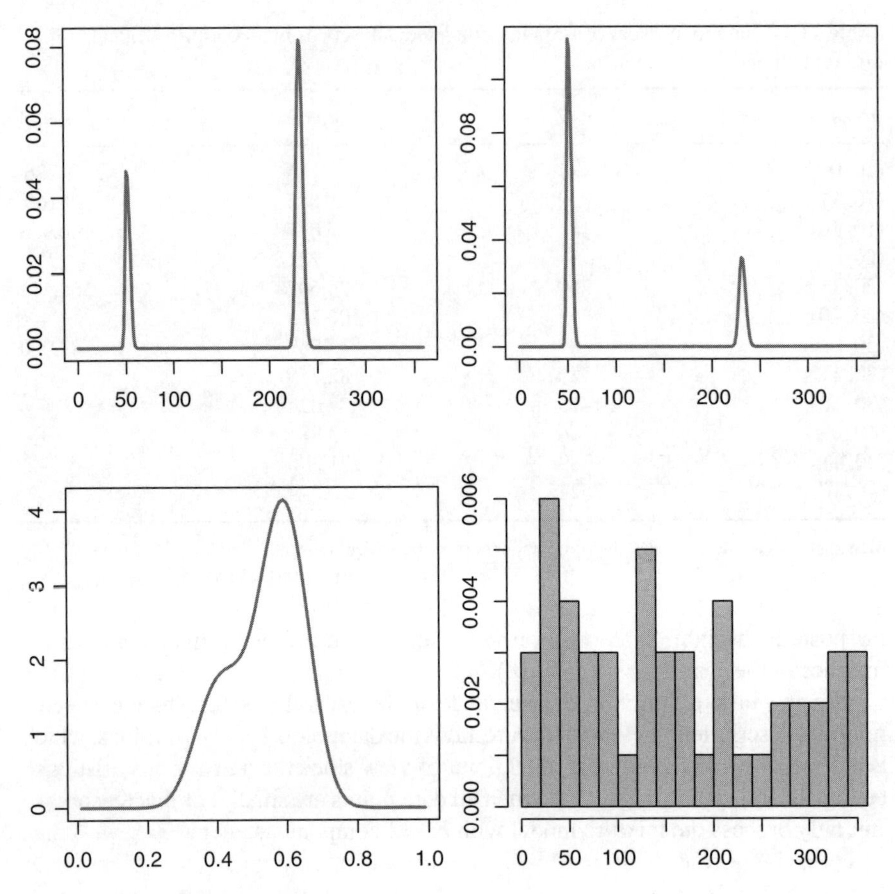

Figure 11.1 Marginal posterior distribution of λ_1 (top left), marginal posterior distribution of λ_2 (top right), marginal posterior distribution of p (bottom left) and histogram of the minnow dataset (bottom right). (The prior parameters are $1/100$ and $1/200$ to remain compatible with the data range.)

If we instead use a completely symmetric prior with identical hyperparameters for λ_1 and λ_2, the output of the algorithm is then also symmetric in both components, as shown by Figure 11.2. The modes of the marginals of λ_1 and λ_2 remain the same, nonetheless.

11.2.4 Multinomial mixtures

The case of a multinomial mixture can be dealt with similarly. If we have n observations $\mathbf{n}_j = (n_{j1}, \ldots, n_{jk})$ from the mixture

$$\mathbf{n}_j \sim p\mathcal{M}_k(d_j; q_{11}, \ldots, q_{1k}) + (1-p)\mathcal{M}_k(d_j; q_{21}, \ldots, q_{2k}),$$

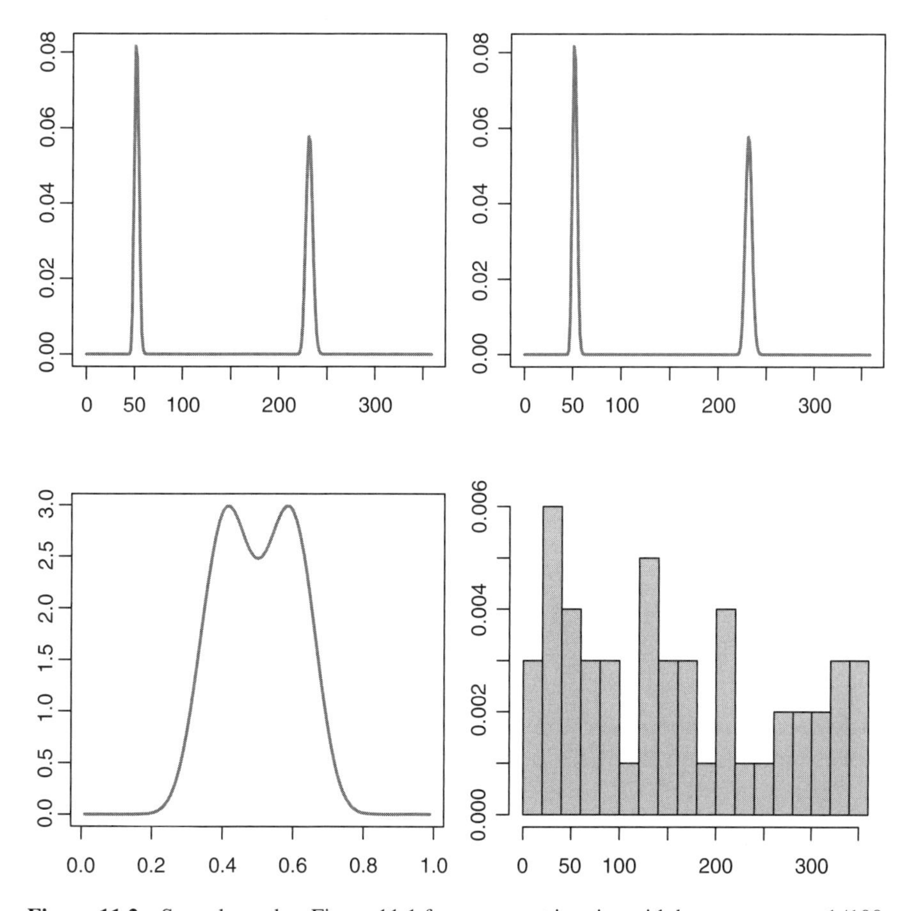

Figure 11.2 Same legend as Figure 11.1 for a symmetric prior with hyperparameter $1/100$.

where $n_{j1} + \cdots + n_{jk} = d_j$ and $q_{11} + \cdots + q_{1k} = q_{21} + \cdots + q_{2k} = 1$, the conjugate priors on the q_{ij}s are Dirichlet distributions ($i = 1, 2$),

$$(q_{i1}, \ldots, q_{ik}) \sim \mathcal{D}(\alpha_{i1}, \ldots, \alpha_{ik}),$$

and we use once again the uniform prior on p. (A default choice for the α_{ij}s is $\alpha_{ij} = 1/2$.) Note that the d_js may differ from observation to observation, since they are irrelevant for the posterior distribution: given a partition $\mathbf{z}$ of the sample, the complete posterior is indeed

$$p^{n_1}(1-p)^{n_2} \prod_{i=1}^{2} \prod_{z_j=i} q_{i1}^{n_{j1}} \cdots q_{ik}^{n_{jk}} \times \prod_{i=1}^{2} \prod_{h=1}^{k} q_{ih}^{-1/2},$$

up to a normalising constant that does not depend on $\mathbf{z}$.

More generally, if we consider a mixture with m components,

$$\mathbf{n}_j \sim \sum_{\ell=1}^{m} p_\ell \mathcal{M}_k(d_j; q_{\ell 1}, \ldots, q_{\ell k}),$$

the complete posterior is also directly available, as

$$\prod_{i=1}^{m} p_i^{n_i} \times \prod_{i=1}^{m} \prod_{z_j=i} q_{i1}^{n_{j1}} \cdots q_{ik}^{n_{jk}} \times \prod_{i=1}^{m} \prod_{h=1}^{k} q_{ih}^{-1/2},$$

once more up to a normalising constant.

The corresponding normalising constant of the Dirichlet distribution being

$$K(\alpha_{i1}, \ldots, \alpha_{ik}) = \frac{\prod_{j=1}^{k} \Gamma(\alpha_{ij})}{\Gamma(\alpha_{i1} + \cdots + \alpha_{ik})},$$

it produces the overall weight of a given partition $\mathbf{z}$ as

$$n_1! n_2! \frac{\prod_{j=1}^{k} \Gamma(\alpha_{1j} + S_{1j})}{\Gamma(\alpha_{11} + \cdots + \alpha_{1k} + S_{1\cdot})} \times \frac{\prod_{j=1}^{k} \Gamma(\alpha_{ij} + S_{2j})}{\Gamma(\alpha_{21} + \cdots + \alpha_{2k} + S_{2\cdot})}, \tag{11.2}$$

where n_i is the number of observations allocated to component i, S_{ij} is the sum of the $n_{\ell j}$s for the observations ℓ allocated to component i and

$$S_{i\cdot} = \sum_j \sum_{z_\ell = i} n_{\ell j}.$$

Given that the posterior distribution only depends on those 'sufficient' statistics S_{ij} and n_i, the same factorisation as in the Poisson case applies, namely that we simply need to count the number of occurrences of a particular local sufficient statistic $(n_1, S_{11}, \ldots, S_{km})$. The bookkeeping algorithm of Fearnhead (2005) applies in this setting as well. (Again, a naïve R program is available from the authors.) As shown in Table 11.3, once again the reduction in the number of cases to be considered is enormous.

11.2.5 Normal mixtures

For a normal mixture, the number of truly different terms in the posterior distribution is much larger than in the previous (discrete) cases, in the sense that only permutations of the members of a given partition within each term of the partition provide the same local sufficient statistics. Therefore, the number of observations that can be handled in an exact analysis is necessarily extremely limited.

As mentioned in Section 11.2.1, the locally conjugate priors for normal mixtures are products of normal $\mu_j | \sigma_j \sim \mathcal{N}(\xi_j, \sigma_j^2 / c_j)$ by inverse gamma $\sigma_j^{-2} \sim \mathcal{G}(a_j/2, b_j/2)$ distributions. For instance, in the case of a two-component

Table 11.3 Number of sufficient statistics $(n_i, S_{ij})_{1\leq i\leq m,1\leq j\leq k}$ corresponding to the 99 % largest posterior, of pairs (n_i, S_i) corresponding to the 99 % largest posterior weights and the total number of statistics for datasets simulated from mixtures of m multinomial $\mathcal{M}_k(d_j; q_1, \ldots, q_k)$ and different parameters.

(n, d_j, k)	$m = 2$	$m = 3$	$m = 4$	$m = 5$
(10,5,3)	33/35	4602/9093	56815/68964	—
(10,5,4)	90/232	3650/21,249	296994/608992	—
(10,5,5)	247/707	247/7857	59195/409600	—
(10,10,2)	19/20	803/885	3703/4800	7267/11 550
(10,10,3)	117/132	1682/1893	48571/60720	—
(10,10,4)	391/514	3022/3510	83757/170864	—
(10,10,5)	287/1008	7031/12960	35531/312320	—
(20,5,2)	129/139	517/1140	26997/45600	947/10626
(20,5,3)	384/424	188703/209736	108545/220320	—
(20,5,4)	3410/6944	819523/1058193	—	—
(20,10,5)	1225/1332	9510/1089990	—	—

Missing terms are due to excessive computational or storage requirements.

normal mixture,

$$x_1, \ldots, x_n \overset{\text{iid}}{\sim} p\mathcal{N}(\mu_1, \sigma_1^2) + (1 - p)\mathcal{N}(\mu_2, \sigma_2^2),$$

we can pick $\mu_1|\sigma_1 \sim \mathcal{N}(0, 10\sigma_1^2)$, $\mu_2|\sigma_2 \sim \mathcal{N}(1, 10\sigma_2^2)$, $\sigma_j^{-2} \sim \mathcal{G}(2, 2/\sigma_0^2)$, if a difference of one between both means is considered likely (meaning of course that the data are previously scaled) and if σ_0^2 is the prior assumption on the variance (possibly deduced from the range of the sample). Obviously, the choice of a gamma distribution with 2 degrees of freedom is open to discussion, as it is not without consequences on the posterior distribution.

The normalising constant of the prior distribution is (up to a true constant)

$$K(a_1, \ldots, a_k, b_1, \ldots, b_k, c_1, \ldots, c_k, \xi_1, \ldots, \xi_k) = \prod_{i=1}^{k} \sqrt{c_j}\,.$$

Indeed, the corresponding posterior is

$$\mu_j|\sigma_j \sim \mathcal{N}\left[(c_j\xi_j + n_j\overline{x}_j, \sigma_j^2/(c_j + n_j)\right]$$

and

$$\sigma_j^{-2} \sim \mathcal{G}\left[\{a_j + n_j\}/2, \{b_j + n_j\hat{\sigma}_j^2 + (\overline{x}_j - \xi_j)^2/(c_j^{-1} + n_j^{-1})\}\right]\,.$$

The number of different sufficient statistics $(n_j, \overline{x}_j, \hat{\sigma}_j^2)$ is thus related to the number of different partitions of the dataset into at most k groups. This is related to the Bell number (Rota, 1964), which grows extremely fast. We therefore do not pursue the example of the normal mixture any further for lack of practical purpose.

References

Casella, G., Robert, C. and Wells, M. (2004) Mixture models, latent variables and partitioned importance sampling. *Statistical Methodology*, **1**, 1–18.

Dempster, A., Laird, N. and Rubin, D. (1977) Maximum likelihood from incomplete data via the EM algorithm (with discussion). *Journal of the Royal Statistical Society, Series B*, **39**, 1–38.

Diebolt, J. and Robert, C. (1990) Estimation des paramètres d'un mélange par échantillonnage bayésien. *Notes aux Comptes–Rendus de l'Académie des Sciences I*, **311**, 653–658.

Diebolt, J. and Robert, C. (1994) Estimation of finite mixture distributions by Bayesian sampling. *Journal of the Royal Statistical Society, Series B*, **56**, 363–375.

Fearnhead, P. (2005) Direct simulation for discrete mixture distributions. *Statistics and Computing*, **15**, 125–133.

Frühwirth-Schnatter, S. (2006) *Finite Mixture and Markov Switching Models*. Springer-Verlag.

Jasra, A., Holmes, C. and Stephens, D. (2005) Markov chain Monte Carlo methods and the label switching problem in Bayesian mixture modeling. *Statistical Science*, **20**, 50–67.

Lee, K., Marin, J. M., Mengersen, K. and Robert, C. (2009) Bayesian inference on mixtures of distributions. In *Perspectives in Mathematical Sciences I: Probability and Statistics* (eds N. N. Sastry, M. Delampady and B. Rajeev), pp. 165–202. World Scientific Singapore.

McLachlan, G. and Peel, D. (2000) *Finite Mixture Models*. John Wiley & Sons, Ltd.

Mengersen, K. and Robert, C. (1996) Testing for mixtures: a Bayesian entropic approach (with discussion). In *Bayesian Statistics 5* (eds J. Berger, J. Bernardo, A. Dawid, D. Lindley and A. Smith), pp. 255–276. Oxford University Press.

Robert, C. (2001) *The Bayesian Choice*, 2nd edn. Springer-Verlag.

Robert, C. and Casella, G. (2004) *Monte Carlo Statistical Methods*, 2nd edn. Springer-Verlag.

Rota, G. C. (1964) The number of partitions of a set. *American Mathematical Monthly*, **71**, 498–504.

Stephens, M. (2000) Dealing with label switching in mixture models. *Journal of the Royal Statistical Society, Series B*, **62**, 795–809.

12

Manifold MCMC for mixtures

Vassilios Stathopoulos and Mark Girolami

12.1 Introduction

Mixture models are useful in describing a wide variety of random phenomena because of their inherent flexibility (Titterington *et al.*, 1985), and therefore in many fields of science they are used to model complex processes and systems. Examples of applications include clustering (McLachlan and Baek, 2010), density estimation (Escobar and West, 1994) and classification (Celeux, 2006). Mixture models still pose several interesting computational and statistical problems, especially in Bayesian inference. In the statistical literature the most prominent approach adopted for Bayesian inference of mixture models is Gibbs sampling; e.g. see Diebolt and Robert (1994), Richardson and Green (1997), McLachlan and Peel (2000) and (Gilks *et al.*, 1999, Chapter 24). Although Gibbs sampling for mixture models is relatively simple to implement it can sometimes be inefficient (Marin *et al.*, 2005).

One problem associated with the inefficiency of the Gibbs sampler is the dependence on a completion constructed by introducing latent indicator variables associating observations with mixture components. This implies first that the dimension of the state space grows linearly with the number of observations. Second, the updates of the parameters and the latent variables are based on their full conditional distributions and thus the length of the proposed moves during simulation may be restricted, resulting in correlated samples (Neal, 1999). Finally, the Gibbs sampler can be very sensitive to initialisation, especially when there are local modes in the vicinity of the starting position. An interesting illustration of these problems can be found in Marin *et al.* (2005).

Mixtures: Estimation and Applications, First Edition. Edited by Kerrie L. Mengersen, Christian P. Robert and D. Michael Titterington.

Some of the previously mentioned problems can be overcome by avoiding such a completion and adopting Metropolis-Hastings sampling, provided that suitable proposals are specified. However, tuning the proposal mechanisms to achieve good mixing and fast convergence is far from straightforward even though some theoretical guidance is provided (Roberts *et al.*, 1997). MCMC methods, such as the Metropolis adjusted Langevin algorithm (MALA) (Roberts and Stramer, 2003) and Hamiltonian Monte Carlo (HMC) (Duane *et al.*, 1987), have also been proposed in the literature and have been shown to be more efficient than random walk Metropolis-Hastings in terms of effective sample size (ESS) and convergence rates on certain problems. However, HMC and MALA also require extensive tuning of the proposal mechanisms; see, for example, Neal (1993) and Roberts and Rosenthal (1998). More recently Girolami and Calderhead (2010) defined HMC and MALA on the Riemann manifold of density functions, thus exploiting the intrinsic geometry of statistical models and thereby providing a principled framework and systematic approach to the proposal design process.

For problems where it is straightforward to sample from the full conditionals, Gibbs sampling is commonly employed, often neglecting the performance issues associated with the method, in terms of both sampling efficiency and computational cost. Given the demonstrated potential of this new class of manifold MCMC algorithms presented in Girolami and Calderhead (2010), it is interesting to investigate their suitability for inference relative to mixture models. In this chapter we aim to provide such a study and suggest viable alternatives to Gibbs sampling for mixture models. We assess manifold MALA and manifold HMC in comparison with their nonmanifold counterparts as well as with Gibbs and Metropolis-Hastings on five mixture models and we present results in terms of sampling efficiency and computational cost. All the algorithms and the scripts used for this study are implemented in Matlab and are available to download from `http://www.ucl.ac.uk/stats/research/manifold_mixtures/`.

Other interesting problems associated with inference for mixture model parameters, which are outside the scope of this chapter, are the label-switching problem (Jasra *et al.*, 2005) and inference on the number of mixture components. Inference for the number of mixture components can be performed using the reversible jump MCMC algorithm (Richardson and Green, 1997) or its population-based extension (Jasra *et al.*, 2007) while for the label-switching problem we refer to Stephens (2000) and Celeux *et al.* (2000). Although we do not deal with the problem of inferring the number of components in this chapter, in Section 12.5 we discuss how the methods discussed here can be extended to handle such problems.

The rest of the chapter is organised as follows. In Section 12.2 we give a brief overview of Riemann manifold MALA and HMC and in Section 12.3 we define the particular family of mixture models that will be used in this study and give the necessary expressions required by the MCMC algorithms. In Section 12.4 we present our experimental framework for tuning and applying the different MCMC algorithms and show results on simulated datasets. Finally, Section 12.5 concludes the chapter and discusses several future directions.

12.2 Markov chain Monte Carlo Methods

In this section we give a brief overview of the MCMC algorithms that we consider in this work. Some familiarity with the concepts of MCMC is required by the reader since an introduction to the subject is beyond the scope of this chapter.

12.2.1 Metropolis-Hastings

For a random vector $\boldsymbol{\theta} \in \mathbb{R}^D$ with density $p(\boldsymbol{\theta})$ the Metropolis-Hastings algorithm employs a proposal mechanism $q(\boldsymbol{\theta}^*|\boldsymbol{\theta}^{t-1})$ and proposed moves are accepted with probability $\min\left\{1, p(\boldsymbol{\theta}^*)q(\boldsymbol{\theta}^{t-1}|\boldsymbol{\theta}^*)/p(\boldsymbol{\theta}^{t-1})q(\boldsymbol{\theta}^*|\boldsymbol{\theta}^{t-1})\right\}$. In the context of Bayesian inference the target density $p(\boldsymbol{\theta})$ corresponds to the posterior distribution of the model parameters. Tuning the Metropolis-Hastings algorithm involves selecting the right proposal mechanism. A common choice is to use a random walk Gaussian proposal of the form $q(\boldsymbol{\theta}^*|\boldsymbol{\theta}^{t-1}) = \mathcal{N}(\boldsymbol{\theta}^*|\boldsymbol{\theta}^{t-1}, \boldsymbol{\Sigma})$, where $\mathcal{N}(\cdot|\boldsymbol{\mu}, \boldsymbol{\Sigma})$ denotes the multivariate normal density with mean $\boldsymbol{\mu}$ and covariance matrix $\boldsymbol{\Sigma}$.

However, selecting the covariance matrix is far from trivial in most cases since knowledge about the target density is required. Therefore a more simplified proposal mechanism is often considered where the covariance matrix is replaced with a diagonal matrix such as $\boldsymbol{\Sigma} = \epsilon \boldsymbol{I}$, where the value of the scale parameter ϵ has to be tuned in order to achieve fast convergence and good mixing. Small values of ϵ imply small transitions and result in high acceptance rates while the mixing of the Markov chain is poor. Large values, on the other hand, allow for large transitions but they result in most of the samples being rejected. Tuning the scale parameter becomes even more difficult in problems where the standard deviations of the marginal posteriors differ substantially, since different scales are required for each dimension, and this is exacerbated when correlations between different variables exist. Adaptive schemes for the Metropolis-Hastings algorithm have also been proposed (Haario *et al.*, 2005) though they should be applied with care (Andrieu and Thoms, 2008).

12.2.2 Gibbs sampling

The Gibbs sampler successively simulates individual or subsets of variables from their full conditional distributions. In cases where these conditionals are not directly available a completion of the original distribution for which it is easier to derive the conditionals can be used. For mixtures, such completion is naturally provided by the missing structure of the problem by introducing indicator variables associating observations with components.

Although the Gibbs sampler is simple to implement and no tuning is required, simulation involves sampling the indicator variables, which are of the order of the observations, therefore increasing the state space considerably. Moreover, sampling the mixture parameters conditioned on indicator variables, and vice versa, implies that the lengths of the transitions are constrained since for a particular value of the indicator variables the mixture parameters are quite concentrated and therefore their new values will not allow for large moves of the indicator variables in the next iteration (Marin *et al.*, 2005). Methods for improving the efficiency of Gibbs sampling include ordered overrelaxation (Neal, 1999), which is applicable when the cumulative and inverse cumulative functions of the conditional distributions can be efficiently calculated.

12.2.3 Manifold Metropolis adjusted Langevin algorithm

With the log of the target density denoted by $\mathcal{L}(\boldsymbol{\theta}) = \log p(\boldsymbol{\theta})$, the manifold MALA (mMALA) method (Girolami and Calderhead, 2010) defines a Langevin diffusion with stationary distribution $p(\boldsymbol{\theta})$ on the Riemann manifold of density functions with metric tensor $\boldsymbol{G}(\boldsymbol{\theta})$. If we employ a first-order Euler integrator to solve the diffusion a proposal mechanism with density $q(\boldsymbol{\theta}^*|\boldsymbol{\theta}^{t-1}) = \mathcal{N}[\boldsymbol{\theta}^*|\boldsymbol{\mu}(\boldsymbol{\theta}^{t-1}, \epsilon), \epsilon^2\boldsymbol{G}^{-1}(\boldsymbol{\theta}^{t-1})]$ is obtained, where ϵ is the integration step size, a parameter that needs to be tuned, and the dth component of the mean function $\boldsymbol{\mu}(\boldsymbol{\theta}, \epsilon)_d$ is

$$\boldsymbol{\mu}(\boldsymbol{\theta}, \epsilon)_d = \boldsymbol{\theta}_d + \frac{\epsilon^2}{2}\left[\boldsymbol{G}^{-1}(\boldsymbol{\theta})\nabla_{\boldsymbol{\theta}}\mathcal{L}(\boldsymbol{\theta})\right]_d - \epsilon^2\sum_{i=1}^{D}\sum_{j=1}^{D}\boldsymbol{G}(\boldsymbol{\theta})^{-1}_{i,j}\Gamma^d_{i,j}, \tag{12.1}$$

where $\Gamma^d_{i,j}$ are the Christoffel symbols of the metric in local coordinates (Kühnel, 2005).

Similarly to MALA (Roberts and Stramer, 2003), due to the discretisation error introduced by the first-order approximation, convergence to the stationary distribution is no longer guaranteed and thus the Metropolis-Hastings ratio is employed to correct this bias. The mMALA algorithm can be simply stated as in Algorithm 1. Details can be found in Girolami and Calderhead (2010).

Algorithm 1: mMALA

Inititialise $\boldsymbol{\theta}^0$
for $t = 1$ to T **do**
  $\boldsymbol{\theta}^* \sim \mathcal{N}[\boldsymbol{\theta}|\boldsymbol{\mu}(\boldsymbol{\theta}^{t-1}, \epsilon), \epsilon^2\boldsymbol{G}^{-1}(\boldsymbol{\theta}^{t-1})]$
  $r = \min\left\{1, p(\boldsymbol{\theta}^*)q(\boldsymbol{\theta}^{t-1}|\boldsymbol{\theta}^*)/p(\boldsymbol{\theta}^{t-1})q(\boldsymbol{\theta}^*|\boldsymbol{\theta}^{t-1})\right\}$
  $u \sim \mathcal{U}_{[0,1]}$
  if $r > u$ **then**
    $\boldsymbol{\theta}^t = \boldsymbol{\theta}^*$
  else
    $\boldsymbol{\theta}^t = \boldsymbol{\theta}^{t-1}$
  end if
end for

We can interpret the proposal mechanism of mMALA as a local Gaussian approximation to the target density similar to the adaptive Metropolis-Hastings of Haario *et al.* (1998). In contrast to Haario *et al.* (1998), however, the effective covariance matrix in mMALA is the inverse of the metric tensor evaluated at the current position and no samples from the chain are required in order to estimate it, thereby avoiding the difficulties of adaptive MCMC discussed in Andrieu and Thoms (2008). Furthermore, a simplified version of the mMALA algorithm can also be derived by assuming a manifold with constant curvature, thus cancelling the last term in Equation (12.1), which depends on the Christoffel symbols. Finally, the mMALA algorithm can be seen as a generalisation of the original MALA (Roberts and Stramer, 2003) since, if the metric tensor $\boldsymbol{G}(\boldsymbol{\theta})$ is equal to the identity matrix corresponding to a Euclidean manifold, then the original algorithm is recovered.

12.2.4 Manifold Hamiltonian Monte Carlo

The Riemann manifold Hamiltonian Monte Carlo (RM-HMC) method defines a Hamiltonian on the Riemann manifold of probability density functions by introducing the auxiliary variables $\boldsymbol{p} \sim \mathcal{N}[\mathbf{0}, \boldsymbol{G}(\boldsymbol{\theta})]$, which are interpreted as the momentum at a particular position $\boldsymbol{\theta}$ and by considering the negative log of the target density as a potential function. More formally, the Hamiltonian defined on the Riemann manifold is

$$H(\boldsymbol{\theta}, \boldsymbol{p}) = -\mathcal{L}(\boldsymbol{\theta}) + \frac{1}{2}\log\left[2\pi|\boldsymbol{G}(\boldsymbol{\theta})|\right] + \frac{1}{2}\boldsymbol{p}^{\mathrm{T}}\boldsymbol{G}(\boldsymbol{\theta})^{-1}\boldsymbol{p}, \tag{12.2}$$

where the terms $-\mathcal{L}(\boldsymbol{\theta}) + \frac{1}{2}\log(2\pi|\boldsymbol{G}(\boldsymbol{\theta})|)$ and $\frac{1}{2}\boldsymbol{p}^{\mathrm{T}}\boldsymbol{G}(\boldsymbol{\theta})^{-1}\boldsymbol{p}$ are the potential energy and kinetic energy terms, respectively. Simulating the Hamiltonian requires a time-reversible and volume-preserving numerical integrator. For this purpose the generalised leapfrog algorithm can be employed and provides a deterministic proposal mechanism for simulating from the conditional distribution, i.e. $\boldsymbol{\theta}^*|\boldsymbol{p} \sim p(\boldsymbol{\theta}^*|\boldsymbol{p})$. More details about the generalised leapfrog integrator can be found in Girolami and Calderhead (2010). To simulate a path across the manifold, the leapfrog integrator is iterated L times, which along with the integration step-size ϵ is a parameter requiring tuning. Again, due to the integration errors on simulating the Hamiltonian, in order to ensure convergence to the stationary distribution the Metropolis-Hastings ratio is applied. Moreover, following the suggestion in Neal (1993) the number of leapfrog iterations L is randomised in order to improve mixing. The RM-HMC algorithm is given in Algorithm 2.

Similarly to the mMALA algorithm, when the metric tensor $\boldsymbol{G}(\boldsymbol{\theta})$ is equal to the identity matrix corresponding to a Euclidean manifold, RM-HMC is equivalent to the HMC algorithm of Duane *et al.* (1987).

12.3 Finite Gaussian mixture models

For this study we consider finite mixtures of univariate Gaussian components of the form

$$p(x_i|\boldsymbol{\theta}) = \sum_{k=1}^{K} \pi_k \mathcal{N}(x_i|\mu_k, \sigma_k^2), \tag{12.3}$$

Algorithm 2: RM-HMC

Inititialise $\boldsymbol{\theta}^0$
for $t = 1$ to T **do**
$\boldsymbol{p}_*^0 \sim \mathcal{N}[\boldsymbol{p}|\boldsymbol{0}, \boldsymbol{G}(\boldsymbol{\theta}^{t-1})]$
$\boldsymbol{\theta}_*^0 = \boldsymbol{\theta}^{t-1}$
$e \sim \mathcal{U}_{[0,1]}$
$N = \text{ceil}(\epsilon L)$
{Simulate the Hamiltonian using a Generalised Leapfrog integrator for N steps}
for $n = 0$ to N **do**

solve $\boldsymbol{p}_*^{n+\frac{1}{2}} = \boldsymbol{p}_*^n - \frac{\epsilon}{2}\nabla_{\boldsymbol{\theta}} H\left(\boldsymbol{\theta}_*^n, \boldsymbol{p}_*^{n+\frac{1}{2}}\right)$

solve $\boldsymbol{\theta}_*^{n+1} = \boldsymbol{\theta}_*^n + \frac{\epsilon}{2}\left[\nabla_{\boldsymbol{p}} H\left(\boldsymbol{\theta}_*^n, \boldsymbol{p}_*^{n+\frac{1}{2}}\right) + \nabla_{\boldsymbol{p}} H\left(\boldsymbol{\theta}_*^{n+1}, \boldsymbol{p}_*^{n+\frac{1}{2}}\right)\right]$

$\boldsymbol{p}_*^{n+1} = \boldsymbol{p}_*^{n+\frac{1}{2}} - \frac{\epsilon}{2}\nabla_{\boldsymbol{\theta}} H\left(\boldsymbol{\theta}_*^{n+1}, \boldsymbol{p}_*^{n+\frac{1}{2}}\right)$

end for
$(\boldsymbol{\theta}^*, \boldsymbol{p}^*) = (\boldsymbol{\theta}_*^{N+1}, \boldsymbol{p}_*^{N+1})$
{Metropolis-Hastings ratio}
$r = \min\left\{1, \exp\left[-H(\boldsymbol{\theta}^*, \boldsymbol{p}^*) + H(\boldsymbol{\theta}^{t-1}, \boldsymbol{p}^{t-1})\right]\right\}$
$u \sim \mathcal{U}_{[0,1]}$
if $r > u$ **then**
$\boldsymbol{\theta}^t = \boldsymbol{\theta}^*$
else
$\boldsymbol{\theta}^t = \boldsymbol{\theta}^{t-1}$
end if
end for

---

where $\boldsymbol{\theta}$ is the vector of size $D = 3K$ containing all the parameters π_k, μ_k and σ_k^2 and $\mathcal{N}(\cdot|\mu, \sigma^2)$ is a Gaussian density with mean μ and variance σ^2. A common choice of prior takes the form

$$p(\boldsymbol{\theta}) = \mathcal{D}(\pi_1, \ldots, \pi_K|\lambda) \prod_{k=1}^{K} \mathcal{N}(\mu_k|m, \beta^{-1}\sigma_k^2)\mathcal{IG}(\sigma_k^2|b, c), \tag{12.4}$$

where $\mathcal{D}(\cdot|\lambda)$ is the symmetric Dirichlet distribution with parameter λ and $\mathcal{IG}(\cdot|b, c)$ is the inverse Gamma distribution with shape parameter b and scale parameter c.

Although the posterior distribution associated with this model is formally explicit it is computationally intractable, since it can be expressed as a sum of K^N terms corresponding to all possible allocations of observations x_i to mixture components (Robert and Casella 2005, Chapter 9). For Bayesian inference of the parameters we can either find a suitable completion of the model which we can efficiently sample using the Gibbs sampler or we can use MCMC algorithms that use the unnormalised log posterior.

12.3.1 Gibbs sampler for mixtures of univariate Gaussians

A natural completion for mixture models that exploits the missing data structure of the problem can be constructed by introducing latent variables $z_{i,k} \in \{0, 1\}$, where $z_i \sim \mathcal{M}(1, \pi_1, \ldots, \pi_K)$ such that if $z_{i,k} = 1$ then the observation x_i is allocated to the kth mixture component. Here $\mathcal{M}(1, p_1, \ldots, p_K)$ is a multinomial distribution for K mutually exclusive events with probabilities $p_1, \ldots, p_K$. The likelihood of the model in Equation (12.3) then becomes

$$p(x_i|\boldsymbol{\theta}, z_i) = \prod_{k=1}^{K} \pi_k^{z_{i,k}} \mathcal{N}(x_i|\mu_k, \sigma_k^2)^{z_{i,k}} \tag{12.5}$$

and under the prior in Equation (12.4) the conditional distributions, which can be used in the Gibbs sampler, follow as

$$z_i|\boldsymbol{X}, \boldsymbol{\theta} \sim \mathcal{M}(1, \rho_{i,1}, \ldots, \rho_{i,K}), \tag{12.6}$$

$$\mu_k|\boldsymbol{X}, \boldsymbol{Z}, \sigma_k^2 \sim \mathcal{N}\left(\frac{m\beta + y_k}{\beta + n_k}, \frac{\sigma_k^2}{\beta + n_k}\right), \tag{12.7}$$

$$\sigma_k^2|\boldsymbol{X}, \boldsymbol{Z}, \mu_k \sim \mathcal{IG}\left(b + \frac{n_k + 1}{2}, c + \frac{s_k + \beta(\mu_k - m)^2}{2}\right), \tag{12.8}$$

$$\boldsymbol{\pi}|\boldsymbol{Z} \sim \mathcal{D}(\lambda + n_1, \ldots, \lambda + n_K), \tag{12.9}$$

where we have defined

$$\rho_{i,k} = \frac{\pi_k \mathcal{N}\left(x_i|\mu_k, \sigma_k^2\right)}{\sum_{k'=1}^{K} \pi_{k'} \mathcal{N}\left(x_i|\mu_{k'}, \sigma_{k'}^2\right)}, \quad s_k = \sum_{i=1}^{N} z_{i,k}(x_i - \mu_k)^2,$$

$$y_k = \sum_{i=1}^{N} z_{i,k} x_i, \quad n_k = \sum_{i=1}^{N} z_{i,k}.$$

12.3.2 Manifold MCMC for mixtures of univariate Gaussians

For the MCMC algorithms described in Section 12.2 we will need a reparameterisation of the mixture model in order to allow the algorithms to operate on an unbounded and unconstrained parameter space. In the model defined in Equation (12.3) the parameters π_k, σ_k^2 are constrained such that $\sum_{k=1}^{K} \pi_k = 1$, $\pi_k > 0$ and $\sigma_k^2 > 0$, thus we reparameterise by using the transformations $\pi_k = \exp(a_k)/\sum_{j=1}^{K} \exp(a_j)$ and $\sigma_k^2 = \exp(\gamma_k)$. We also use an equivalent prior of independent gamma distributions on $\exp(a_k)$ instead of the symmetric Dirichlet prior in Equation (12.4). Finally, the Jacobians of the transformations are used to scale the prior when calculating the Metropolis-Hastings ratio in order to ensure that we are sampling from the correct distribution. The final form of the transformed joint log-likelihood including the Jacobian terms is then

$$\mathcal{L}(\boldsymbol{\theta}) = \sum_{i=1}^{N} \log\left(\sum_{k=1}^{K} \pi_k \mathcal{N}(x|\mu_k, \sigma_k^2)\right) + \sum_{k=1}^{K} \log \mathcal{G}[\exp(a_k)|\lambda, 1] + a_k$$
$$+ \sum_{k=1}^{K} \log \mathcal{N}\left(\mu_k|m, \beta\sigma_k^2\right) + \sum_{k=1}^{K} \log \mathcal{IG}\left(\sigma_k^2|b, c\right) + \gamma_k \qquad (12.10)$$

The partial derivatives of the log-likelihood and the log-prior required for calculating the gradient can be found in the Appendix.

12.3.3 Metric tensor

The last quantities required are the metric tensor and its partial derivatives with respect to the parameters $\boldsymbol{\theta}$. For density functions the natural metric tensor is the expected Fisher Information, $\boldsymbol{I}(\boldsymbol{\theta})$ (Amari and Nagaoka, 2000), which is nonanalytical for mixture models. In this work an estimate of the Fisher information, the empirical Fisher information as defined in McLachlan and Peel (2000, Chapter 2) is used as the metric tensor, and its form is given in Equation (12.11), where we have defined the $N \times D$ score matrix $\boldsymbol{S}$ with elements $S_{i,d} = [\partial \log p(x_i|\boldsymbol{\theta})]/\partial\theta_d$ and $\boldsymbol{s} = \sum_{i=1}^{N} \boldsymbol{S}_{i,\cdot}^{\mathrm{T}}$. The derivatives of the empirical Fisher information are also easily computed as they require calculation of the second derivatives of the log-likelihood and their general form is given in Equation (12.12):

$$\boldsymbol{G}(\boldsymbol{\theta}) = \boldsymbol{S}^{\mathrm{T}}\boldsymbol{S} - \frac{1}{N}\boldsymbol{s}\boldsymbol{s}^{\mathrm{T}}, \qquad (12.11)$$

$$\frac{\partial \boldsymbol{G}(\boldsymbol{\theta})}{\partial\theta_d} = \left(\frac{\partial \boldsymbol{S}^{\mathrm{T}}}{\partial\theta_d}\boldsymbol{S} + \boldsymbol{S}^{\mathrm{T}}\frac{\partial \boldsymbol{S}}{\partial\theta_d}\right) - \frac{1}{N}\left(\frac{\partial \boldsymbol{s}}{\partial\theta_d}\boldsymbol{s}^{\mathrm{T}} + \boldsymbol{s}\frac{\partial \boldsymbol{s}^{\mathrm{T}}}{\partial\theta_d}\right). \qquad (12.12)$$

For the sake of completeness, the second-order partial derivatives of the log-likelihood of the transformed model required in Equation (12.12) are also given in the Appendix.

It is interesting to note here that the metric tensor is defined by the choice of the distance function between two densities. The Fisher information can be derived by taking a first-order expansion of the symmetric Kullback–Leibler (KL) divergence between two densities; the Hellinger distance yields the same metric as it provides a bound on the KL divergence. The choice of distance function therefore dictates the form of the metric tensor and we can see that the analytical intractability of the expected Fisher information for mixture models is 'inherited' from the Kullback–Leibler divergence. However, by considering a different divergence we can define a metric tensor that has an analytical form. In fact, if we consider the L_2 distance between two densities given by

$$\int_{\mathcal{X}} |p(x|\boldsymbol{\theta} + \delta\boldsymbol{\theta}) - p(x|\boldsymbol{\theta})|^2 \,\mathrm{d}x$$

and take a first-order expansion, the metric tensor under the L_2 metric is

$$\int_{\mathcal{X}} \nabla_{\boldsymbol{\theta}} p(x|\boldsymbol{\theta}) \nabla_{\boldsymbol{\theta}}^T p(x|\boldsymbol{\theta}) \mathrm{d}x,$$

where for mixtures of Gaussian densities the integral can be evaluated explicitly. The L_2 metric is a special case of the power divergence (Basu *et al.*, 1998) and has been used in robust estimation of parametric models in Scott (2001). Although we do not explore these ideas further in this chapter we believe that this is an interesting topic for further research.

12.3.4 An illustrative example

To illustrate the differences between Gibbs, MALA, HMC and their Riemann manifold counterparts we use a simple example for which it is easy to visualise the gradient fields and the paths of the Markov chains. For this example we use a mixture model of the form

$$p(x|\mu, \sigma^2) = 0.7 \times \mathcal{N}(x|0, \sigma^2) + 0.3 \times \mathcal{N}(x|\mu, \sigma^2), \tag{12.13}$$

where the variance σ^2 is the same for both components. We use the same reparameterisation for the variance term, namely $\sigma^2 = \exp(\gamma)$, and priors as discussed in Section 12.3. We generated a synthetic dataset of 500 random samples from this model with true parameters set at $\mu = 2.5$ and $\sigma^2 = 1$ and we draw 10 000 samples of the parameters from the posterior using the Gibbs sampler, MALA, HMC, mMALA and RM-HMC as well as the simplified version of mMALA discussed in Section 12.2. For all examples the same starting position was used.

In Figure 12.1 we illustrate the contravariant form of the gradients and the contravariant metric tensor for the simple model in Equation (12.13). The ellipses correspond to the contravariant metric tensor evaluated at different locations and plotted on top of the joint log-likelihood surface. We can see how the contravariant metric tensor reflects the local geometry and note especially the ellipses at the bottom left of the figure, which are elongated along the axis where the target density changes less rapidly, i.e. μ_2. For the mMALA algorithm the contravariant metric tensor corresponds to the covariance matrix of the proposal mechanism and we can see how the proposal mechanism is adapted optimally, allowing for large transitions in areas where the target density is flat and smaller transitions in areas where the target density is steep, preventing the algorithm from 'overshooting'.

In Figure 12.2 we illustrate the differences between MALA and mMALA by comparing the first 1000 samples from their chains. To ensure a fair comparison we have tuned the step-size parameter for both algorithms such that the acceptance rate remains between 40 % and 60 %, as discussed in Christensen *et al.* (2005) and Girolami and Calderhead (2010). For MALA, achieving such an acceptance rate at stationarity was not trivial since for step-sizes above 0.000 424 the algorithm failed to converge in 10 000 samples by rejecting all proposed moves. In Christensen *et al.* (2005) such behaviour for the transient phase of MALA is also reported and a different scaling is suggested. We have experimented with both scalings without

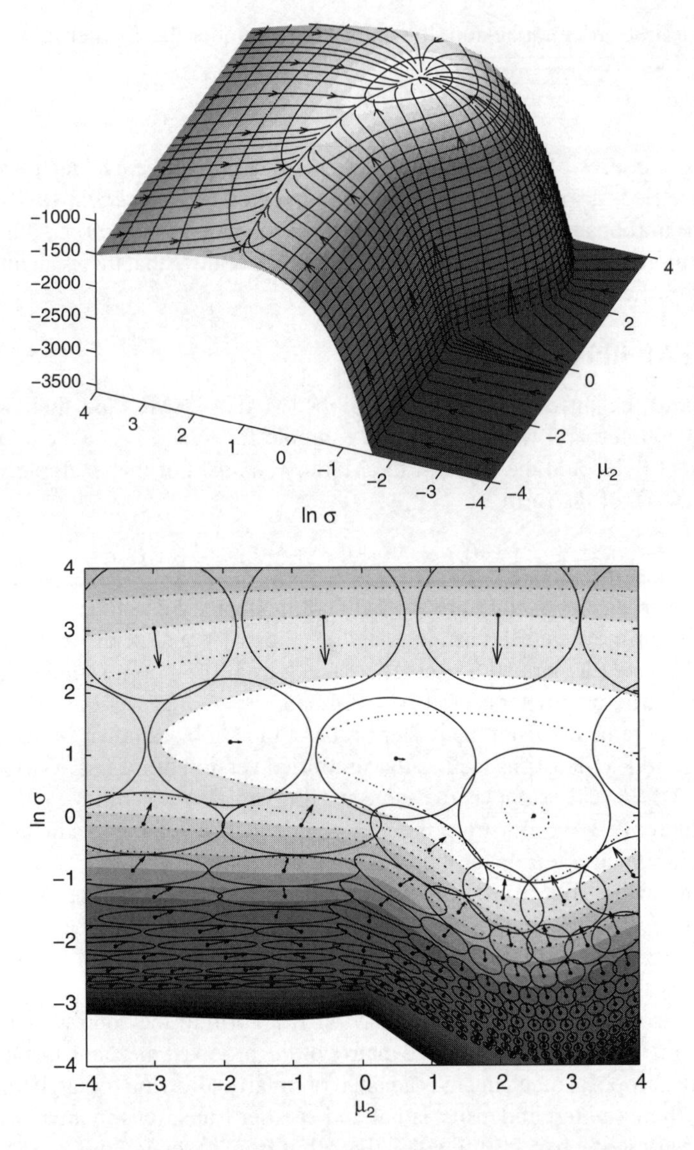

Figure 12.1 Surface of the joint log-likelihood (top) and contravariant form of the gradients and the contravariant metric tensor evaluated at different locations (bottom). Arrows correspond to the contravariant form of the gradients and ellipses to the contravariant metric tensor. Dashed lines are the isocontours of the joint log-likelihood.

any significant difference for this particular case. We argue that the problem lies in the large discrepancy between the gradient magnitudes at the starting position and around the mode. From Figure 12.1 we can see that at the starting position a small step-size is required in order not to 'overshoot' due to the large gradient. On the

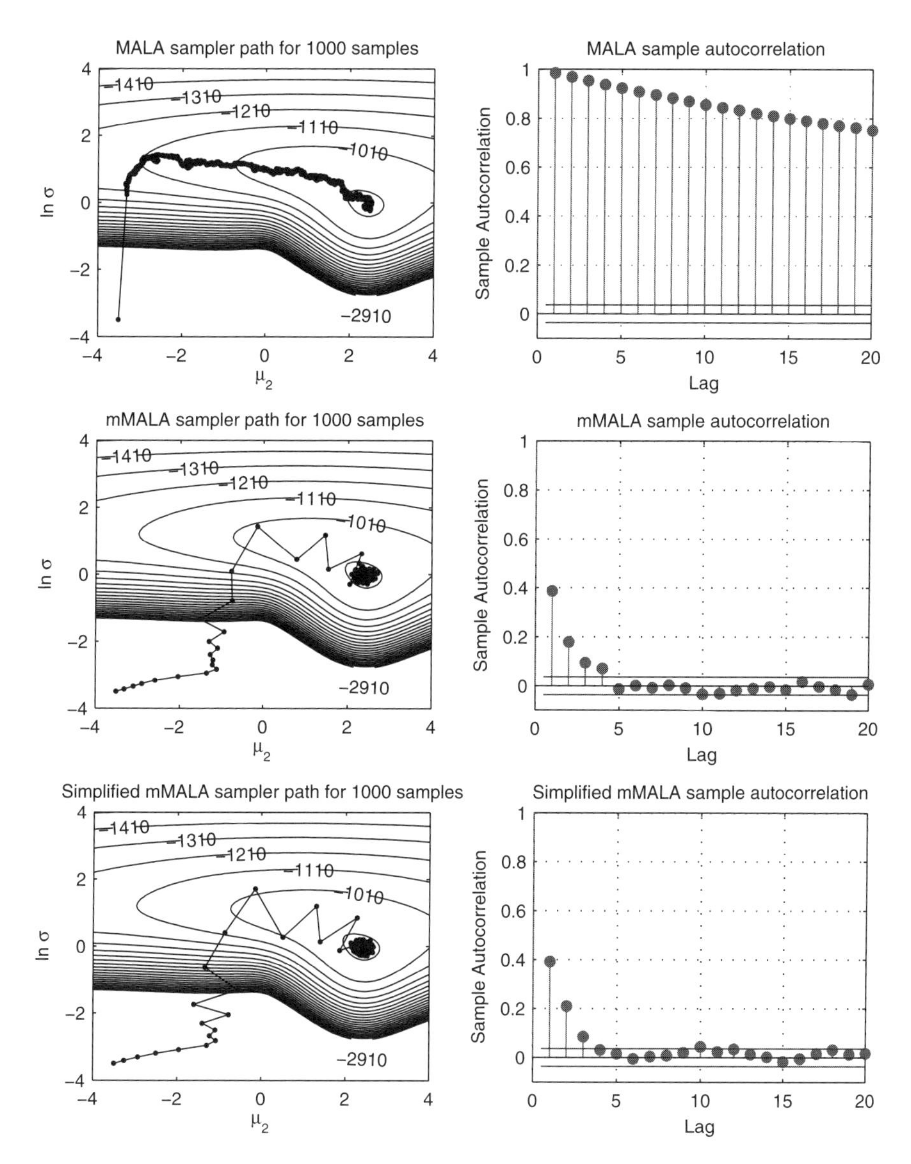

Figure 12.2 Comparison of MALA (top), mMALA (middle) and simplified mMALA (bottom) convergence paths and autocorrelation plots. Autocorrelation plots are from the stationary chains, i.e. once the chains have converged to the stationary distribution.

contrary, around the mode the gradient magnitudes are small, suggesting a larger step-size in order to obtain less-correlated samples.

The sample autocorrelation plots from the stationary chains, i.e. after the chains have converged to the stationary distribution, for both algorithms are also shown in Figure 12.2. We can see that the mMALA algorithm converges rapidly and, if it has converged, the correlation between samples is very low. We have seen similar behaviour in our experiments with different starting positions while retaining the same value for the integration step-size parameter. Moreover, in Figure 12.2 we show the convergence path and autocorrelation for simplified mMALA and we can see that, despite the simplifying assumption of a constant manifold, the algorithm remains efficient. In fact we calculated the values of the last term in Equation (12.1), which depends on the Christoffel symbols for several positions around the mode of the target density, and we found that the values are very close to zero, indicating that the manifold has almost constant curvature around the mode.

Similarly we perform the same experiments with HMC and RM-HMC, where the step-size parameters were selected such that the acceptance rate was above 70 % and the number of leapfrog iterations was fixed at 10 for both algorithms. The results are presented in Figure 12.3. We can again see a behaviour similar to that in the previous example with MALA and mMALA. The HMC method needs several iterations to reach the mode due to the small step-size required to escape from the low potential region at the starting position, while RM-HMC rapidly converges to the mode and the autocorrelation of the samples remains low. We also note that the RM-HMC algorithm seems more efficient than mMALA, something that is also reported in Girolami and Calderhead (2010) and relates to the fact that for mMALA a single discretisation step of the diffusion is used to propose a new sample, while for RM-HMC the Hamiltonian is simulated for several iterations. Finally, similar plots for the Gibbs sampler are presented in Figure 12.3 where we can see that the Gibbs sampler using the full conditional distributions follows a path along the gradients, although no gradient information is used explicitly. Despite its fast convergence, however, the chain mixing is not as good as for RM-HMC, as indicated by the autocorrelation function.

12.4 Experiments

In this section we describe a set of experiments designed to evaluate the MCMC algorithms in more realistic scenarios than the one presented in Section 12.3.4. We considered five different simulated datasets randomly generated from the densities in Table 12.1 taken from McLachlan and Peel (2000). The densities were selected in order to assess the performance of the methods under different conditions. For example, in the ‘bimodal’ density the modes are well separated and the posterior standard deviations are of the same order. On the other hand, the ‘claw’ density has overlapping components of different variances and the posterior standard deviations have large differences. Plots of the densities are shown in Figure 12.4 while the number of parameters is also given in Table 12.1. For the datasets ‘kurtotic’, ‘bimodal’ and ‘skewed bimodal’ we generated 2000 random samples while for the datasets ‘trimodal’ and ‘claw’ we generated 4000 and 5000 samples, respectively. Different samples were used in order to guarantee that the different components are well represented in the dataset.

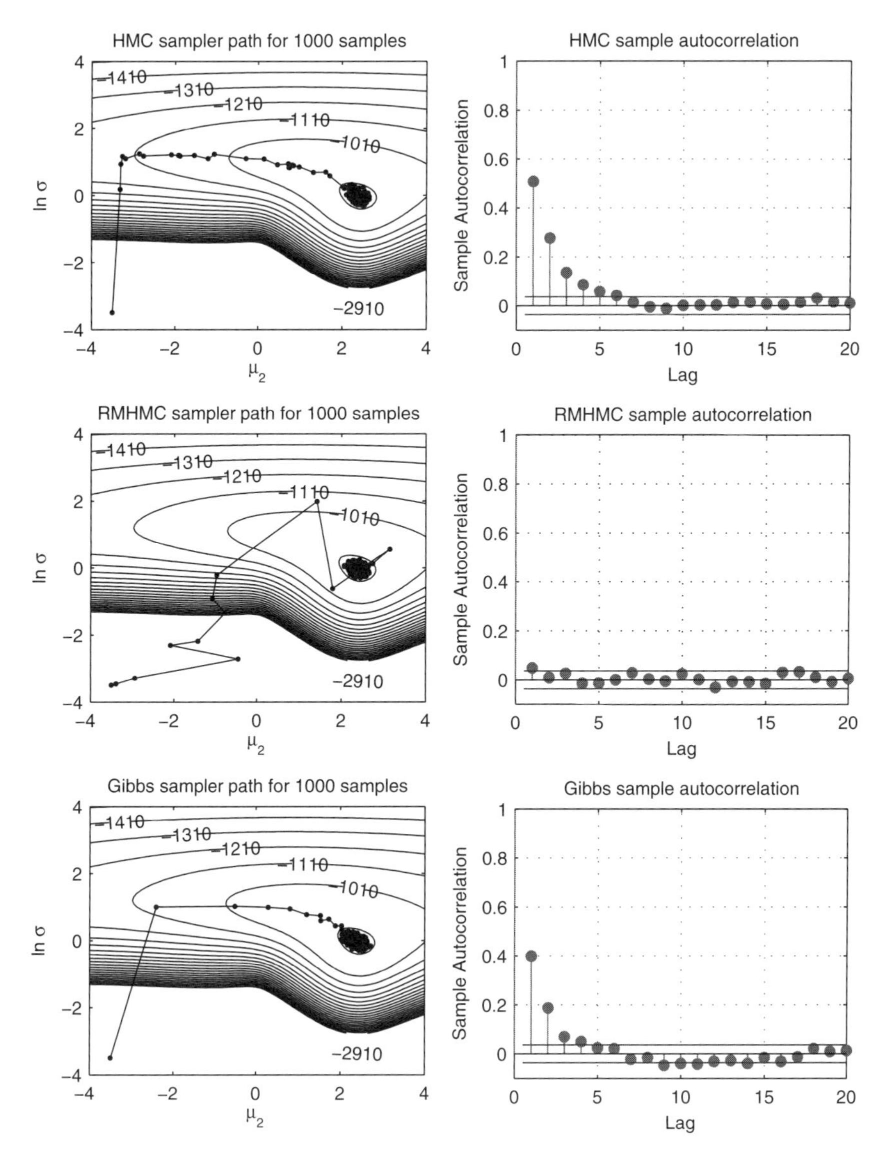

Figure 12.3 Comparison of HMC (top), RM-HMC (middle) and Gibbs (bottom) convergence paths and autocorrelation plots. Autocorrelation plots are from the stationary chains, i.e. once the chains have converged to the stationary distribution.

To ensure fair comparison, all algorithms were tuned by following the suggestions in Roberts *et al.* (1997), Christensen *et al.* (2005) and Neal (1993). More precisely, we tuned the Metropolis-Hastings scale parameter such that the acceptance rate was between 20 % and 30 %. The scale parameter for MALA was set such that the acceptance rate was between 40 % and 60 %. For HMC we used as

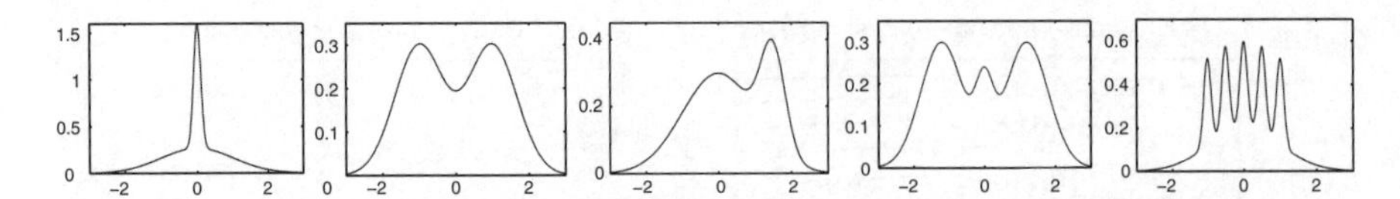

Figure 12.4 Densities used to generate synthetic datasets. From left to right the densities are in the same order as in Table 12.1. The densities are taken from McLachlan and Peel (2000).

the step-size parameter ϵ, the smallest standard deviation of the marginal posterior, and we set the number of leapfrog iterations L such that ϵL was above the largest standard deviation. Of course such knowledge is not known a priori and pilot runs to obtain the marginal posterior are needed when HMC is applied in practice. For the RM-HMC, mMALA and simplified mMALA, this knowledge of the target density is not required, and we follow Girolami and Calderhead (2010) and tune the step-size parameters such that the acceptance rate was above 70 % and the number of leapfrog steps for RM-HMC was kept fixed at five.

It is interesting to note here that the suggestions for tuning the Metropolis-Hastings and the MALA algorithms (Christensen *et al.*, 2005; Roberts *et al.*, 1997) assume stationarity. Therefore, tuning those algorithms requires several pilot runs where the chains are simulated until convergence and the acceptance rate is measured to ensure that it lies between the suggested values. Similarly, tuning HMC requires pilot runs in order to obtain the posterior standard deviations. In contrast, the tuning of mMALA and RM-HMC does not require such pilot runs; instead the first set of parameters that achieved an acceptance rate above 70 % was used to set the step-size. The acceptance rate during the transient and stationary phases of mMALA and RM-HMC were at the same levels across all our experiments.

We run all algorithms for 100 000 iterations and we keep only the last 10 000 samples as samples from the posterior, provided that the chains have converged. Convergence was assessed by inspection of the trace plots. We have also compared the posterior means with the true values provided in Table 12.1 to ensure that the

Table 12.1 Densities used for the generation of synthetic datasets.

Dataset name	Density function	Number of parameters
Kurtotic	$\frac{2}{3}\mathcal{N}(x\|0,1)+\frac{1}{3}\mathcal{N}\left(x\|0,\left(\frac{1}{10}\right)^2\right)$	6
Bimodal	$\frac{1}{2}\mathcal{N}\left(x\|-1,\left(\frac{2}{3}\right)^2\right)+\frac{1}{2}\mathcal{N}\left(x\|1,\left(\frac{2}{3}\right)^2\right)$	6
Skewed bimodal	$\frac{3}{4}\mathcal{N}(x\|0,1)+\frac{1}{4}\mathcal{N}\left(x\|\frac{3}{2},\left(\frac{1}{3}\right)^2\right)$	6
Trimodal	$\frac{9}{20}\mathcal{N}\left(x\|-\frac{6}{5},\left(\frac{3}{5}\right)^2\right)+\frac{9}{20}\mathcal{N}\left(x\|\frac{6}{5},\left(\frac{3}{5}\right)^2\right)+\frac{1}{10}\mathcal{N}\left(x\|0,\left(\frac{1}{4}\right)^2\right)$	9
Claw	$\frac{1}{2}\mathcal{N}(x\|0,1)+\sum_{i=0}^{4}\frac{1}{10}\mathcal{N}\left(x\|\frac{i}{2}-1,\left(\frac{1}{10}\right)^2\right)$	18

chains have converged to the correct mode. In the experiments presented here, we have not encountered the label-switching problem in any of the algorithms or datasets. This is due to the careful tuning of samplers to achieve relatively high acceptance rates. In our pilot runs, however, where larger step-sizes and scales were used, we have seen some label-switching from Metropolis-Hastings, HMC and RM-HMC, although the acceptance rates were below the suggested values. The lack of label-switching due to different tuning of parameters is also reported in Marin *et al.* (2005). From the output of each algorithm we measure the minimum, median and maximum effective sample size (ESS) (Geyer, 1992) across parameters and the number of samples per second. The experiments were repeated 10 times and all results presented in Table 12.2 are averages over the 10 different runs.

From Table 12.2 we can immediately see that, in terms of raw ESS, mMALA and simplified mMALA were always better than MALA and Metropolis-Hastings. Similarly, RM-HMC was in all cases better than HMC. This highlights the fact that by exploiting the intrinsic geometry of the model in MCMC we can achieve superior mixing. Another interesting observation is that the correlation of samples across different parameters (minESS and max ESS) is almost the same for RM-HMC and mMALA while for all other algorithms it varies significantly. This is related to the difficulties in tuning the scale and step-size parameters of MALA and HMC. When the standard deviations of the marginal posterior exhibit very large differences in scale it is very difficult to tune MALA and HMC to achieve good mixing across all parameters. For RM-HMC and mMALA the gradients are appropriately scaled by the metric tensor and therefore a smaller variation across the parameters is expected, making the algorithms less sensitive to tuning parameters. This argument is also supported by the results of Metropolis-Hastings, MALA and HMC for the 'bimodal' dataset where the difference between minESS and maxESS is small, suggesting a small difference in scale of the posterior's standard deviation. Indeed, the minimum and maximum standard deviations of the posterior for the 'bimodal' dataset were found to be 0.0402 and 0.0822, respectively. Finally, it is interesting to note that in terms of raw ESS the Gibbs sampler was only better on 2 out of 5 times compared to mMALA and simplified mMALA, while it was always worse than RM-HMC.

In practice, however, raw ESS is not very informative about the practical significance of an MCMC algorithm since less computationally demanding algorithms can obtain the required number of effectively independent samples faster than a more demanding but more efficient (in terms of mixing) algorithm. For that reason we also report the number of effectively uncorrelated samples per second (minESS/s). The results suggest that, for the most difficult examples, such as the 'trimodal', 'claw' and 'kurtotic' datasets, the RM-HMC algorithm is always better while the less computationally demanding simplified mMALA performs almost equally well in some cases. Interestingly, the Gibbs sampler, which is the most widely used algorithm for inference in mixture models, is always substantially the worst when compared to some of the other MCMC methods, with the exception of the claw dataset.

Table 12.2 Evaluation of MCMC algorithms for univariate mixtures of normals.

Dataset	Algorithm	Min ESS	Med ESS	Max ESS	minESS/s
	M-H	37 ± 10	133 ± 24	1347 ± 58	4.5 ± 1.2
	MALA	19 ± 7	78 ± 10	5172 ± 304	0.7 ± 0.2
	mMALA	158 ± 13	193 ± 13	235 ± 15	2.6 ± 0.1
Kurtotic	simp. mMALA	173 ± 15	206 ± 16	231 ± 12	4.5 ± 0.4
	HMC	892 ± 42	4167 ± 412	10000 ± 0	4.6 ± 0.2
	RM-HMC	9759 ± 56	9961 ± 89	10000 ± 0	$\mathbf{28.9 \pm 0.6}$
	Gibbs	1728 ± 173	3102 ± 73	9733 ± 230	9.0 ± 0.8
	M-H	83 ± 13	103 ± 17	119 ± 23	10.2 ± 1.7
	MALA	141 ± 26	169 ± 31	187 ± 35	5.4 ± 1.0
	mMALA	695 ± 20	798 ± 53	869 ± 55	9.02 ± 0.3
Bimodal	simp. mMALA	774 ± 24	817 ± 27	937 ± 86	20.4 ± 0.7
	HMC	1509 ± 149	1675 ± 144	1747 ± 142	$\mathbf{24.4 \pm 2.4}$
	RM-HMC	4593 ± 128	4920 ± 184	5215 ± 152	13.7 ± 0.3
	Gibbs	473 ± 97	504 ± 93	574 ± 97	2.5 ± 0.5
	M H	50 ± 9	68 ± 17	259 ± 102	6.2 ± 1.2
	MALA	77 ± 9	113 ± 19	312 ± 51	2.9 ± 0.3
	mMALA	437 ± 59	511 ± 63	669 ± 91	5.6 ± 0.7
Skewed	simp. mMALA	537 ± 86	587 ± 77	703 ± 96	$\mathbf{14.2 \pm 2.3}$
	HMC	1491 ± 57	1849 ± 93	3613 ± 187	$\mathbf{14.1 \pm 0.5}$
	RM-HMC	4793 ± 639	5152 ± 704	6969 ± 690	$\mathbf{14.7 \pm 1.5}$
	Gibbs	407 ± 40	469 ± 53	1032 ± 83	2.1 ± 0.2
	M-H	10 ± 3	42 ± 17	136 ± 30	0.6 ± 0.1
	MALA	20 ± 4	75 ± 23	312 ± 11	0.3 ± 0.07
	mMALA	209 ± 44	272 ± 21	342 ± 24	0.9 ± 0.2
Trimodal	simp. mMALA	224 ± 13	272 ± 15	319 ± 37	$\mathbf{2.5 \pm 0.1}$
	HMC	582 ± 79	1713 ± 262	6851 ± 363	1.9 ± 0.2
	RM-HMC	2066 ± 117	2369 ± 176	2622 ± 175	$\mathbf{2.3 \pm 0.1}$
	Gibbs	205 ± 31	381 ± 40	621 ± 29	0.5 ± 0.08
	M-H	—	—	—	—
	MALA	—	—	—	—
	mMALA	79 ± 11	145 ± 15	222 ± 28	0.09 ± 0.01
Claw	simp. mMALA	95 ± 13	154 ± 7	242 ± 30	0.4 ± 0.05
	HMC	923 ± 215	1739 ± 140	6529 ± 70	$\mathbf{0.5 \pm 0.1}$
	RM-HMC	1756 ± 427	2075 ± 503	2825 ± 815	0.3 ± 0.06
	Gibbs	238 ± 71	966 ± 51	4197 ± 367	$\mathbf{0.5 \pm 0.1}$

ESS is estimated using 10 000 samples from the posterior and minESS/s denotes the number of uncorrelated samples per second. The Metropolis-Hastings and MALA algorithms failed to converge after 100 000 samples in almost all of the 10 runs for the 'claw' dataset.

Table 12.3 Posterior distribution of the kurtotic dataset. True values are depicted by vertical lines or the plus sign.

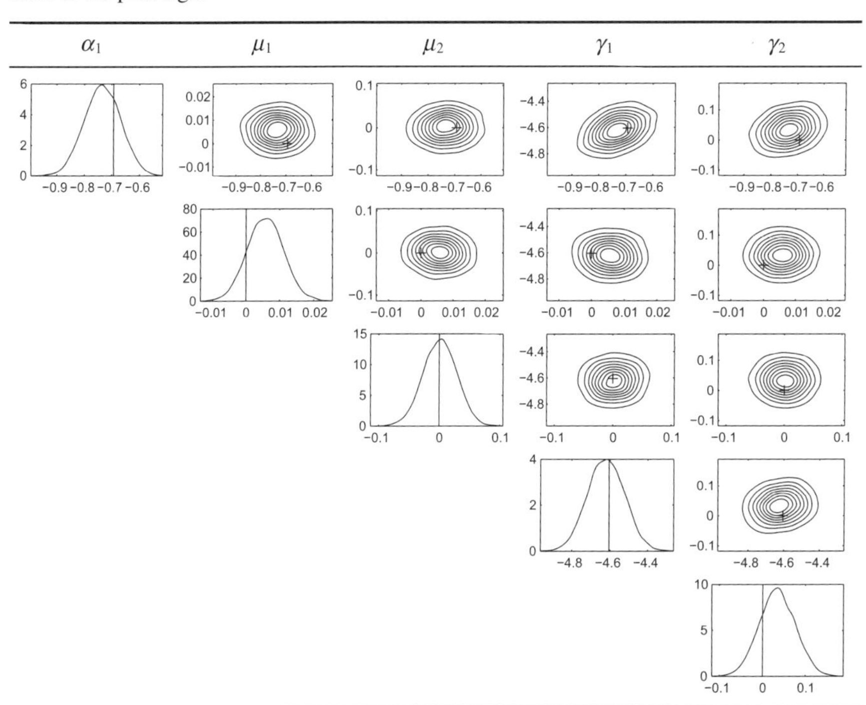

In Table 12.3 the marginal univariate and bivariate posteriors for the 'kurtotic' dataset are shown. Note the large difference in variance for μ_1 compared to all other parameters as well as the correlation between α_1 and γ_1,γ_2. Parameter γ_1 corresponds to the log of the variance of the second component of the 'kurtotic' density in Table 12.1. Its density is depicted in Figure 12.4, and the second component corresponds to the peak around 0. When the variances of both components increase the observations around 0 have higher likelihood under the second component, thus increasing the value of its mixing coefficient and therefore explaining the correlation in Table 12.3.

The datasets used in this study have relatively large sizes and since a finite-sample estimate of the Fisher information is used it is unknown if asymptotic behaviour is affecting our comparisons. For this reason we conducted experiments using the 'kurtotic' density in which we have created datasets of different sizes and compared the minimum effective sample size of Gibbs and RM-HMC samplers. For these experiments we use 10 000 burn-in samples and measure ESS in 1000 posterior samples provided the chains have converged. All experiments are repeated for 10 runs and results are presented in Figure 12.5. The minimum ESS for RM-HMC is increasing with the size of the dataset and after 500 observations we can see that it remains almost constant. In comparison, the Gibbs sampler is independent of

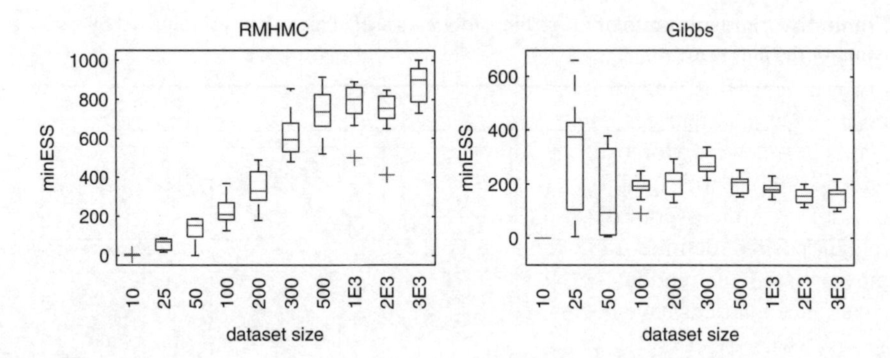

Figure 12.5 Comparison of minESS for the RM-HMC (left) and Gibbs (right) samplers for different dataset sizes. For datasets of size 10 both algorithms failed to converge after 10 000 iterations.

the sample size, as was expected. Moreover, we can see that even for small datasets RM-HMC has superior mixing properties when compared to the Gibbs sampler.

12.5 Discussion

In this chapter we have studied the application of Riemann manifold Metropolis adjusted Langevin and Hamiltonian Monte Carlo algorithms on mixtures of univariate Gaussian distributions. We have assessed their efficiency in terms of effective sample size and computational time and then compared them against MALA, HMC, Metropolis-Hastings and Gibbs sampling in five different simulated datasets. We provided some additional insight regarding the geometrical concepts exploited by mMALA and RM-HMC with an illustrative example. Our results indicated that RM-HMC and mMALA are always better in terms of raw ESS when compared to MALA and HMC while only limited tuning is required. Moreover, we showed that despite its simplicity the Gibbs sampler is not the most efficient algorithm, in terms of both ESS and computational time, and due to this observation the alternative manifold-based methods could be considered, subject to further experimental evaluation. Finally, we showed that exploiting the natural Riemann manifold in MCMC is also possible when exact analytical solutions for the natural metric are unavailable and finite sample estimates are used.

In our experiments none of the Markov chains has visited any of the symmetric modes associated with the invariance of the log-likelihood to label switching, casting doubts about the performance of the MCMC algorithms considered in Section 12.2 in problems where the posterior density is multimodal. However, we have shown that mMALA and RM-HMC are very efficient for exploring unimodal posteriors. Integration of these algorithms in a population MCMC scheme is trivial and therefore they can accommodate multimodal posteriors.

For the future, we believe that the extension to multivariate densities as well as inferring the unknown number of components are natural extensions of our work.

For multivariate mixtures of Gaussians the main difficulty is the reparameterisation of the components' covariance matrices in order to ensure that the proposal mechanisms of RM-HMC and mMALA always propose moves in the space of symmetric positive definite matrices. Such reparameterisations have been studied in Pinheiro and Bates (1996). Moreover, introducing reversible jump steps (Richardson and Green, 1997) is not trivial for multivariate normals since the split and merge moves have to leave the overall dispersion constant and the covariance matrices have to remain positive definite. In Dellaportas and Papageorgiou (2006) split and merge moves based on random permutation of the eigenvectors of the components' covariance matrices have been used with success.

Acknowledgements

This research was funded by the Engineering and Physical Sciences Council (EPSRC) grant number EP/H024875/1.

Appendix

The scores for each observation x_i under the reparameterisations introduced in Section 12.3.2 are

$$\frac{\partial \log p(x_i|\boldsymbol{\theta})}{\partial a_k} = \rho_{i,k} - N\pi_k,$$

$$\frac{\partial \log p(x_i|\boldsymbol{\theta})}{\partial \mu_k} = \rho_{i,k}\frac{x_i - \mu_k}{\sigma_k^2},$$

$$\frac{\partial \log p(x_i|\boldsymbol{\theta})}{\partial \gamma_k} = \rho_{i,k}\left(\frac{(x_i - \mu_k)^2}{2\sigma_k^2} - \frac{1}{2}\right).$$

Similarly, the partial derivatives of the prior are

$$\frac{\partial \log p(\boldsymbol{\theta})}{\partial a_k} = \lambda - 1 - \exp(a_k),$$

$$\frac{\partial \log p(\boldsymbol{\theta})}{\partial \mu_k} = -\frac{\mu_k - m}{\sigma_k^2}\beta,$$

$$\frac{\partial \log p(\boldsymbol{\theta})}{\partial \gamma_k} = \frac{(\mu_k - m)^2}{2\sigma_k^2}\beta - \frac{1}{2} - b - 1 + \frac{c}{\sigma_k^2}.$$

Finally, second partial derivatives of the log-likelihood with respect to the parameters required for estimating the partial derivatives of the empirical Fisher information in Section 12.3.2 are

$$\frac{\partial^2 \log p(x_i|\boldsymbol{\theta})}{\partial a_k \partial a_j} = -\rho_{i,j}\rho_{i,k} + \pi_j \pi_k + \delta_{k,j}\left(\rho_{i,k} - \pi_k\right),$$

$$\frac{\partial^2 \log p(x_i|\boldsymbol{\theta})}{\partial a_k \partial \mu_j} = (\delta_{k,j}\rho_{i,k} - \rho_{i,j}\rho_{i,k})\left(\frac{x_i - \mu_j}{\sigma_j^2}\right),$$

$$\frac{\partial^2 \log p(x_i|\boldsymbol{\theta})}{\partial a_k \partial \sigma_j^2} = (\delta_{k,j}\rho_{i,k} - \rho_{i,j}\rho_{i,k})\left[\frac{(x_i - \mu_j)^2}{2\sigma_j^2} - \frac{1}{2}\right];$$

$$\frac{\partial^2 \log p(x_i|\boldsymbol{\theta})}{\partial \mu_k \partial \mu_j} = -\rho_{i,j}\rho_{i,k}\left(\frac{x_i - \mu_j}{\sigma_j^2}\right)\left(\frac{x_i - \mu_k}{\sigma_k^2}\right) + \delta_{j,k}\rho_{i,k}\left[\left(\frac{x_i - \mu_k}{\sigma_k^2}\right)^2 - \frac{1}{\sigma_k^2}\right],$$

$$\frac{\partial^2 \log p(x_i|\boldsymbol{\theta})}{\partial \mu_k \partial \sigma_j^2} = -\rho_{i,j}\rho_{i,k}\left[\frac{(x_i - \mu_j)^2}{2\sigma_j^2} - \frac{1}{2}\right]\left(\frac{x_i - \mu_k}{\sigma_k^2}\right)$$
$$+\delta_{j,k}\rho_{i,k}\left(\frac{x_i - \mu_k}{\sigma_k^2}\right)\left[\frac{(x_i - \mu_j)^2}{2\sigma_j^2} - \frac{1}{2} - 1\right],$$

$$\frac{\partial^2 \log p(x_i|\boldsymbol{\theta})}{\partial \sigma_k^2 \partial \sigma_j^2} = -\rho_{i,j}\rho_{i,k}\left[\frac{(x_i - \mu_j)^2}{2\sigma_j^2} - \frac{1}{2}\right]\left[\frac{(x_i - \mu_k)^2}{2\sigma_k^2} - \frac{1}{2}\right]$$
$$+\delta_{j,k}\rho_{i,k}\left\{\left[\frac{(x_i - \mu_k)^2}{2\sigma_k^2} - \frac{1}{2}\right]^2 - \frac{(x_i - \mu_k)^2}{2\sigma_k^2}\right\},$$

where $\delta_{i,j}$ is the Kronecker delta function.

References

Amari, S.-I. and Nagaoka, H. (2000) *Methods of Information Geometry.* Translations of Mathematical Monographs. American Mathematical Society.

Andrieu, C. and Thoms, J. (2008) A tutorial on adaptive MCMC. *Statistics and Computing*, **18**, 343–373.

Basu, A., Harris, I. R., Hjort, N. L. and Jones, M. C. (1998) Robust and efficient estimation by minimising a density power divergence. *Biometrika*, **85**, 549–559.

Celeux, G. (2006) Mixture models for classification. In *Advances in Data Analysis, Proceedings of the 30th Annual Conference of the Gesellschaft für Klassifikation e.V., Freie Universität Berlin*, pp. 3–14.

Celeux, G., Hurn, M. and Robert, C. P. (2000) Computational and inferential difficulties with mixture posterior distributions. *Journal of the American Statistical Association*, **95**, 957–970.

Christensen, O. F., Roberts, G. O. and Rosenthal, J. S. (2005) Scaling limits for the transient phase of local Metropolis-Hastings algorithms. *Journal of The Royal Statistical Society, Series B*, **67**, 253–268.

Dellaportas, P. and Papageorgiou, I. (2006) Multivariate mixtures of normals with unknown number of components. *Statistics and Computing*, **16**, 57–68.

Diebolt, J. and Robert, C. P. (1994) Estimation of finite mixture distributions through Bayesian sampling. *Journal of the Royal Statistical Society, Series B*, **56**, 363–375.

Duane, S., Kennedy, A. B., Pendleton, J. B. and Roweth, D. (1987) Hybrid Monte Carlo. *Physics Letters B*, **195**, 216–222.

Escobar, M. D. and West, M. (1994) Bayesian density estimation and inference using mixtures. *Journal of the American Statistical Association*, **90**, 577–588.

Geyer, C. J. (1992) Practical Markov chain Monte Carlo. *Statistical Science*, **7**, 473–483.

Gilks, W. R., Richardson, S. and Spiegelhalter, D. J. (1999) *Markov chain Monte Carlo In Practice*. Chapman & Hall/CRC.

Girolami, M. and Calderhead, B. (2010) Riemann manifold Langevin and Hamiltonian Monte Carlo (with discussion). *Journal of the Royal Statistical Society, Series B*, (to appear).

Haario, H., Saksman, E. and Tamminen, E. (2005) Componentwise adaptation for high dimensional MCMC. *Computational Statistics*, **20**, 265–273.

Haario, H., Saksman, E. and Tamminen, J. (1998) An adaptive Metropolis algorithm. *Bernoulli*, **7**, 223–242.

Jasra, A., Holmes, C. C. and Stephens, D. A. (2005) MCMC and the label switching problem in Bayesian mixture models. *Statistical Science*, **20**, 50–67.

Jasra, A., Stephens, D. A. and Holmes, C. C. (2007) Population-based reversible jump Markov chain Monte Carlo. *Biometrika*, **94**, 787–807.

Kühnel, W. (2005) *Differential Geometry: Curves – Surfaces – Manifolds*, Vol. 16 of Student Mathematical Library. AMS.

McLachlan, G. J. and Baek, J. (2010) Clustering of high-dimensional data via finite mixture models. In *Advances in Data Analysis, Data Handling and Business Intelligence*, pp. 33–44.

McLachlan, G. and Peel, D. (2000) *Finite Mixture Models*. John Wiley & Sons, Ltd.

Marin, M. J., Mengersen, K. and Robert, C. P. (2005) Bayesian modelling and inference on mixtures of distributions. In *Handbook of Statistics*, pp. 15840–15845.

Neal, R. M. (1993) Probabilistic inference using Markov chain Monte Carlo methods. Technical Report CRG-TR-93-1, Department of Computer Science, University of Toronto.

Neal, R. M. (1999) Suppressing random walks in Markov chain Monte Carlo using ordered overrelaxation. In *Learning in Graphical Models* (ed. M. Jordan), pp. 205–228. MIT Press.

Pinheiro, J. C. and Bates, D. M. (1996) Unconstrained parametrizations for variance–covariance matrices. *Statistics and Computing*, **6**, 289–296.

Richardson, S. and Green, P. J. (1997) On Bayesian analysis of mixtures with an unknown number of components (with discussion). *Journal of the Royal Statistical Society, Series B*, **59**, 731–792.

Robert, C. P. and Casella, G. (2005) *Monte Carlo Statistical Methods*, Springer-Verlag.

Roberts, G. O. and Rosenthal, J. S. (1998) Optimal scaling of discrete approximations to Langevin diffusions. *Journal of the Royal Statistical Society, Series* B **60**, 255–268.

Roberts, G. O. and Stramer, O. (2003) Langevin diffusions and Metropolis-Hastings algorithms. *Methodology and Computing in Applied Probability*, **4**, 337–357.

Roberts, G. O., Gelman, A. and Gilks, W. R. (1997) Weak convergence and optimal scaling of random walk Metropolis algorithms. *Annals of Applied Probability*, **7**, 110–120.

Scott, D. W. (2001) Parametric statistical modeling by minimum integrated square error. *Technometrics*, **43**, 274–285.

Stephens, M. (2000) Dealing with label switching in mixture models. *Journal of the Royal Statistical Society, Series B*, **62**, 795–809.

Titterington, D. M., Smith, A. F. M. and Makov, U. E. (1985) *Statistical Analysis of Finite Mixture Distributions*. John Wiley & Sons, Ltd.

13

How many components in a finite mixture?

Murray Aitkin

13.1 Introduction

In Aitkin (2001), I compared several Bayesian analyses of the galaxy recession velocity data set described by Roeder (1990). These analyses diverged widely in their conclusions about the number of normal components in the fitted mixture needed to represent the velocity distribution. I asked rhetorically why these analyses were so diverse, and concluded as follows:

> The complexity of the prior structures needed for Bayesian analysis, the obscurity of their interaction with the likelihood, and the widely different conclusions they lead to over different specifications, leave the user completely unclear about 'what the data say' from a Bayesian point of view about the number of components in the mixture. For this mixture parameter at least, the Bayesian approach seems unsatisfactory in its present state of development.

Bayesians of whom I have asked this question directly have generally responded, as did some of the authors of these papers: 'The results are different, because of the different priors used.' This restatement of the problem does not throw light on its solution. In this chapter I revisit the issue and suggest that the diversity is caused, not specifically by the different priors used but by the settings of the hyperparameters of these priors. In the galaxy dataset these settings lead to unbelievable numbers of components in several of the analyses.

Mixtures: Estimation and Applications, First Edition. Edited by Kerrie L. Mengersen, Christian P. Robert and D. Michael Titterington.

This difficulty is endemic in the comparison of models by integrated likelihoods, and I show that an alternative approach using the full posterior distributions of the likelihoods under each model does not suffer from this difficulty and leads to a reasonable conclusion about the number of components.

13.2 The galaxy data

The galaxy data published by Roeder (1990) are the recession velocities, in units of 10^3 km/s, of 82 galaxies from six well-separated conic sections of space; the tabled velocities are said by astronomers to be in error by less than 0.05 units. Roeder (1990) notes that the distribution of velocities is important, as it bears on the question of 'clumping' of galaxies: if galaxies are clumped by gravitational attraction, the distribution of velocities would be multimodal; conversely, if there is no clumping effect, the distribution would increase initially and then gradually tail off.

We do not analyse separately the data from the six regions, following all authors including the astronomers Postman *et al.* (1986) who gave the full data by region. The individual regions have very small datasets, from which not much can be learned about clustering among or within regions. The data are reproduced below, ordered from smallest to largest and scaled by a factor of 1000 as in Roeder (1990) and Richardson and Green (1997).

```
       Recession velocities (/1000) of 82 galaxies

 9.172  9.350  9.483  9.558  9.775 10.227 10.406 16.084
16.170 18.419 18.552 18.600 18.927 19.052 19.070 19.330
19.343 19.343 19.440 19.473 19.529 19.541 19.547 19.663
19.846 19.856 19.863 19.914 19.918 19.973 19.989 20.166
20.175 20.179 20.196 20.215 20.221 20.415 20.629 20.795
20.821 20.846 20.875 20.986 21.137 21.492 21.701 21.814
21.921 21.960 22.185 22.209 22.242 22.249 22.314 22.374
22.495 22.746 22.747 22.888 22.914 23.206 23.241 23.263
23.484 23.538 23.542 23.666 23.706 23.711 24.129 24.285
24.289 24.366 24.717 24.990 25.633 26.960 26.995 32.065
32.789 34.279
```

The empirical cumulative distribution function (CDF) of the velocities is shown in Figure 13.1 (left), together with the fitted normal CDF.

It is immediately clear that the normal distribution does not fit, with a gap or jump between the seven smallest observations around 10 and the large central body of observations between 16 and 26, and another gap between 27 and 32 for the three largest observations. We might therefore expect to find at least three 'clumps' or mixture components in the data.

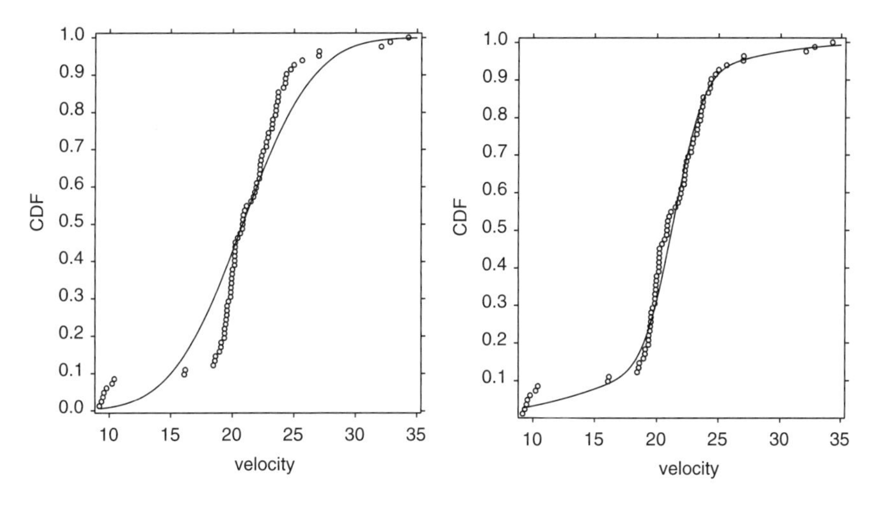

Figure 13.1 Normal distribution fit to the galaxy data (left); two-component mixture fit (right).

13.3 The normal mixture model

The general model for a K-component normal mixture has different means μ_k and variances σ_k^2 in each component:

$$f(y) = \sum_{k=1}^{K} \pi_k f(y|\mu_k, \sigma_k),$$

where

$$f(y|\mu_k, \sigma_k) = \frac{1}{\sqrt{2\pi}\sigma_k} \exp\left[-\frac{1}{2\sigma_k^2}(y - \mu_k)^2\right]$$

and the π_k are positive with $\sum_{k=1}^{K} \pi_k = 1$. Given a sample $y_1, ..., y_n$ from $f(y)$, the likelihood is

$$L(\theta) = \prod_{i=1}^{n} f(y_i),$$

where $\theta = (\pi_1, ..., \pi_{K-1}, \mu_1, ..., \mu_K, \sigma_1, ..., \sigma_K)$.

Frequentist and Bayesian analyses are straightforward using the EM algorithm and Gibbs sampling, respectively, provided the component standard deviations are bounded below to prevent single observations defining components with zero variances.

Figures 13.1 (right) and 13.2 show the maximum-likelihood fitted models and the data for two, three and four components. The estimated three-component model gives a close fit to the empirical CDF, except in the interval 18–23 where a fourth

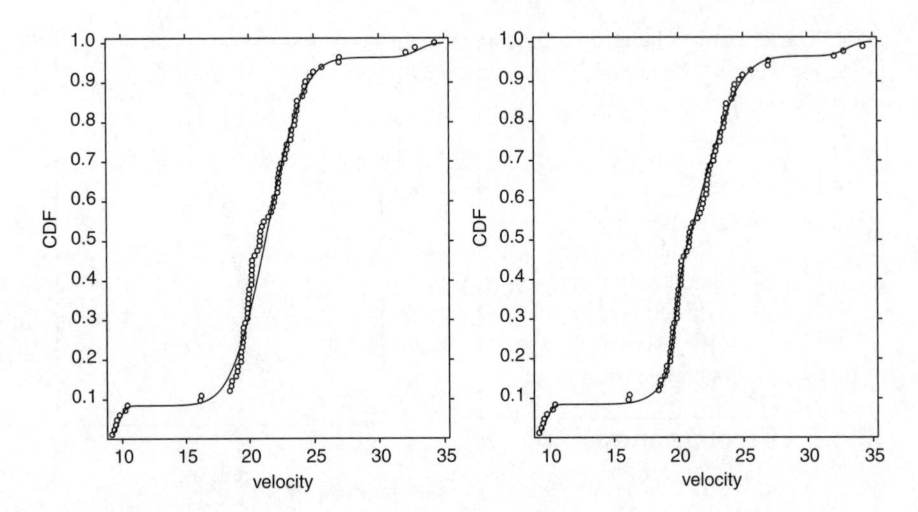

Figure 13.2 Three-component (left) and four-component (right) fits.

component may be needed. The estimated four-component model gives a very close fit to the empirical CDF.

13.4 Bayesian analyses

The Bayesian analyses of these data, and the choice of priors and their hyperparameters, were discussed in detail in Aitkin (2001) for the analyses by Escobar and West (1995), Carlin and Chib (1995), Phillips and Smith (1996), Roeder and Wasserman (1997) and Richardson and Green (1997); full discussions are given in these references. I have added the discussion by Stephens (2000), which was not included in Aitkin (2001).

Most of the analyses took the number of components K as fixed over a grid, used MCMC with prior specifications regarded as appropriate and set hyperparameters to reasonable values, in terms of their tuning to the data. After convergence of MCMC was established for each K, the likelihood was integrated over the proper priors, and the integrated likelihoods for all K were converted to posterior model probabilities, incorporating the prior for K.

Here we focus on the effects of integrating the likelihoods for a subset of the approaches. We exclude the Carlin and Chib (1995) analysis as their model had equal component variances, unlike all the others. We also exclude the Nobile analysis (Nobile, 2004); his prior for K was unspecified.

Some analyses used flat priors for K and others used informative priors; this makes it difficult to compare 'what the data say' about the number of components. To allow direct comparability among analyses we convert the posteriors for the informative priors to the corresponding posteriors for a flat prior. The effect of this is shown by giving both analyses, with the authors' original priors and the flat prior. The effect of changing the prior is easily assessed: if the original prior and posterior were $\pi_1(K)$ and $\pi_1(K \mid \mathbf{y})$, and the new prior and posterior are $\pi_2(K)$ and $\pi_2(K \mid \mathbf{y})$, then

$$\pi_2(K \mid \mathbf{y}) = c\pi_1(K \mid \mathbf{y}) \cdot \pi_2(K)/\pi_1(K),$$

where c scales the new posterior to sum to 1. The reversible jump MCMC (RJMCMC) used by Richardson and Green (1997) did not use integrated likelihoods directly but included K in the parameter space with a flat prior, so the posterior for K was obtained directly from the MCMC output.

We give the Bayesian analyses below in some detail, to show both the diversity of priors and the choice of hyperparameters in each analysis.

13.4.1 Escobar and West

Escobar and West (1995) (EW) used a Dirichlet process prior for the mixture means and variances depending on the sample size n and a hyperparameter α; this induces a prior on K depending strongly on n, and stochastically increasing with α. For the galaxy sample size of $n = 82$, the prior distribution of K for the chosen $\alpha = 1$ is given in Table 13.1 (as EW) with the priors used by the other authors, identified by their initials.

The prior mean distribution of the process prior was normal, with normal priors for the component means, $\mu_k \sim N(m, \tau\sigma_k^2)$ conditional on the component variances and inverse gamma distributions for the variances depending on shape $s/2$ and scale $S/2$ hyperparameters. A diffuse reference prior was used for m and a diffuse inverse gamma prior for τ with hyperparameters 1/2 and 50. For the variances, the prior parameters s and S were taken as 2 and 1.

The posterior distribution of K for this choice of priors is given in Table 13.2. In Table 13.3 we rescale it to be comparable with the posteriors found by the other authors using a flat prior for K.

13.4.2 Phillips and Smith

Phillips and Smith (1996) (PS) used a truncated Poisson (3) prior for $K \geq 1$, a uniform prior on π, independent $G(\gamma, \delta)$ priors on the inverse component variances and independent $N(\mu, \tau^2)$ priors on the component means. The hyperparameters were set at $\mu = 20$, $\tau^2 = 100$, $\gamma = 2$ and $\delta = 0.5$.

Table 13.1 Prior distributions for K.

K	1	2	3	4	5	6	7	8	9	10
EW	0.01	0.06	0.14	0.21	0.21	0.17	0.11	0.06	0.02	
PS	0.16	0.24	0.24	0.18	0.10	0.05	0.02	0.01		
S	0.58	0.29	0.10	0.02	0.004	0.001	—	—	—	
RW	0.10	0.10	0.10	0.10	0.10	0.10	0.10	0.10	0.10	0.10
RG	0.03	0.03	0.03	0.03	0.03	0.03	0.03	0.03	0.03	0.03...

Table 13.2 Posterior distributions for K.

K	3	4	5	6	7	8	9	10	11	12	13
EW	—	0.03	0.11	0.22	0.26	0.20	0.11	0.05	0.02	—	—
PS	—	—	—	0.03	0.39	0.32	0.22	0.04	—	—	—
S	0.55	0.34	0.09	0.01	—	—	—	—	—	—	—
RW	0.999	0.00	—	—	—	—	—	—	—	—	—
RG	0.06	0.13	0.18	0.20	0.16	0.11	0.07	0.04	0.02	0.01	0.01

The posterior distribution for K was relatively concentrated, with a mode at 7 and posterior probability 0.93 for $7 \leq K \leq 9$. In Table 13.3 we rescale it to be comparable with the posteriors found by the other authors using a flat prior for K.

13.4.3 Roeder and Wasserman

Roeder and Wasserman (1997) (RW) used a uniform prior for K, for $K \leq 10$. The prior for π was the symmetric Dirichlet (1,...,1) and the priors for σ_k/A were independent scaled inverse-χ_ν distributions, with the hyperparameter A having a log-uniform reference prior for a scale parameter and the hyperparameter ν being set equal to 1. Conditional on σ_k, (the ordered) μ_k had a truncated Markov prior with mean μ_{k-1} and variance $2M^2/(1/\sigma_k^2 + 1/\sigma_{k-1}^2)$, depending on a hyperparameter M, which was set equal to 5.

Since the full prior contains the improper reference component for A, the integrated likelihood is undefined; Roeder and Wasserman (1997) dealt with this problem by estimating the integrated likelihood from the Schwartz or Bayesian information criterion (BIC). They approximated the maximised likelihood in the BIC by substituting the posterior means of the parameters for the maximum-likelihood estimates.

13.4.4 Richardson and Green

Richardson and Green (1997) (RG) used a uniform prior for K for $K \leq 30$. The prior for π was the symmetric Dirichlet $(\delta, \delta, ...\delta)$ with δ set equal to 1. The priors for μ_k and σ_k^{-2} were independent $N(\xi, \kappa^{-1})$ and $\Gamma(\alpha, \beta)$. The prior parameters for the μ_k distribution were set to $\xi = 21.73$, the sample mean velocity, and

Table 13.3 Posterior distributions for K (flat prior).

K	3	4	5	6	7	8	9	10	11	12	13
EW		0.01	0.03	0.07	0.13	0.18	0.30	0.28	?	?	?
PS				0.00	0.10	0.21	0.43	0.26	?	?	?
S	0.10	0.25	0.35	0.29	?	?	?	?	?	?	?
RW	>0.999	<0.001									
RG	0.06	0.13	0.18	0.20	0.16	0.11	0.07	0.04	0.02	0.01	0.01

$\kappa^{-1} = R^2 = 25.11^2$ ($\kappa = 0.0016$), where R is the sample range of the data. The value of α was set to 2 and β was given a hyperprior gamma distribution with parameters g and h; these were set to $g = 0.2$ and $h = 0.573$.

13.4.5 Stephens

Stephens (2000) (S) used a truncated Poisson (λ) prior with several different values for the mean λ. His other priors were the same as those used by Richardson and Green (1997). He used a marked point birth-and-death process for K when this was treated as variable and included in the parameter space, as an alternative to Richardson and Green's reversible jump approach. He noted the considerable sensitivity of the posterior for K to the priors for both K and the other component parameters and gave results for both fixed and variable K. Numerical values for the posterior for K were given only for the case of fixed K and $\lambda = 1$, though histograms were given for other values of λ, showing the great variability in the posterior for K.

13.5 Posterior distributions for K (for flat prior)

Table 13.2 gives the posterior distribution for K from each of the analyses reported above using the authors' priors for K and Table 13.3 gives the posteriors standardised for a flat prior for K. For the rescaled analyses by Escobar and West (1995), Phillips and Smith (1996) and Stephens (2000), the posterior probabilities for extreme values of K could not be computed from the limited precision given in the available results and are represented by question marks. All these posterior distributions were decreasing beyond the last value given, and we assume they continue to do so with increasing K. The unknown tail values have been ignored in rescaling the posteriors to sum to 1.

It is immediately striking that the posteriors for the EW and PS analyses have modes at $K = 9$, with high probability also for $K = 10$, while that for RG has its mode at 6, is very diffuse and does not rule out $K = 9$ or $K = 3$ or 4. The posterior for S has a mode at $K = 5$ and a slightly lower value at $K = 6$. The RW distribution is almost a spike at $K = 3$. The PS posterior rules out $K \leq 6$.

It is hard to imagine a more diverse, and inconsistent, set of posterior conclusions about a parameter. In the discussion of Aitkin (2001) and in Stephens (2000), this difference is obscured by the strongly informative priors for K used by EW, PS and S, which almost eliminate the possibility of $K \geq 9$. If we have no prior view about the number of mixture components, which conclusions are believable?

There is an obvious difficulty with the EW and PS conclusions, and that is the sample size relative to the number of model parameters. With seven components in the mixture, the *average* sample size per parameter is only four. More importantly, the well-separated sets of seven and three observations at the extremes of the velocity range clearly define single components, so the remaining central group of 72 observations are being modelled by five components. This is clear from the maximum likelihood analyses described in Aitkin (2001), from which the ML estimates of μ_k, σ_k and π_k for $K = 7$ are shown below:

K	k	mean	sd	prop
7	1	33.04	0.92	0.037
	2	26.98	0.018	0.024
	3	23.42	0.99	0.300
	4	22.13	0.25	0.085
	5	19.89	0.73	0.444
	6	16.13	0.043	0.024
	7	9.71	0.42	0.085
deviance			363.04	

Components 2 and 6 each contain only two very close observations (2/82 = 0.024), giving very small standard deviations, while components 3 and 4 differ in mean by only 1.3 component 3 standard deviations.

For $K \geq 9$, the two extreme groups are further split into subgroups with single observations, for which standard deviations cannot be estimated – the ML analysis of the model breaks down at this point. It therefore seems unbelievable that the Bayesian analysis can give nine or 10 components with high probability.

Since the prior forms used by EW, PS, S and RG are very similar, apart from those for K, the explanation of the unbelievable integrated likelihoods must lie in the effects of the hyperparameters on the integrated likelihoods for the EW and PS analyses. The hyperparameter values were chosen ('tuned to the data'), as is the case for most MCMC analyses, to ensure that the chain starts from a region of reasonable likelihood. Badly chosen random values from diffuse priors may lead to the chain starting from a region of flat likelihood and converging extremely slowly, if at all. This is an important issue, but it leads to the determination of the integrated likelihood in terms of these hyperparameter values: Stephens (2000) noted the great sensitivity of the posterior conclusions about K to *all* the prior assumptions. As is well known (Kass and Raftery, 1995), the prior values have a marked effect on the integrated likelihood, which does not dissipate with increasing sample size.

The RW analysis leads to a dramatically opposite conclusion: that there are, with almost certainty, only three components in the mixture (their prior only allowed for 2, 3 or 4). RW used a truncated Markov prior for the means, and their analysis depended on three hyperparameter values. They avoided the difficulty of integrating the likelihood by using the Bayesian information criterion (BIC), which depends on the data only through the maximised -2 log-likelihood (the frequentist deviance) and the number of model parameters. The maximised likelihood for each model was approximated by the likelihood evaluated at the posterior means of the parameters.

Since the BIC can be evaluated from a maximised-likelihood analysis without any prior specification or MCMC analysis, it is of interest to compare the RW results with those from the ML analysis. The latter BIC conclusions are summarised

Table 13.4 BIC for $K = 2, ..., 7$.

K	p	Deviance	$p \log n$	BIC	Posterior probability
2	5	440.72	22.03	462.75	0.000
3	8	406.96	35.26	442.22	0.229
4	11	395.43	48.48	443.91	0.098
5	14	392.27	61.70	453.97	0.000
6	17	365.15	74.92	440.07	0.670
7	20	363.04	88.14	451.18	0.003

in Table 13.4, using the frequentist deviances from Aitkin (2001). If we restrict consideration to at most four components, BIC gives a preference to $K = 3$ (0.70) over $K = 4$ (0.30), nothing like the 0.999 given by RW. However, for the range up to $K = 7$, the six-component model has a smaller BIC than the three-component one, giving a posterior probability of 0.67 for the six-component model and 0.23 for the three-component model. Two, five and seven components have negligible posterior probability.

13.6 Conclusions from the Bayesian analyses

The EW and PS results, supporting strongly models with nine or 10 components, are apparently determined by their choices of hyperparameter values for the MCMC initialisation. They are inconsistent with the ML analysis, which leads to single-observation components with variances held at the limit for these models, and frequentist deviances which depend on this limit. The prior specifications for the variances will not prevent the posteriors for these variances being strongly peaked near zero. It is clear that the mixture model cannot support so many components with this small dataset.

The RW analysis, identifying almost certainly three components with the BIC, is inconsistent with the same frequentist analysis, which gives appreciable probability to four components in the restricted set considered by these authors and highest posterior probability to six components when up to seven are allowed.

The RJMCMC analysis of RG gives a very diffuse posterior: a 90% central credible interval for K goes from 4 to 10 (credible intervals for EW, PS and S could not be determined because of the uncertainty in the upper tail). The S posterior provides limited information, giving appreciable probability to four to six components and possibly more.

The quote from my paper of 2001 on the inability of the Bayesian approach, in its present state of development, to provide 'what the data say' about the number of components in the mixture of normals, seems if anything stronger from this detailed look at the five Bayesian analyses above.

An alternative approach is provided by the use of the full posterior distribution of the likelihoods, or the deviances, from each model. This approach is set out in detail in Aitkin (2010), from which the following discussion of the galaxy analysis is extended.

13.7 Posterior distributions of the model deviances

The use of the posterior distribution of the likelihood, and of the deviance, was pioneered by Dempster (1974, reprinted in 1997) and extended and generalised by Aitkin (1997) and Aitkin *et al.* (2005). Recent applications can be found in Liu and Aitkin (2008) and Aitkin *et al.* (2009): similar applications are given by Fox (2005), and by Congdon (2005, 2006) under the name 'parallel sampling'. A book-length treatment is given in Aitkin (2010).

Posterior deviance distributions for a finite mixture problem (not the galaxy data) were given by Richardson and Green (1997). They used kernel density estimates for each K that overlapped considerably, making identification of the best-supported model difficult. The use of cumulative distribution functions instead of densities greatly simplifies and clarifies the use of posterior deviances for model comparison.

The computation of the posterior distribution of the likelihood is quite simple. For a given model $f_k(y \mid \theta_k)$ with likelihood $L_k(\theta_k)$, we make M independent draws $\theta_k^{[m]}$ from the posterior of θ_k and substitute them into the model likelihood to give M independent likelihood draws $L_k^{[m]} = L_k(\theta_k^{[m]})$, and equivalently M independent deviance draws $D_k^{[m]} = -2 \log L_k^{[m]}$. These are ordered to form the empirical CDF of the likelihood, or the deviance. As shown below, the deviance distribution is much better behaved, and we use this generally for inference.

13.8 Asymptotic distributions

For regular models $f(y \mid \theta)$ with flat priors, giving an MLE $\hat{\theta}$ internal to the parameter space, the second-order Taylor expansion of the deviance $D(\theta) = -2 \log L(\theta) = -2\ell(\theta)$ about $\hat{\theta}$ gives

$$\begin{aligned}
-2\ell(\theta) &\doteq -2\ell(\hat{\theta}) - 2(\theta - \hat{\theta})'\ell'(\hat{\theta}) - (\theta - \hat{\theta})'\ell''(\hat{\theta})(\theta - \hat{\theta}) \\
&= -2\ell(\hat{\theta}) + (\theta - \hat{\theta})' I(\hat{\theta})(\theta - \hat{\theta}), \\
L(\theta) &\doteq L(\hat{\theta}) \exp[-(\theta - \hat{\theta})' I(\hat{\theta})(\theta - \hat{\theta})/2], \\
\pi(\theta|\mathbf{y}) &\doteq c \exp[-(\theta - \hat{\theta})' I(\hat{\theta})(\theta - \hat{\theta})/2].
\end{aligned}$$

Therefore, asymptotically, given the data $\mathbf{y}$, we have the posterior distributions

$$\begin{aligned}
\theta &\sim N(\hat{\theta}, I(\hat{\theta})^{-1}), \\
(\theta - \hat{\theta})' I(\hat{\theta})(\theta - \hat{\theta}) &\sim \chi_p^2, \\
D(\theta) &\sim D(\hat{\theta}) + \chi_p^2, \\
L(\theta) &\sim L(\hat{\theta}) \exp(-\chi_p^2/2).
\end{aligned}$$

The likelihood $L(\theta)$ has a *scaled* $\exp(-\chi_p^2/2)$ distribution, while the deviance $D(\theta)$ has a *shifted* χ_p^2 distribution, shifted by the frequentist deviance $D(\hat{\theta})$, where p is the dimension of θ. The frequentist deviance is an origin parameter for the posterior deviance distribution: no random draw can give a smaller deviance value than the frequentist deviance.

13.9 Posterior deviances for the galaxy data

For each K we report the 10 000 deviance draws from Celeux *et al.* (2006) using a diffuse Dirichlet prior on the component proportions π_k and diffuse conjugate priors on the means μ_k and inverse variances σ_k^2. (The deviances were kindly supplied by Christian Robert.)

These provide a *reference* analysis, in a sense similar to that of Berger and Bernardo (1989): no hyperparameter is needed in the prior specification, but they can be introduced if desired and their effects assessed relative to the diffuse prior. Since no integration over the prior is used, the effect of informative priors on the posterior distributions of the parameters, and therefore of the deviance, is greatly reduced.

Afer convergence 10 000 values were sampled from the thinned posterior distributions and the K-component mixture likelihoods computed for each parameter set. Figures 13.3 to 13.6 show the deviance distributions for $K = 1, 2, ..., 7$ (solid), with the asymptotic χ^2_{3K-1} distribution (dotted). The deviance draw distribution agrees almost exactly with its asymptotic form (χ^2_2) for $K = 1$. As K increases, two different phenomena are visible:

- The frequentist deviance *moves further away* from the deviance draw distribution:

  The frequentist deviance is *increasingly unrepresentative of the deviance draw distribution minimum.*

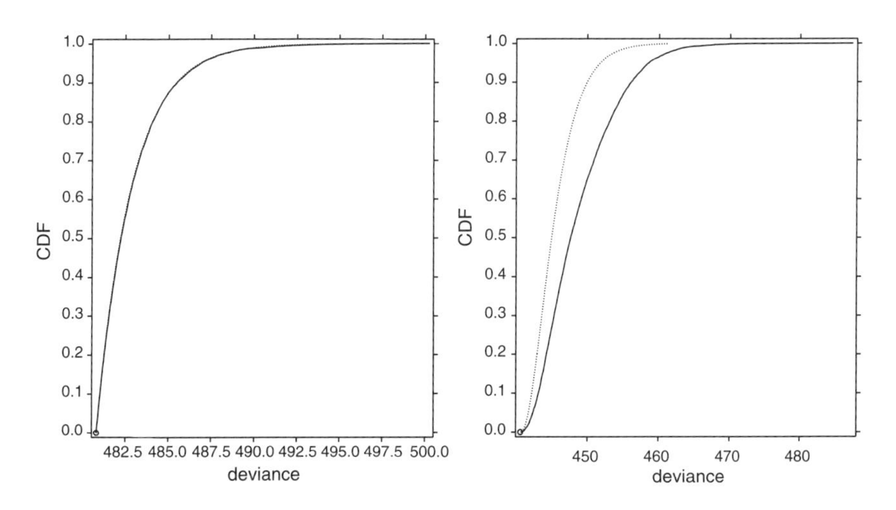

Figure 13.3 One component (left) and two components (right).

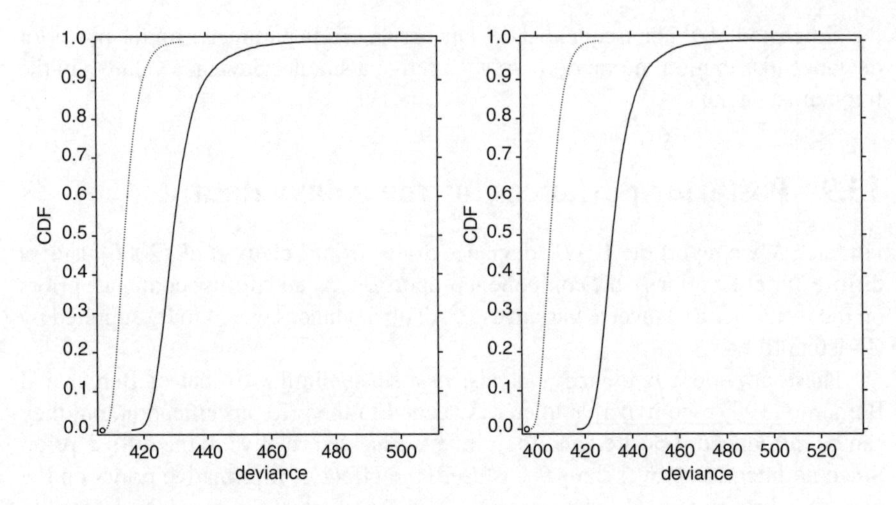

Figure 13.4 Three components (left) and four components (right).

- The deviance draw distribution is *increasingly more diffuse* than the asymptotic distribution:

  The data are spread *increasingly thinly* over the increasing number of parameters, so all posteriors are more diffuse, as is the deviance posterior.

We show in Figure 13.6 (right) the deviance distributions on the same scale. The interpretation of this figure can be simply summarised as follows:

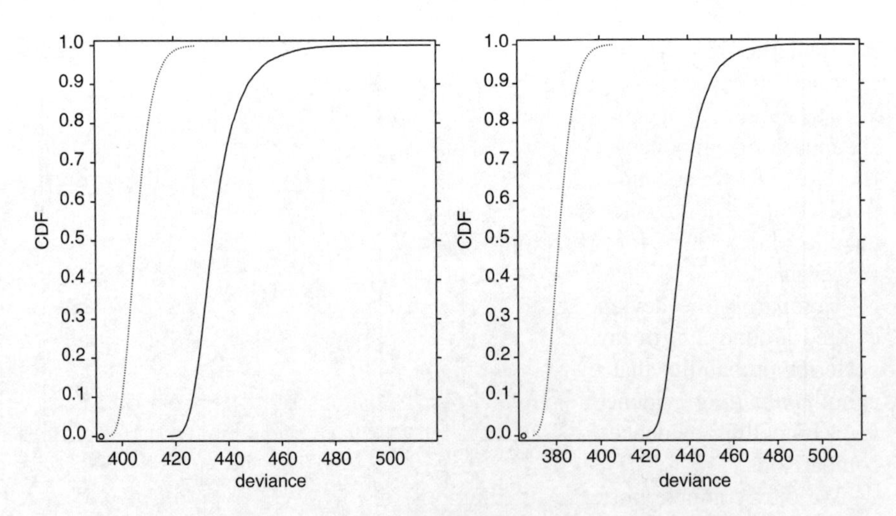

Figure 13.5 Five components (left) and six components (right).

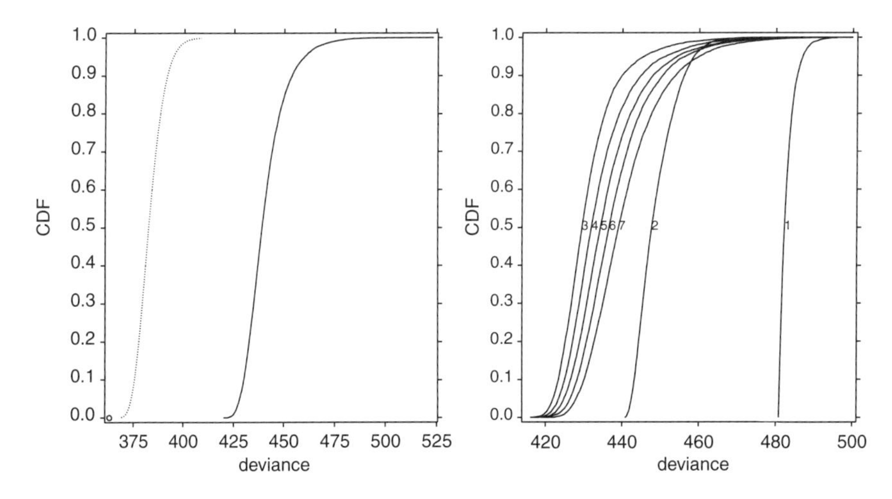

Figure 13.6 Seven components (left) and deviances for one to seven components (right).

- The deviance distribution for $K = 2$ greatly improves on that for the single normal.
- The improvement continues for $K = 3$.
- As the number of components increases beyond three the deviance distributions move steadily to the right, to larger values (lower likelihoods).
- They also become more *diffuse*, with increasing slope.
- It is clear that components beyond three *add only noise*.

These comparisons can be formalised by examining the distributions of deviance *differences* between the three-component model and those with more components. We randomly pair the draws from the three-component deviance with those from the other models and construct the empirical distributions for the differences.

The number of negative deviance differences, in the 10000 draws, for two, four, six and seven components compared to three, are 9620 for two, 6098 for four, 6921 for five, 6591 for six and 8226 for seven. Thus, the empirical probabilities that the three-component deviance is the smaller of the two being compared are (to three decimal places) 0.962 for two, 0.610 for four, 0.692 for five, 0.659 for six and 0.823 for seven.

Comparing the deviances for three and four components, the difference is centred around 2.5, in favour of the three-component model, and the empirical posterior probability that the three-component deviance is smaller is 0.61. This is not *compelling* evidence for three rather than four components, but it is much *less* compelling evidence for four rather than three, and similarly for the other comparisons.

We may compute posterior distributions for the posterior probability of each number of components K (up to seven), by drawing $L_K^{[m]}$ from each likelihood

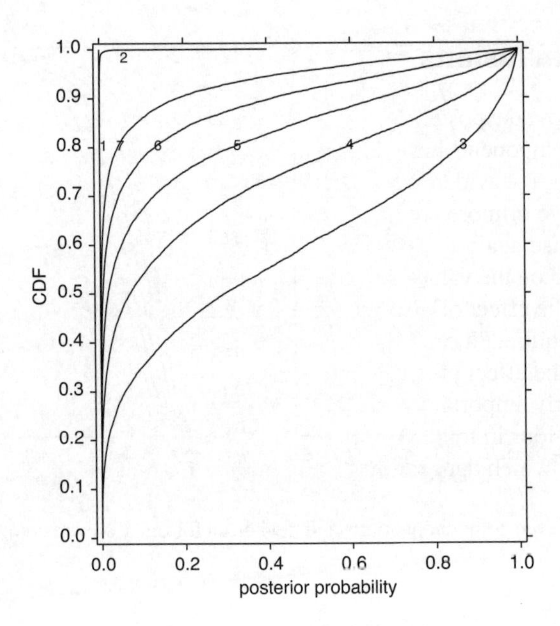

Figure 13.7 Posterior probabilities for one to seven components.

distribution, converting to the posterior probabilities,

$$pp_K^{[m]} = L_K^{[m]} / \sum_K L_K^{[m]},$$

and ordering to give the CDFs for each K. The resulting graph is shown in Figure 13.7.

At approximately the 65th percentile of these distributions, the posterior probabilities for 1, 2, 3, 4, 5, 6 and 7 components are 0, 0, 0.64, 0.24, 0.08, 0.03 and 0.01. An alternative is to summarise the posterior model probabilities using the *posterior median* likelihoods. The median deviances, and corresponding posterior probabilities, are given in Table 13.5: to two decimal places they are 0, 0, 0.75, 0.17, 0.05, 0.02 and 0.01. These are different because the probability and deviance scales are nonlinearly related.

Table 13.5 Posterior median deviances and model probabilities.

K	1	2	3	4	5	6	7
Deviance	482.2	448.9	431.0	433.9	436.2	438.3	440.8
Probability	0	0	0.745	0.175	0.055	0.019	0.006

13.10 Conclusions

We conclude from the posterior likelihood approach that *what the data say about K* are that three components have high posterior probability, though four components are not ruled out (the evidence for them was visible in the fit of the three-component mixture), but five or more components have very low probability, less than 0.1.

We emphasise again that this analysis does not require informative priors and does not depend on the values of hyperparameters. It provides a *reference analysis* against which the effect of informative priors can be assessed, but, since the latter priors are not required, it can provide a *standard analysis*; no sensitivity assessment is required for the effect of varying the priors.

A particularly important point is the divergence of the likelihood from log-quadratic behaviour in this example for K as small as 2. This invalidates reliance on AIC or BIC, which depend on this asymptotic form of the likelihood.

References

Aitkin, M. (1997) The calibration of P-values, posterior Bayes factors and the AIC from the posterior distribution of the likelihood (with discussion). *Statistics and Computing*, **7**, 253–272.

Aitkin, M. (2001) Likelihood and Bayesian analysis of mixtures. *Statistical Modelling*, **1**, 287–304.

Aitkin, M. (2010) *Statistical Inference: An Integrated Bayesian/Likelihood Approach*. Chapman & Hall.

Aitkin, M., Boys, R. J. and Chadwick, T. (2005) Bayesian point null hypothesis testing via the posterior likelihood ratio. *Statistics and Computing*, **15**, 217–230.

Aitkin, M., Liu, C. C. and Chadwick, T. (2009) Bayesian model comparison and model averaging for small-area estimation. *Annals of Applied Statistics*, **3**, 199–221.

Berger, J. O. and Bernardo, J. M. (1989) Estimating a product of means: Bayesian analysis with reference priors. *Journal of the American Statististical Association*, **84**, 200–207.

Carlin, B. P. and Chib, S. (1995) Bayesian model choice via Markov chain Monte Carlo methods. *Journal of the Royal Statistical Society, Series B*, **57**, 473–484.

Celeux, G., Forbes, F., Robert, C. P. and Titterington, D. M. (2006) Deviance information criteria for missing data models. *Bayesian Analysis*, **1**, 651–674.

Congdon, P. (2005) Bayesian predictive model comparison via parallel sampling. *Computational Statistics and Data Analysis*, **48**, 735–753.

Congdon, P. (2006) Bayesian model comparison via parallel model output. *Journal of Statistical Computation and Simulation*, **76**, 149–165.

Dempster, A. P. (1974) The direct use of likelihood in significance testing. In *Proceedings of the Conference on Foundational Questions in Statistical Inference* (eds O. Barndorff-Nielsen, P. Blaesild and G. Sihon), pp. 335–352.

Dempster, A. P. (1997) The direct use of likelihood in significance testing. *Statistics and Computing*, **7**, 247–252.

Escobar, M. D. and West, M. (1995) Bayesian density estimation and inference using mixtures. *Journal of the American Statistical Association*, **90**, 577–588.

Fox, J. -P. (2005) Multilevel IRT using dichotomous and polytomous response data. *British Journal of Mathematical Statistical Psychology*, **58**, 145–172.

Kass, R. E. and Raftery, A. E. (1995) Bayes factors. *Journal of the American Statistical Association*, **90**, 773–795.

Liu, C. C. and Aitkin, M. (2008) Bayes factors: prior sensitivity and model generalizability. *Journal of Mathematical Psychology*, **52**, 362–375.

Nobile, A. (2004) On the posterior distribution of the number of components in a finite mixture. *Annals of Statistics*, **32**, 2044–2073.

Phillips, D. B. and Smith, A. F. M. (1996) Bayesian model comparison via jump diffusions. In *Markov Chain Monte Carlo in Practice* (eds W.R. Gilks, S. Richardson and D. J. Spiegelhalter). Chapman & Hall.

Postman, M., Huchra, J. P. and Geller, M. J. (1986) Probes of large-scale structures in the Corona Borealis region. *The Astronomical Journal*, **92**, 1238–1247.

Richardson, S. and Green, P. J. (1997) On Bayesian analysis of mixtures with an unknown number of components (with discussion). *Journal of the Royal Statistical Society, Series B*, **59**, 731–792.

Roeder, K. (1990) Density estimation with confidence sets exemplified by superclusters and voids in the galaxies. *Journal of the American Statistical Association*, **85**, 617–624.

Roeder, K. and Wasserman, L. (1997) Practical Bayesian density estimation using mixtures of normals. *Journal of the American Statistical Association* **92**, 894–902.

Stephens, M. (2000) Bayesian analysis of mixtures with an unknown number of components – an alternative to reversible jump methods. *Annals of Statistics*, **28**, 40–74.

14

Bayesian mixture models: a blood-free dissection of a sheep

Clair L. Alston, Kerrie L. Mengersen and Graham E. Gardner

14.1 Introduction

The use of computed tomography (CT) scanning to measure attributes of tissue composition in animal experiments has grown steadily since the early 1990s. This technology is used on a range of experiments, such as nutrition trials for live animals, as well as on carcases after slaughter.

A CT scan returns measurements averaged over a pixel area that represent the denseness of the tissue. This tissue denseness is related to tissue type, with fat being generally less dense then muscle and bone being the most dense tissue we study. However, tissue denseness is not well separated, leading to a large overlap on the boundaries between types.

Normal mixture models have proved to be an efficient analytical technique for estimating the proportion of tissue types in individual CT scans, with MCMC output providing measures of variability that are unavailable in the standard cut-point modelling approach. These models are then used in conjunction with integration techniques to estimate the tissue volumes within a carcase.

In this paper we initially model individual scan data using a hierarchical mixture model, where skewed tissue densities are represented by the addition of two or more components.The mixture model is then extended to account for some of the spatial information using a Markov random field represented by a Potts model in

Mixtures: Estimation and Applications, First Edition. Edited by Kerrie L. Mengersen, Christian P. Robert and D. Michael Titterington.

terms of the allocation vector. A scheme for choosing starting values for component parameters is presented. The paper concludes with the use of the Cavalieri approach to combine individual scan estimates in order to estimate the carcase volume.

14.2 Mixture models

14.2.1 Hierarchical normal mixture

Figure 14.1 illustrates the log transformation of the Hounsfield units of the image on the left, represented in histogram form. This representation clearly indicates the suitability of such data to analysis using a normal mixture model for the following reasons:

- Tissue components are clearly 'mixed' at the boundary between fat and muscle, at around the level 5.4. Bone is more separated in the region around 6.
- The tissue densities appear to be skewed and hence representation via the addition of two or more components is appropriate.

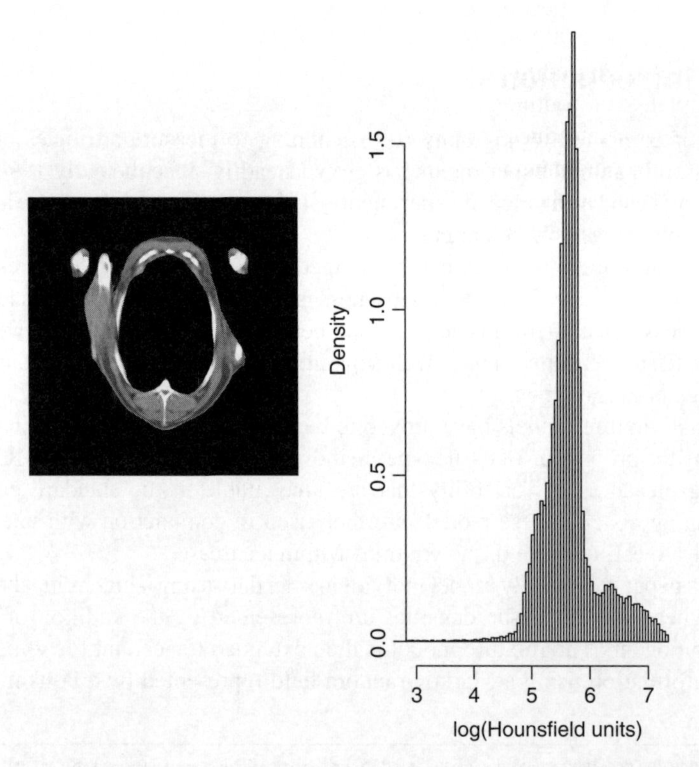

Figure 14.1 Image from CT scanning a live sheep (left) and representation of these data in histogram form (right).

- In addition, viewing the histograms from multiple scans Alston *et al.* (2009) indicate that the denseness of fat and muscle differs over carcase location. This attribute is readily accommodated in the mixture model analysis.

The normal mixture model is represented by the likelihood

$$p(y \mid \mu, \sigma^2, \lambda) = \prod_{i=1}^{N} \sum_{j=1}^{k} \lambda_j \frac{1}{\sqrt{2\pi\sigma_j^2}} \exp\left[-\frac{1}{2}\left(\frac{y_i - \mu_j}{\sigma_j}\right)^2\right],$$

where N is the sample size, k is the number of components in the model, λ_j is the weight of component j and μ_j and σ_j^2 are the mean and variance for the component.

In the Bayesian setting, it is usual to allocate a prior distribution to the unknown parameters of interest, and estimate the posterior distribution for these parameters according to Bayes, rule:

$$p(\mu, \sigma^2, \lambda \mid y) \propto p(y \mid \mu, \sigma^2, \lambda) p(\mu, \sigma^2, \lambda).$$

Hence, upon allocation of appropriate prior distributions, an explicit estimator for the Gaussian mixture model is available. However, the posterior distribution given by Equation (14.2.1) in Robert and Casella (1999, page 433), is the sum of k^N terms making computations, such as posterior expectations, too intensive to contemplate as a routine analytical method, even with moderate sample sizes. Instead, we now introduce a vector, z, that represents the missing observations associated with component membership; this, along with the use of conjugate priors, allows us to analyse the data using this model with a standard Gibbs sampling algorithm. The conjugate priors used were

$$\begin{aligned} p(\lambda) &\propto \text{Dir}(\alpha_1, \alpha_2, \cdots, \alpha_k), \\ p(\mu_j \mid \sigma_j^2) &\propto \text{N}(\xi_j, \sigma_j^2/m_j), \\ p(\sigma_j^2) &\propto \text{IG}(\nu_j/2, s_j^2/2). \end{aligned}$$

These priors were used as they were computationally convenient as well as being an adequate representation of the biology as known by scientists in the field.

The two-stage Gibbs sampler for estimating the parameters is as follows:

- *Stage 1*
  - Allocate pixels (y_i) to a component (z_i) using the conditional distribution $p(z_i \mid y_i, \mu, \sigma^2, \lambda) \sim \text{multinomial}(1; \gamma_1, \gamma_2, \cdots, \gamma_k)$, where

$$\gamma_j = \frac{\lambda_j(\sqrt{2\pi\sigma_j^2})^{-1} \exp\left[-\frac{1}{2}\left(\frac{y_i-\mu_j}{\sigma_j}\right)^2\right]}{\sum_{l=1}^{k} \lambda_l(\sqrt{2\pi\sigma_l^2})^{-1} \exp\left[-\frac{1}{2}\left(\frac{y_i-\mu_l}{\sigma_l}\right)^2\right]}.$$

- *Stage 2*
  - Using the allocations z_i from stage 1, update the estimates of μ, σ^2 and λ as follows:
  - Update μ_j from the conditional posterior,

$$p(\mu_j \mid y, z, \sigma_j^2) \sim \mathrm{N}\left(\frac{n_j\bar{y}_j + m_j\xi_j}{n_j + m_j}, \frac{\sigma_j^2}{n_j + m_j}\right),$$

where $n_j = \sum_{i=1}^{N} z_{ij}$ and $\bar{y}_j = \sum_{i=1}^{N} z_{ij} y_i / n_j$. Here $z_{ij} = 1$ if pixel i is allocated to component j, and 0 otherwise.
  - Update σ_j^2 from the marginal posterior distribution,

$$p(\sigma_j^2 \mid y) \sim \mathrm{IG}\left\{\frac{n_j + v_j + 1}{2}, \frac{1}{2}\left[s_j^2 + \hat{s}_j^2 + \frac{n_j m_j}{n_j + m_j}(\bar{y}_j - \xi_j)^2\right]\right\}$$

where $\hat{s}_j^2 = \sum_{i=1}^{N} z_{ij} \left(y_i - \bar{y}_j\right)^2$.
  - Update λ from the conditional posterior distribution,

$$p(\lambda \mid z) \sim \mathrm{Dir}(n_1 + \alpha_1, \ldots, n_k + \alpha_k)$$

The skewness of the density of each tissue type (Figure 14.1) meant that the number of components in the model was unknown, other than being at least three (fat, muscle and bone) or greater (to account for the skewness). The model was extended to incorporate reversible jump (RJMCMC) to estimate component numbers (Richardson and Green, 1997). However, when naively implemented, this algorithm was unsuitable as the posterior estimate of k did not necessarily give resulting estimates from a single model with k components. As a requirement of this analysis was for each component to be allocated to a single tissue group, we instead use the BIC to compare models by sequentially adding components.

This model was implemented in Alston *et al.* (2004). Results in that paper indicated that the model was a good representation of the underlying tissue group. In particular, it was shown that fat and muscle proportions were more accurately estimated than by a simple boundary approach and that the mixture model had an added advantage of providing measures of variability associated with these estimates.

14.3 Altering dimensions of the mixture model

The changes in tissue composition, both over a carcase and between animals, ensures that different numbers of components will be required to represent the skewed densities of these tissues between scans. We found starting the mixture model with $k = 3$ components and sequentially adding components, using the BIC statistic to assess model fit, to be quite successful.

To compute these models efficiently we require a method of simulating new starting values for the mean and variance of the component that is to be added to the current model. The nature of these data does not lend itself to simple schemes, such as allowing the starting values of the component means (μ_j) to be equally spaced over the range of the data. Having fitted a k-component model, we propose to incorporate a new component using the method outlined in Alston *et al.* (2007). This algorithm works by calculating the discrepancy between the observed and fitted density of the data:

$$d(y) = \tilde{g}(y) - \hat{g}(y).$$

In this case we take the observed density, $\tilde{g}(y)$, to be the number of pixels of equal Hounsfield units divided by the total number of pixels. The estimated density, $\hat{g}(y)$, is the estimated density from the k-component mixture model.

This disparity measure is used to identify areas between current component means where the model is underfitting the data and, as such, areas that may benefit from the addition of a component. We refer to this algorithm as the targeted addition of components (TAC).

This difference, $d(y)$, which is observed in Figure 14.2(a) as the discrepancy between the histogram values and the model density line, is graphed against the

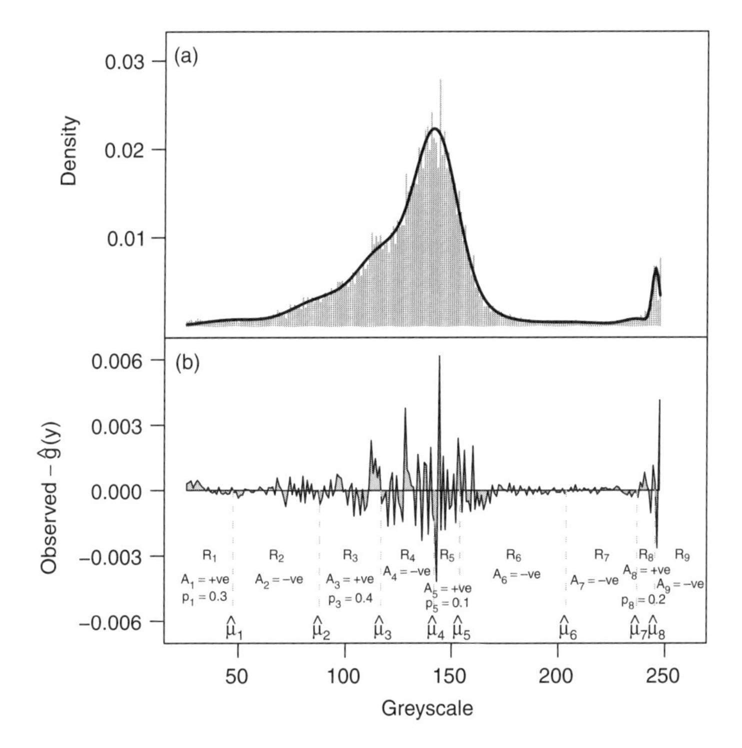

Figure 14.2 Diagrammatic representation of the TAC algorithm. The top figure is the data represented by the histogram with the fitted model overlaid (solid line). The bottom figure represents the discrepancy between the data and the model fit. Current estimates of parameter means are marked (vertical lines) and areas of under/overfit are calculated.

respective greyscale values in Figure 14.2(b) and partitioned into $k+1$ regions. These regions, R_j, which are allocated between the estimated component means (Figure 14.2(b)) of the current k-component mixture model, represent potential sites for the addition of a new component. Within each region R_j, we calculate the area (A_j) between $d(y)$ and the zero axis. Then A_j represents the over- or underestimation of the current mixture model in terms of the raw data for region R_j. If A_j is positive (i.e. more grey shading above the zero axis than below), this is an indication that the observed density is generally greater than the modelled density, and, as such, region j may benefit from the addition of a component. In the example given in Figure 14.2(b), there are nine regions of which four have an overall positive value for their respective areas (R_1, R_3, R_5 and R_8). The region R_j in which the new component is to be added is drawn from a multinomial distribution,

$$R_j \sim \text{multinomial}(1; p_1, p_2, \ldots, p_{k+1})$$

where $p_j = a_j / \sum a_t$ and $a_j = A_j$ if $A_j > 0$, and otherwise $a_j = 0$. These probabilities, rounded to one decimal place, are supplied for the regions with positive area in our example graph (Figure 14.2(b)). Then a new component is drawn from the region R_j. The starting value for the mean of the new component is simulated as

$$\mu_{\text{new}} \sim \text{N}\left[\hat{\mu}_{j-1} + \frac{\hat{\mu}_j - \hat{\mu}_{j-1}}{2}, \left(\frac{\hat{\mu}_j - \hat{\mu}_{j-1}}{2 \times 3}\right)^2\right],$$

in order to gain a new component mean that has an expected value equal to the midpoint of the chosen region and a standard deviation that will ensure approximately 99.9 % of drawn values belong to the interval $\{\hat{\mu}_{j-1}, \hat{\mu}_j\}$.

Similarly, to obtain a realistic starting value for the variance for this new component, we simulate a new value from a normal distribution that is truncated at zero:

$$\sigma_{\text{new}} \sim \text{N}\left[\frac{\hat{\mu}_j - \hat{\mu}_{j-1}}{2 \times 3}, \left(\frac{\hat{\mu}_j - \hat{\mu}_{j-1}}{2 \times 3 \times 5}\right)^2\right].$$

Although we have used the measure $d(y)$ in our scheme to add and delete components to the model, alternative measures such as the likelihood could be employed in a similar manner.

14.4 Bayesian mixture model incorporating spatial information

Figure 14.3 illustrates a typical draw of the allocation vector, z, from a normal mixture model resulting from a five-component model applied to the data in Figure 14.1. It was speculated from images such as these that the overall fit of

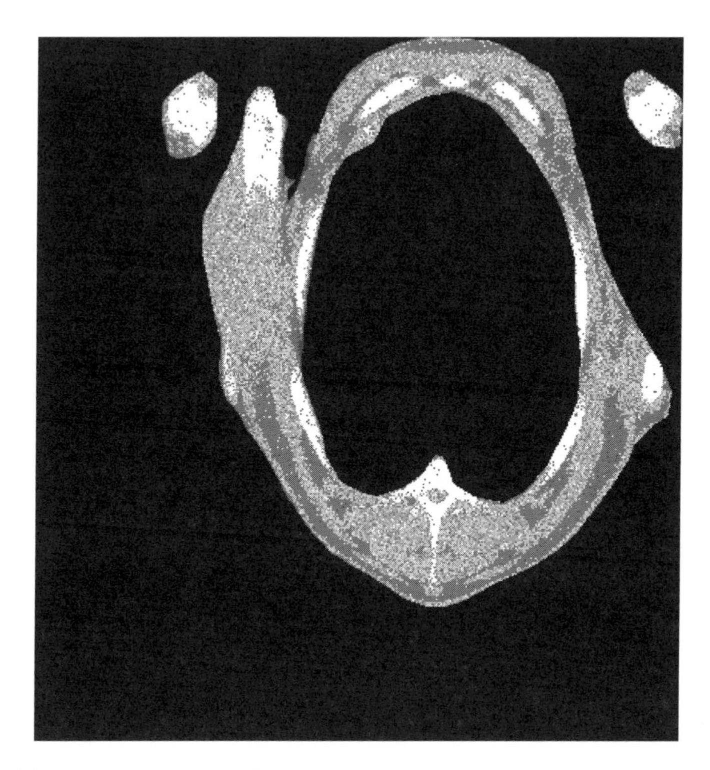

Figure 14.3 Typical allocation of variable z during stage 1 of the Gibbs sampler for the five-component mixture model.

the image was preserving and adequately representing the main features of the data. However, the speckled appearance of the image indicated that the allocation step of the model may be improved by using the information about denseness contained in the neighbouring pixels. Improvement of the allocation step would also improve the estimates of component means, variances and weights.

In this model, the assumption of independence for the allocation variable, z_i, is replaced by considering z_i as being drawn from a Markov random field (MRF), which involves the first-order neighbours of pixel i. Under the MRF model, the independence assumption on the observed allocation vector is removed, and information supplied by neighbouring pixels can influence the allocation of the origin pixel into a component group (McLachlan and Peel, 2000, Chapter 13).

Other papers that use similar models include Stanford (1999), who modelled PET scan, aerial and satellite image data using a normal mixture model in conjunction with an MRF and computed via the EM algorithm, and Fernández and Green (2002), who modelled spatially indexed disease data using a Poisson mixture with spatially correlated weights. In their approach they use RJMCMC (Green, 1995) to estimate parameters and determine component numbers.

In our model we assume that the joint distribution for the unobserved component membership, z, can be represented by a Potts model (Potts, 1952; Winkler, 2003):

$$p(z \mid \beta) = C(\beta)^{-1} \exp\left(\beta \sum_{i \sim j} z_i z_j\right),$$

where $C(\beta)$ is a normalising constant, $i \sim j$ indicates neighbouring pixels, $z_i z_j = 1$ if $z_i = z_j$, and otherwise $z_i z_j = 0$. The parameter β estimates the level of spatial homogeneity in component membership between neighbouring pixels in the image. In this case we would expect β to be positive in value as pixels from the same tissue group are clustered together.

The model is implemented using the Gibbs sampler to generate realisations of the parameters μ, σ^2, β and z from their full conditional distributions:

$$p(z_i \mid z_{\partial i}, y, \beta, \mu, \sigma) \propto p(y_i \mid z_i, \mu, \sigma) p(z_i \mid z_{\partial i}, \beta), \tag{14.1}$$

$$p(\mu, \sigma^2 \mid \beta, z, y) \propto \left[\prod_{i=1}^{N} p(y_i \mid z_i, \mu, \sigma)\right] p(\mu, \sigma),$$

$$p(\beta \mid \mu, \sigma, z, y) \propto p(z \mid \beta) p(\beta). \tag{14.2}$$

Updating the current estimates for the unobserved component membership (Equation (14.1)) is a straightforward simulation from a multinomial distribution:

$$p(z_i \mid z_{\partial i}, y, \beta, \mu, \sigma) \sim \text{multinomial}(1; \omega_{i1}, \omega_{i2}, \cdots, \omega_{ik}),$$

with the posterior probability of y_i belonging to component j being

$$\omega_{ij} = \frac{\left(\sqrt{2\pi\sigma_j^2}\right)^{-1} \exp\left[-\frac{1}{2}\left(\frac{y_i - \mu_j}{\sigma_j}\right)^2 + \beta U\left(z_{\partial i}, j\right)\right]}{\sum_{t=1}^{k}\left(\sqrt{2\pi\sigma_t^2}\right)^{-1} \exp\left[-\frac{1}{2}\left(\frac{y_i - \mu_t}{\sigma_t}\right)^2 + \beta U\left(z_{\partial i}, t\right)\right]},$$

where $z_{\partial i}$ are the first-order neighbours of pixel y_i and $U(z_{\partial i}, j)$ is the number of pixels in the neighbourhood of z_i currently allocated to component j.

Estimation of the component weights, λ, is no longer required in this model. However, for interest, they can be estimated when the allocation of z is finalised as

$$\lambda_j = \frac{\sum_{i=1}^{N} \omega_{ij}}{\sum_{i=1}^{N} \sum_{t=1}^{k} \omega_{it}}.$$

The component means (μ_j) and variances (σ_j^2) are simulated as in the independent mixture model.

Generating realisations of the spatial parameter, β, is complicated by the inclusion of the normalising constant in the joint distribution $p(z \mid \beta)$. We use the methodology of Rydén and Titterington (1998) and replace the joint distribution in

Equation (14.2) with the psuedolikelihood,

$$p_{PL}(z \mid \beta) = \prod_{i=1}^{N} p(z_i \mid z_{\partial i}, \beta)$$
$$= \prod_{i=1}^{N} \frac{\exp[\beta \mathrm{U}(z_{\partial i}, z_i)]}{\sum_{j=1}^{k} \exp[\beta \mathrm{U}(z_{\partial i}, j)]}.$$

The prior for β is taken as Un(ζ, δ) and β is updated at each stage of the Gibbs sampler using a Metropolis-Hastings step:

- Draw $\beta_{new} \sim$ Un(ζ, δ).
- Calculate the ratio $\psi = p(\hat{\beta}_{new} \mid \mu, \sigma, z, y)/p(\hat{\beta}_{current} \mid \mu, \sigma, z, y)$.
- If $\psi \geq 1$ accept β_{new}, otherwise:
  - Draw $u \sim$ Un$(0, 1)$.
  - If $u \leq \psi$ accept β_{new}, otherwise,
  - Reject β_{new} and retain $\beta_{current}$.

As we are working with images, and as such we expect pixels to be similar to their neighbours, we set the prior distribution for the spatial parameter as $p(\beta) \sim$ Un$(0, 3)$. The hyperparameters on the component means (μ_j) and variances (σ_j^2) are unchanged from the independent model.

14.4.1 Results

The parameters of the spatial mixture model were estimated using the Gibbs sampler and using TAC and the BIC statistic to determine the number of components in the model. The number of components in the spatial mixture was estimated as eight, which is higher than the five estimated under the independence of pixels model. The additional components are feasible in a spatial mixture model due to the enhancements in the allocation step. As neighbouring pixels are now more likely to be allocated to the same components, a reduction in variance of components is observed; see Alston *et al.* (2005) for full examples, which in turn requires the addition of several components to model the likelihood adequately.

The use of neighbouring information in the allocation of pixels to components in the spatial mixture model proved to be particularly beneficial in the boundary regions between tissue types. As a consequence, the estimated proportion of each tissue type has become more plausible and the associated credible intervals for the tissue proportions are much tighter in the spatial mixture model estimates (Table 14.1).

Figure 14.4 illustrates a typical allocation of the pixels y to components z under the eight-component spatial mixture model. The speckled appearance characteristic of the independent allocation (Figure 14.3) is no longer evident, indicating that the inclusion of neighbouring information has made pixel allocations in regions of high component mixing more stable. The estimate of the spatial parameter $\hat{\beta} = 1.42$ [1.38, 1.45] is confirmation of the spatial cohesion within an image.

Table 14.1 Comparison of estimated proportions using the independent mixture model and the spatial mixture model for data given in Figure 14.1.

	Estimated proportion			
Tissue	Mixture	Independent [95 % CI]	Spatial mixture	[95 % CI]
Fat	0.3866	[0.3631, 0.4100]	0.3050	[0.2933, 0.3096]
Muscle	0.4245	[0.4068, 0.4431]	0.5734	[0.5683, 0.5792]
Bone	0.1887	[0.1790, 0.1993]	0.1216	[0.1190, 0.1231]

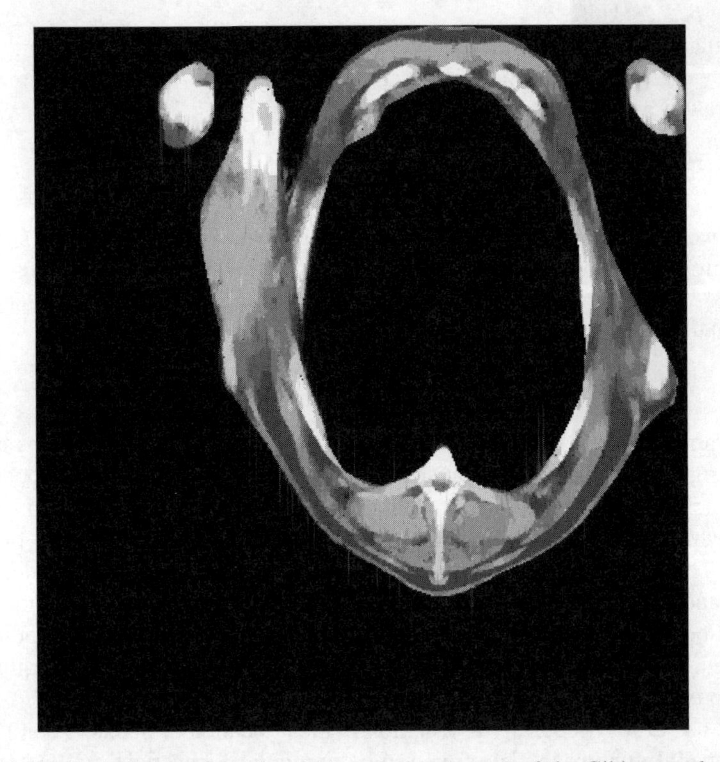

Figure 14.4 Typical allocation of variable z during stage 1 of the Gibbs sampler for the eight-component spatial mixture model.

14.5 Volume calculation

Estimation of whole carcase tissue weight or volume is achieved by taking CT scans at fixed distances along the animal's body and using the Cavalieri principle to estimate the whole mass. In this example, the scans are taken at 4 m intervals, resulting in the set of images shown in Figure 14.5. As can be observed in these

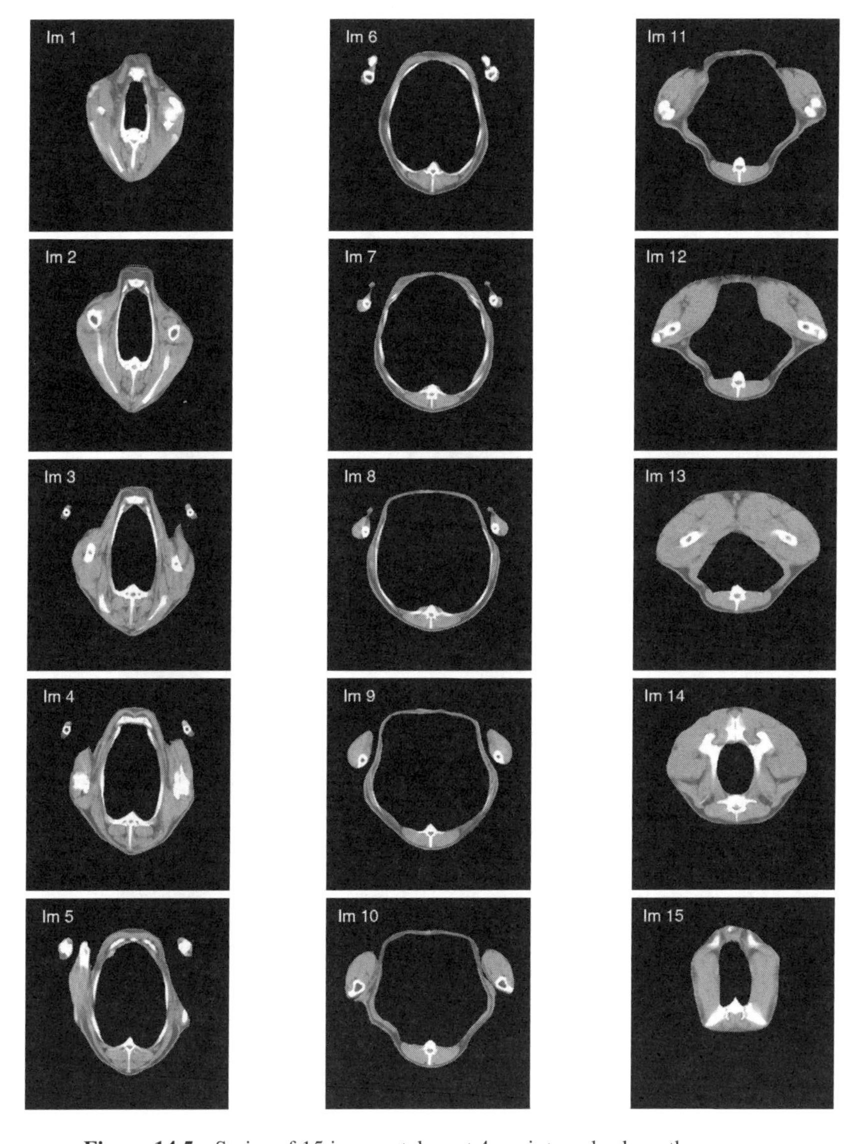

Figure 14.5 Series of 15 images taken at 4 cm intervals along the carcase.

images, the proportion of each tissue type changes according to location on the carcase. This is better illustrated by the histogram representation of the data given in Figure 14.6. The histograms also illustrate that the denseness of tissue types changes with location, indicated by the different areas displaying peaks within known tissue regions, and this biological feature favours the use of the mixture model over the fixed boundary method, such as those displayed in Figure 14.6.

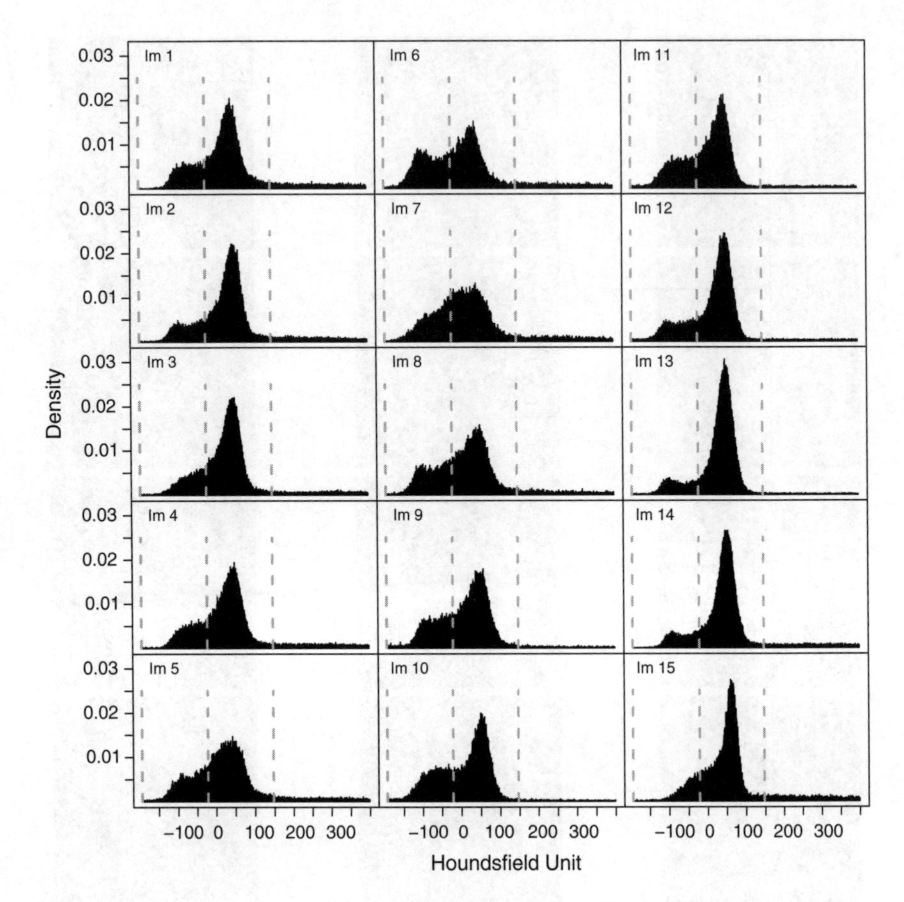

Figure 14.6 Histograms of data from the 15 images in Figure 14.5. Grey lines indicate boundaries commonly used in sheep studies (Kvame and Vangen, 2007) to allocate pixels to fat ($y \leq -23$), muscle ($-22 \geq y \leq 146$) and bone ($y > 146$) components.

The estimates of the proportions for each tissue type resulting from the separate analysis of each of the 15 scans is shown in Figure 14.7. It can be seen that the transition of proportions between scans is mostly smooth, as would be expected. Exceptions to this are in the regions of images 13 and 14, where the bone content increases dramatically (with an equally large decrease in fat). This is plausible as image 14 is taken in the pelvic region of the carcase, where bone content is comparatively high.

The volume of tissue is calculated using the Cavalieri method as it has proven to be reliable in most situations (Gundersen *et al.*, 1988; Gundersen and Jensen, 1987). The volume is calculated as

$$V_{\text{Cav}} = d \times \sum_{g=1}^{m} \text{area}_g - t \times \text{area}_{max}, \tag{14.3}$$

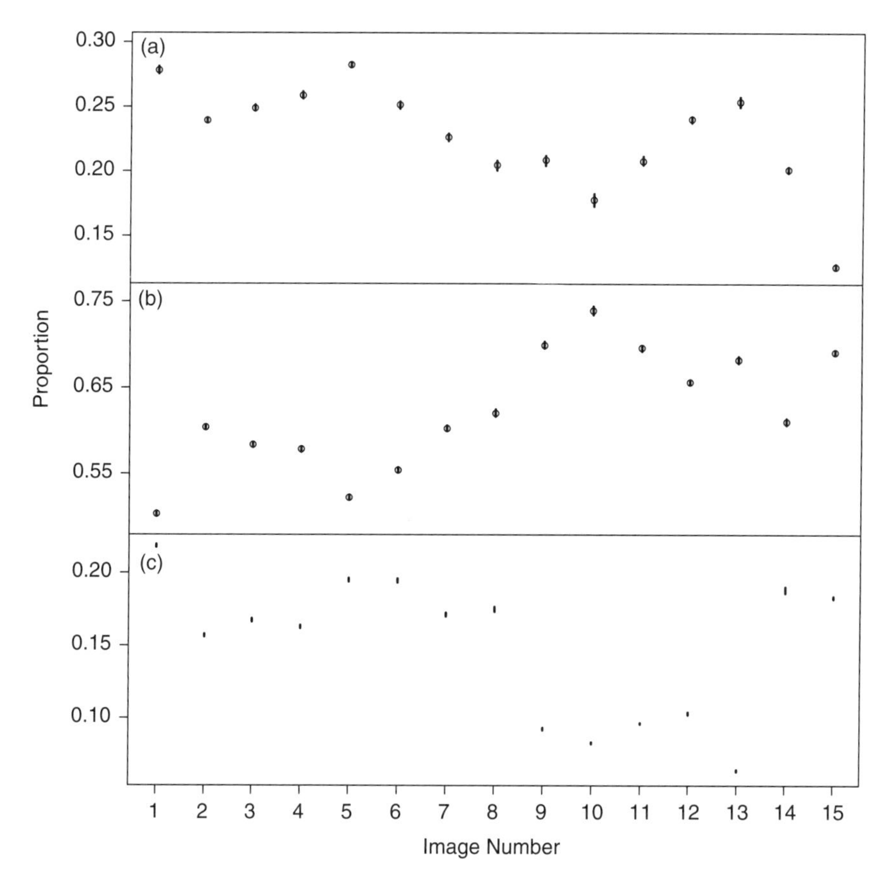

Figure 14.7 Estimated proportion and 95 % credible intervals for each scan using the spatial mixture model: (a) fat, (b) muscle, (c) bone.

where m is the number of CT scans taken (15) and d is the distance the CT scans are apart, in this case 4 cm. The value of t is the thickness of each slice (g), in this example 1 cm, and area_{max} is the maximum area of any of the m scans.

In some experiments the tissue weight is of interest. Calculating tissue weight is straightforward. We use the established conversion to grams per cubic centimetre given by Fullerton (1980):

$$\text{Established density}(\text{g/cm}^3) = 1.0062 + \text{Hu} \times 0.00106. \tag{14.4}$$

To estimate the weight of each tissue we use the equation

$$\widehat{\text{Tissue weight}} = \left[\left(1.0062 + \overline{\text{Hu}}_{tissue} \times 0.00106\right) \times \text{NumPix}_{tissue} \times \left(\text{pixels/cm}^3\right)^{-1} \right] / 1000,$$

Table 14.2 Mixture model estimates with 95 % credible intervals for the percentage and weight of tissue components in the carcase. The boundary method provides point estimates only and are defined as fat: $y \leq -23$, muscle: $-22 \geq y \leq 146$ and bone: $y > 146$.

	Cavalieri		
	Mixture		Boundary
Percentage (%)			
Fat	23.03	[22.94, 23.10]	32.38
Muscle	61.89	[61.80, 61.98]	42.78
Bone	14.75	[14.72, 14.79]	24.84
Weight (kg)			
Carcase	23.77	[23.771, 23.776]	24.30
Fat	4.77	[4.75, 4.78]	7.87
Muscle	14.35	[14.33, 14.38]	10.39
Bone	4.64	[4.63, 4.65]	6.04

where $\overline{\text{Hu}}$ is the mean Hounsfield number (Hu) for pixels allocated to the tissue of interest and NumPix_{tissue} is the number of pixels allocated to the tissue type. The number of pixels/cm^3 is 113.78 in this dataset.

The mixture model is well suited to volume calculation, as parameter estimation using the Gibbs sampler results in the quantities required to calculate volume (Equation (14.3)) and weight (Equation (14.4)) being estimated at each iteration. Therefore, the estimation based on all scans can be calculated at each iteration, after a suitable burn-in period, resulting in posterior quantities such as means and credible intervals. The resulting mean and credible intervals for proportions and weights of this carcase are given in Table 14.2, along with the results from a simple boundary approach.

Based on the mixture model results, the Cavalieri method yields estimates of tissue proportions that are in close agreement with the findings of an extensive study published by Thompson and Atkins (1980). Additionally, a chemical analysis of this carcase indicated a fat percentage of 20.2 % (4.47 kg), which is also in close agreement with these estimates. However, the results obtained using the boundary approach indicate that thc proportions of fat and bone in this carcase are being overestimated quite substantially.

While it is of less scientific interest, the weight of the carcase and individual tissues therein were also estimated (Table 14.2). A weight measurement of the carcase at slaughter was 25.4 kg, which again shows a close agreement with the estimates obtained using the mixture model approach, particularly given the error that is associated with manually editing images to extract the unwanted viscera from the live animal CT scans.

14.6 Discussion

In this paper we have summarised our previous publications (Alston *et al.*, 2004, 2005, 2007, 2009), which focused on the mixture model analysis for individual CT scans and how these results may be used to calculate whole carcase volume of tissue types in live sheep using the Cavalieri integration technique.

The use of the Bayesian normal mixture model to estimate tissue volume and allocate pixels to each of the three tissue types is shown to be advantageous relative to the standard practice of imposing fixed boundaries to achieve these estimates. The change in mean Hounsfield units for the allocated tissues along carcase regions, along with the illustrated change in histogram densities, provides clear evidence that information may be misinterpreted if assessment is made using a standard boundary model.

In current work we are examining the possibility of calibrating the fixed boundaries for factors such as breed, sex, carcase region and age, rather then applying a standard set to all sheep in general. However, as the main disadvantage of the mixture model is the computing burden, we are also working in the area of advanced computing, such as parallel processing of the allocation step and the use of GPU for these models (Carson *et al.*, 2009).

An additional benefit of the mixture model over the standard boundary technique is the ability to calculate measures of uncertainty around estimates, such as credible intervals. The boundary method provides a point estimate only, hence giving no means of assessing uncertainty.

References

Alston, C. L., Mengersen, K. L., Thompson, J. M., Littlefield, P. J., Perry, D. and Ball, A. J. (2004) Statistical analysis of sheep CAT scan images using a Bayesian mixture model. *Australian Journal of Agricultural Research*, **55**, 57–68.

Alston, C. L., Mengersen, K. L., Thompson, J. M., Littlefield, P. J., Perry, D. and Ball, A. J. (2005) Extending the Bayesian mixture model to incorporate spatial information in analysing sheep CAT scan images. *Australian Journal of Agricultural Research*, **56**, 373–388.

Alston, C. L., Mengersen, K. L., Robert, C. P., Thompson, J. M., Littlefield, P. J., Perry, D. and Ball, A. J. (2007) Bayesian mixture models in a longitudinal setting for analysing sheep cat scan images. *Computational Statistical Data Analysis*, **51**, 4282–4296.

Alston, C. L. Mengersen, K. L. and Gardner, G. E. (2009) A new method for calculating the volume of primary tissue types in live sheep using computed technology (CT) scanning. *Animal Production Science*, **49**, 1035–1042.

Carson, B., Murison, R. and Alston, C. (2009) Parallel computing of a spatial model for density component estimates, (in preparation).

Fernández, C. and Green, P. J. (2002) Modelling spatially correlated data via mixtures: a Bayesian approach. *Journal of the Royal Statistical Society, Series B*, **64**, 805–826.

Fullerton, G. D. (1980) Fundamentals of CT tissue characterization. In *Medical Physics of CT and Ultrasound: Tissue Imaging and Characterization*, (eds G. D. Fullerton and

J. Zagzebski), pp. 125–162. Medical Physics of CT and Medical Physics Monograph 6, American Institute of Physics.

Green, P. J. (1995) Reversible jump Markov chain Monte Carlo computation and Bayesian model determination. *Biometrika*, **82**, 711–732.

Gundersen, H. J. G. and Jensen, E. B. (1987) The efficiency of systematic sampling in sterology and its prediction. *Journal of Microscopy*, **147**, 229–263.

Gundersen, H. J. G., Bendtsen, T. F., Korbo, L., Marcussen, N., Møller, A., Nielsen, K., Nyengaard, J. R., Pakkenberg, B., Sørensen, F. B., Vesterby, A. and West, M. J. (1988) Some new, simple and eficient stereological methods and their use in pathological research and diagnosis. *APMIS. Acta Pathologica, Microbiologica et Immunologica Scandinavica*, **96**, 379–394.

Kvame, T. and Vangen, O. (2007) Selection for lean weight based on ultrasound and CT in a meat line of sheep. *Livestock Science*, **106**, 232–242.

McLachlan, G. and Peel, D. (2000) *Finite Mixture Models*. John Wiley & Sons, Ltd.

Potts, R. B. (1952) Some generalized order-disorder transitions. *Proceedings of the Cambridge Philosophical Society*, **48**, 106–109.

Richardson, S. and Green, P. J. (1997) On Bayesian analysis of mixtures with an unknown number of components (with discussion). *Journal of the Royal Statistical Society, Series B*, **59**, 731–792.

Robert, C. P. and Casella, G. (1999) *Monte Carlo Statistical Methods*. Springer-Verlag.

Rydén, T. and Titterington, D. M. (1998) Computational Bayesian analysis of hidden Markov models. *Journal of Computational Graphical Statistics*, **7**, 194–211.

Stanford, D. C. (1999) Fast automatic unsupervised image segmentation and curve detection in spatial point patterns. PhD Thesis, University of Washington.

Thompson, J. M. and Atkins, K. D. (1980) Use of carcase measurements to predict percentage carcase composition in crossbred lambs. *Australian Journal of Agricultural Research and Animal Husbandry*, **20**, 144–150.

Winkler, G. (2003) *Image Analysis, Random Fields and Markov Chain Monte Carlo Methods: A Mathematical Introduction*, 2nd edn. Springer-Verlag.

Index

acceptance rate, 257, 263, 266–269
Akaike information criterion (AIC), 291
  Monte Carlo (AICM), 115, 116
assumed density filtering, 22
autocorrelation, 265–267
averaging, 42, 44

Bayesian
  cross-validation, 129
  inference, 142, 164, 172, 213, 215, 216, 224, 255, 260
  information criterion (BIC), 109, 115, 199, 206, 222, 237, 282, 284, 285, 296, 301
  Lasso, 233–234
  model choice, 137, 226
  nonparametrics, 148, 150
  paradigm, 113
  rose tree model, 165
  variable selection, 124, 127
Bell number, 253
Benter's model, 108
book-keeping algorithm, 252
bootstrap, parametric, 82
boundary, 294, 296, 301, 303, 306, 307

carcase, 293
Cavalieri integration, 304, 307
chi-square distribution, 85
Chinese restaurant process, 164, 178
clumping of galaxies, 278
clustering, 101–104, 190, 225, 226
  greedy agglomeration, 163
components
  allocation, 242, 244
  heavy-tailed, 124
  latent, 130
  membership, 295
  number of, 124, 132, 256, 272, 277, 301
confidence region, 78
covariate, 101, 103, 106
  concomitant, 101
  influential, 109
  informative, 103
credible interval, 285
CT scan, 293

deviance, 286
  asymptotic distribution, 286
Dirichlet prior, 216, 221
  stick-breaking representation, 222
Dirichlet process, 132, 145
  hierarchical, 146, 148
  IBP compound, 145, 150–153
  mixture model, 145
  mixture of Gaussian process, 172
  prior, 216

effective sample size (ESS), 256, 269, 271
elliptically symmetric distribution, 202
EM algorithm, 1–7, 36, 109, 123, 190, 194, 196, 206, 279
  alternative, conditional (AECM), 197
  initial value, 209
  online, *see* online EM algorithm
  partial iteration, 56
EM test, 55
entropy, 111
Expectation Maximization Algorithm, *see* EM Algorithm

Mixtures: Estimation and Applications, First Edition. Edited by Kerrie L. Mengersen, Christian P. Robert and D. Michael Titterington.

galaxy data, 277, 278, 287
  deviance distribution for, 289
gamma-Poisson process, 155
gating network model, 105
Gibbs sampling, 19, 215, 255–258, 272, 279, 295
graphical model, 105
greedy agglomerative clustering, 174

Hamiltonian, 256, 259
  Monte Carlo algorithm, 256, 257–259
heterogeneity, 194, 231–237
hidden Markov model, 7, 18–20
hierarchical
  clustering, 162, 165, 176
  prior, 233
hyperparameter, 280
  settings, 277
hyperspherical coordinate, 86

identifiability, 215, 233, 246
  asymptotic, 90
  degenerate, 91
  labelling, 78, 81, 90, 91
  topological, 91
identification, *see* identifiability
incremental EM, 48
Indian buffet process, 146, 150, 159
integrated likelihood, 278, 280

k-means clustering, 214, 217, 219, 226

label switching, 215, 216, 219, 245, 246
large sample property, 56
latent position cluster model (LPCN), 112, 113, 115, 118
leapfrog algorithm, 259, 266
likelihood, 222
  complete, 243, 247
  confidence region, 79
  marginal, 132, 176, 233
  profile, 78
likelihood ratio test, 55, 199
  modified, 56
log predictive density score, 132
logistic regression, 102, 105, 108, 113

machine learning, 123
manifold
  constant, 266
  Euclidean, 256, 258, 259
  MALA, 258
  MCMC, 261
  Riemann, 256, 263
marginal likelihood, 222
Markov chain Monte Carlo (MCMC), 105, 124, 132, 172, 213, 215, 218, 224–228, 231, 233, 235, 237, 256, 257, 259, 261, 266, 280, 281, 284
  birth-and-death, 283
  reversible, 132, 256
Markov random field (MRF), 299
Markov switching model, 219
maximum a posteriori (MAP), 41
maximum likelihood, 104, 109
  estimator, 79, 164
Metropolis adjusted Langevin algorithm (MALA), 256
  Riemann manifold, 256, 258, 259, 263, 272
Metropolis algorithm, 124
Metropolis-within-Gibbs algorithm, 113, 129, 130
missing data, 241, 242, 261
mixed membership model, 145, 146, 148
mixture, 90
  Bayesian rose tree, 161
  complexity, 244, 248
  Dirichlet process, 158
  hierarchical, 293
  likelihood, 170
  of beta distributions, 18
  of common factor analysers (MCFA), 193–198
  of Dirichlet distributions, 26
  of experts model, 17, 102–118, 123
    gating network, 106
  of exponential family distributions, 34
  of factor analysers, 189
  of Gaussian distributions, 4, 16, 137, 272, 295
    multivariate, 198, 203, 233, 273
  of known densities, 26
  of linear submodels, 192
  of multinomial distributions, 242, 243

of Poisson distributions, 242, 243, 246, 299
of principal component analysers, 191
of Student's t distributions, 17
of two known densities, 4, 26
order of, 55
overfitting, 220, 221
smooth, 123–125
modal simulation, 78, 82–89
model selection, 163, 172, 183, 222

nonparametrics, Bayesian, 148, 150
number of components
estimation, 220–230
number of factors, 199

odds ratio, 110
online EM, 40–50
convergence, 41
for mixtures, 41, 46
Titterington's algorithm, 50
use for MAP estimation, 41
use of mini-batches, 45
order of a mixture, 55

partition, 165, 251
number, 244, 245, 253
Plackett-Luce model, 108
point-process representation, 214–220, 228, 230, 233, 237
Polyak–Ruppert averaging, *see* averaging, 36
posterior
deviance, 286
distribution, 242, 244, 245, 248, 250
decomposition, 248
of likelihoods, 285
of the deviance, 286
estimate, 296
mode, 1, 2, 8
probability of a model, 285, 289
Potts model, 293
preference data, 107
prior
conjugate, 242, 243
Dirichlet, 242, 243
prior distribution
choice of, 281
probabilistic
editor, 25
PCA, 34
product partition model, 176
profile likelihood, 80

quasi-Bayes, 25

Rand index, 199
rank data, 108
Robbins–Monro algorithm, 40
rose tree, 163, 165
mixture, 172, 175

shrinkage prior, 233, 235, 237
skewed response variable, 124
social network, 112
sparse learning, 154, 155
spatial correlation, 299
spike and slab model, 153, 155
star-shaped set, 86
nonexistence, 94
stick-breaking representation, 222
stochastic approximation, 40
sufficient statistic, 243

tightness, 62
tissue denseness, 293
topic model, 146, 147
focused, 153
sparse, 155
tree consistent partition, 166, 167

variable selection, 125, 233
visualisation, 93
voting bloc, 102

Wald confidence
region, 94
set, 98
what the data say?, 277, 280, 285, 291